ANNALS OF THE NEW YORK ACADEMY OF SCIENCES

Volume 744

CELLULAR GENERATION, TRANSPORT, AND EFFECTS OF EICOSANOIDS

BIOLOGICAL ROLES AND PHARMACOLOGICAL INTERVENTION

ANNALS OF THE NEW YORK ACADEMY OF SCIENCES
Volume 744

CELLULAR GENERATION, TRANSPORT, AND EFFECTS OF EICOSANOIDS

BIOLOGICAL ROLES AND PHARMACOLOGICAL INTERVENTION

Edited by Edward J. Goetzl, Robert A. Lewis, and Marek Rola-Pleszczynski

The New York Academy of Sciences
New York, New York
1994

The diagram on the printed cover is of the structure of soybean lipoxygenase-1. It shows regions of α-helix as cylinders and regions of β-sheet as arrows. Colors are used to highlight important amino acid side chains, which are given in space-filling representation without hydrogen atoms. Side chains lining cavity I are blue; those lining cavity II are red except for the two (L480 and M341) that block the entrance to this cavity, which are turquoise. The magenta side chains are the sulfhydryls; the iron atom is yellow. The diagram was prepared using the program SETOR, courtesy of S. Evans (1993, *J. Mol. Graphics* 11: 134–138). A black-and-white version of this diagram has been published in B. J. Gaffney, J. C. Boyington, L. M. Amzel, K. S. Doctor, S. T. Prigge & S. M. Yuan, 1994, Lipozygenase structure and mechanism, in *Advances in Prostaglandin, Thromboxane and Leukotriene Research: Proceedings of the 9th International Conference on Prostaglandins and Related Compounds,* Raven Press, New York, NY.

Library of Congress Cataloging-in-Publication Data

Cellular generation, transport, and effects of eicosanoids :
biological roles and pharmacological intervention / edited by Edward
J. Goetzl, Robert A. Lewis, and Marek Rola-Pleszczynski.
p. cm. — (Annals of the New York Academy of Sciences, ISSN
0077-8923 ; v. 744)
Proceedings of a conference held November 14-17, 1993, in
Stockholm, Sweden, and sponsored by the New York Academy of
Sciences.
Includes bibliographical references and index.
ISBN 0-89766-874-X (cloth : alk. paper). — ISBN 0-89766-875-8
(pbk. : alk. paper)
1. Prostaglandins—Physiological effect—Congresses.
2. Eicosanoic acid—Derivatives—Physiological effect—Congresses.
3. Cellular signal transduction—Congresses. I. Goetzl, Edward J.
II. Lewis, Robert A. (Robert Allen) III. Rola-Pleszczynski, Marek.
IV. New York Academy of Sciences. V. Series.
[DNLM: 1. Eicosanoids—congresses. 2. Receptors, Eicosanoid-
congresses. W1 AN626YL v. 744 1994 / QU 90 C3935 1994]
Q11.N5 vol. 744
[QP801.P68]
500 s—dc20
[615.7]
DNLM/DLC
for Library of Congress 94-40352
CIP

PCP

Printed in the United States of America

ISBN 0-89766-874-X (cloth)
ISBN 0-89766-875-8 (paper)
ISSN 0077-8923

ANNALS OF THE NEW YORK ACADEMY OF SCIENCES

Volume 744
November 15, 1994

CELLULAR GENERATION, TRANSPORT, AND EFFECTS OF EICOSANOIDS

BIOLOGICAL ROLES AND PHARMACOLOGICAL INTERVENTION[a]

Editors
EDWARD J. GOETZL, ROBERT A. LEWIS, AND MAREK ROLA-PLESZCZYNSKI

Conference Organizers
K. FRANK AUSTEN, EDWARD J. GOETZL, ANTHONY FORD-HUTCHINSON, SVEN HAMMARSTROM, ROBERT D. KRELL, ROBERT A. LEWIS, ROBERT C. MURPHY, JOHN A. OATES, PETER W. RAMWELL, MAREK ROLA-PLESZCZYNSKI, AND BENGT SAMUELSSON

CONTENTS

[a] This volume is the result of a conference entitled ''Cellular Generation, Transport, and Effects of Eicosanoids: Biological Roles and Pharmacological Intervention,'' which was held on November 14–17, 1993 in Stockholm, Sweden and sponsored by the New York Academy of Sciences.

Part II. Enzymatic and Other Pathways of Generation, Transport, and Metabolism

Part III. Cellular Recognition and Responses

Part IV. Biological Effects and Experimental Models

Part V. Pathogenetic Roles and Pharmacological Intervention in Human Diseases

Poster Papers

Financial assistance was received from:

Supporters

- ABBOTT LABORATORIES
- SYNTEX DISCOVERY RESEARCH
- TAKEDA CHEMICAL INDUSTRIES, LTD.
- ZENECA PHARMACEUTICALS GROUP

Contributors

- ASTRA DRACO
- BOEHRINGER INGELHEIM PHARMACEUTICALS, INC.
- CIBA-GEIGY LIMITED
- THE DU PONT MERCK PHARMACEUTICAL COMPANY
- GENETICS INSTITUTE, INC.
- HOFFMANN-LA ROCHE INC.
- THE R. W. JOHNSON PHARMACEUTICAL RESEARCH INSTITUTE
- MERCK FROSST CENTRE FOR THERAPEUTIC RESEARCH
- MERCK RESEARCH LABORATORIES
- MILES INC.
- MONSANTO COMPANY
- NATIONAL INSTITUTE OF ALLERGY AND INFECTIOUS DISEASES, NATIONAL INSTITUTES OF HEALTH
- PFIZER CENTRAL RESEARCH
- SMITHKLINE BEECHAM PHARMACEUTICALS
- WARNER-LAMBERT/PARKE-DAVIS

Stockholm, November 15, 1993

DEAR COLLEAGUES,

Once again an eicosanoid meeting comes to Stockholm.

Since the '50s the Karolinska Institute has been the place of inspiration and guidance for anyone interested in this new field of biology and medicine.

This has become generally appreciated with the First International Meeting held in this city in 1967 under the direction of Professor Bengt Samuelsson and Professor Sune Bergstrom, followed by a symposium in Italy devoted to the new class of leukotrienes.

It is of interest to remember how the eicosanoid discovery was built over many years as a combined effort of intuition, hard work, and new methodology.

For the first time a new series of cell regulators were discovered, and the compounds found had such a low concentration that they defied the existing methods of evaluation. This key problem–among many others–was solved brilliantly by the Swedish group who developed mass spectrometry combined with gas chromatography, a procedure refined by Ragnar Ryhage, a close colleague of Sune Bergstrom. This new procedure had by itself a significant effect on the further development of pharmacokinetics in animals and man and in toxicology.

At the present time the eicosanoid field has reached maturity: the cloning of the most relevant receptors; the rapid development of new regulatory drugs, which may soon reach clinical use; the widespread appreciation that this lipid class of compounds plays a fundamental role in cell metabolism, in immunology, and in normal and pathologic growth are examples of the successful, cooperative work of many groups around the world. Most of the more significant representatives of this research are present here today.

Unlike other fields of research this area has remained united–under the keen eye of Bengt Samuelsson–over the last decades, with a remarkable blending of continuity and innovation.

This also reminds us of Stockholm, where prostaglandins were born–a city where nothing seems to change and where so much innovative work goes on.

Ladies and gentlemen, I wish each of you and all of us a pleasant, successful symposium among the new treasures that the eicosanoid field has to offer, in this traditional academic atmosphere that made so much possible.

PROFESSOR RODOLFO PAOLETTI

Introduction

EDWARD J. GOETZL,[a] ROBERT A. LEWIS,[b] AND MAREK ROLA-PLESZCZYNSKI[c]

[a]*University of California*
San Francisco, California 94143

[b]*Discovery Research*
Syntex Corporation
Palo Alto, California 94304

[c]*University of Sherbrooke*
Sherbrooke, Québec, Canada J1H 5N4

More than 15 years since completion of the structural characterization of the principal products of oxygenation of arachidonic acid, many normal and pathophysiologic aspects of the cellular generation, recognition, and effects of these potent mediators remain unsolved mysteries. Eighty authorities in the field from ten different countries met in Stockholm on November 14-17, 1993 to present new findings, discuss current concepts and controversies, and outline future directions of research. The impact of modern methods of cellular and molecular biology were evident in the sophistication of the present approaches, and advances in pharmacology were continuing to provide new specific tools, as well as hopes for more effective treatments of relevant diseases.

Knowledge of the cellular biology and enzymology of phospholipases and diverse oxygenases has progressed rapidly. Distinctive forms of some of the enzymes, most notably cyclooxygenase, are activatable by growth factors and other cytokines. In some instances, such as 5-lipoxygenase, other cellular proteins are required for optimal and effectively rapid activation by specific stimuli. The nuclear localization of a portion of some oxygenases and activating proteins as well as the expression of novel mechanisms for cellular transport refocus attention on earlier theories suggesting a role for oxidized fatty acids as intracellular mediators.

The nearly ubiquitous distribution of pathways for generation of many of the eicosanoid mediators earlier had placed great emphasis on the times and amounts of production to explain biological specificity. The recent cloning of receptors for many of the eicosanoids, all of which belong to the superfamily of G protein-associated cell membrane receptors, has permitted an understanding of several unique mechanisms for specificity of action of the eicosanoid mediators. As one example, the three major subsets of receptors for prostaglandin E2 not only exhibit differences in structure and affinity of binding of a series of related prostaglandins, but also show distinctive patterns of tissue distribution and pathways of signal transduction as a result of selective association with one or more G proteins.

New insights into involvement of eicosanoids in host defense and diseases came from results of studies of regulatory activities for nucleic acid biosynthesis, growth and differentiation of normal and tumor cells, and cytokine production and activities in the immune system. The participation of eicosanoid mediators in asthma, allergic diseases, and inflammation was the primary focus of the clinical studies and discussions of new drugs capable of selectively inhibiting generation and activities of some eicosanoids.

The accelerating pace of research in most disciplines of the field suggest that the normal homeostatic and disease-related roles of eicosanoids will soon be known, and useful treatments will be available. The contributions in this volume reflect both the excellence of the ongoing science and optimism for rapid progress in the next decade.

Regulation of Prostaglandin E_2 Production in P388D_1 Macrophage-Like Cells[a]

RETO ASMIS [b,c] AND EDWARD A. DENNIS [d,e,f]

[b]*Institute for Biochemistry and Molecular Biology*
University of Berne
CH-3012 Berne, Switzerland

[d]*Department of Chemistry*
University of California at San Diego
La Jolla, California 92093-0601

INTRODUCTION

Eicosanoids play a central and often synergistic role in numerous aspects of inflammatory responses and host defense.[1,2] They are synthesized by most if not all cells via either the cyclooxygenase, lipoxygenase, or epoxygenase pathways. While these different pathways have been elucidated and the enzymes involved are well characterized, the release of arachidonic acid and its regulation in intact cells are still not fully understood. The availability of arachidonic acid appears to be the rate-limiting event in the biosynthesis of eicosanoids. It is believed to be controlled by a phospholipase-type enzyme, the simplest mechanism being the release of arachidonic acid from the *sn*-2 position of phospholipids by a phospholipase A_2 (PLA_2; EC 3.1.1.4).[3]

Two families of mammalian PLA_2s that are structurally unrelated have been well characterized. The first family consists of the low molecular mass PLA_2s (14 kDa), which are subdivided based on conserved disulfide bonds into group I (pancreatic type), group II (nonpancreatic type) and group III.[4] The recently described high molecular mass (85 kDa) cytosolic PLA_2 comprises the second family.[5,6] Although the cytosolic PLA_2, which preferentially hydrolyzes arachidonic acid-containing phospholipids at submicromolar Ca^{2+} concentrations,[5,7,8] has repeatedly been suggested to be the most likely candidate for the stimulus-coupled release of arachidonic acid,[9–14] recent evidence suggests that the low molecular mass PLA_2 might also play an important role in eicosanoid biosynthesis.[11,15–17] The signaling pathways, however, by which these PLA_2s are activated and regulated as well as their physiological functions remain to be elucidated.

[a]This work was supported by National Institutes of Health Grant HD26171.

[c]Supported in part by the Swiss National Fund and the American Heart Association, California Affiliate Postdoctoral Fellowship.

[e]Member, Center for Molecular Genetics.

[f]To whom correspondence should be addressed.

Several PLA_2 activities have been demonstrated and characterized in the $P388D_1$ macrophage-like cell line.[18–20] Inhibitor studies on an isolated Ca^{2+}-dependent, membrane-associated PLA_2 *in vitro* and in intact cells correlated enzyme activity, arachidonic acid release, and the formation of prostaglandin E_2 in these cells.[20–22] Hence, $P388D_1$ cells provide a unique model system to study the signaling events that lead to arachidonic acid release and eicosanoid biosynthesis in macrophages, which play an important role in the pathogenesis of inflammation and other diseases.

STIMULATION OF $P388D_1$ CELLS BY PAF

$P388D_1$ cells synthesize eicosanoids, predominantly the prostaglandins E_2 (PGE_2) and D_2 (PGD_2), in response to various stimuli.[21,23] The lipid autocoid, platelet activating factor (PAF), is a potent activator of arachidonic acid release and prostanoid synthesis in these cells.[21] However, at physiological concentrations PAF stimulates the release of PGE_2 only if the $P388D_1$ cells have been primed with bacterial lipopolysaccharide (LPS) prior to stimulation[22] (see below). In order to better understand the mechanisms involved in the regulation of PLA_2 activity in intact cells, we have investigated the role of inositol-1,4,5-trisphosphate [(1,4,5)IP_3], Ca^{2+}, and cAMP in the PAF-stimulated release of PGE_2 in LPS-primed $P388D_1$ cells. Our results are summarized in FIGURE 1. PAF stimulates $P388D_1$ cells via a high-affinity PAF receptor,[24] the structure of which has recently been elucidated,[25] resulting in the rapid formation of (1,4,5)IP_3.[26] The rapid catabolism of (1,4,5)IP_3 appears to be mediated by a cAMP-dependent protein kinase[33] (PKA, see below).

Both the release of (1,4,5)IP_3 and the subsequent increase in the intracellular free Ca^{2+} concentration ($[Ca^{2+}]_i$) are necessary but not sufficient for PAF to stimulate PGE_2 formation. Although LPS priming is necessary for PAF to stimulate PGE_2 formation, LPS priming has no effect on either (1,4,5)IP_3 levels or $[Ca^{2+}]_i$, nor is it required for the PAF-induced release of (1,4,5)IP_3 and increase in $[Ca^{2+}]_i$.[26] (1,4,5)IP_3 formation, the Ca^{2+} transient, and PGE_2 synthesis stimulated by PAF, however, are inhibited if $P388D_1$ cells are treated with pertussis toxin, suggesting the involvement of a pertussis toxin-sensitive G protein in this signaling sequence.[26] Although the type of the various G proteins and their sensitivity to pertussis toxin or cholera toxin varies in other macrophages and monocytic cell lines,[27] cell activation by PAF appears to follow the general scheme: PAF $\rightarrow$ PAF receptor $\rightarrow$ G protein $\rightarrow$ PLC $\rightarrow$ (1,4,5)IP_3 $\rightarrow$ Ca^{2+} signal.

Our study on the role of Ca^{2+} in this signaling sequence has demonstrated that the removal of extracellular Ca^{2+} prevented the influx of Ca^{2+}, but did not affect the initial Ca^{2+} transient nor PGE_2 formation induced by PAF. Under the same conditions, the Ca^{2+} ionophore- (ionomycin) induced PGE_2 formation was completely inhibited even though ionomycin stimulated an identical Ca^{2+} transient compared to PAF. PGE_2 formation, however, can be rescued by the subsequent addition of PAF even though no further Ca^{2+} signal is observed. PAF, therefore, appears to trigger two signals: the rise in $[Ca^{2+}]_i$ (STEP 1) and a second signal (STEP 2) that is not Ca^{2+} mediated (FIG. 1). Intracellular forms of PLA_2 possess Ca^{2+}-sensitive translocation domains.[5] One attractive possibility is that the second step corresponds to the translocation of the enzyme once it has bound Ca^{2+}. Translocation would then be induced or facilitated

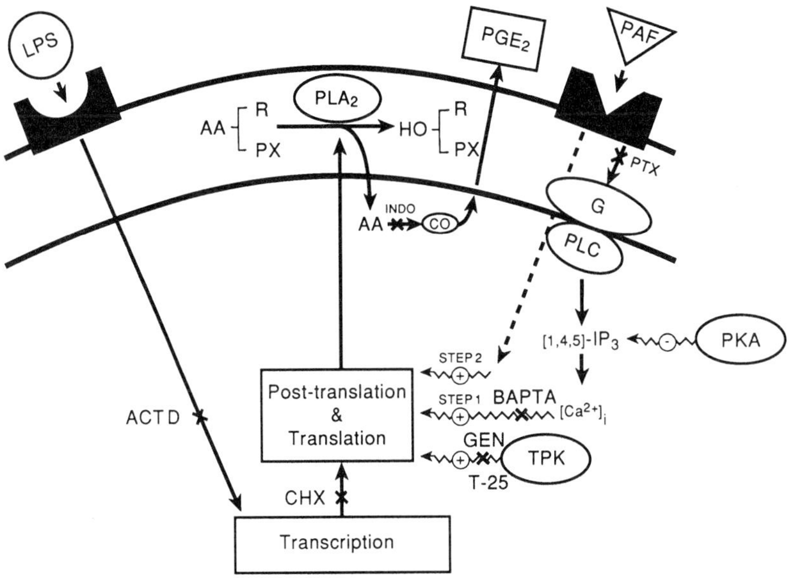

FIGURE 1. Signal transduction model for LPS priming and PAF-stimulated PGE_2 formation in $P388D_1$ cells. LPS priming and the PAF-stimulated release of arachidonic acid (AA) by phospholipase A_2 (PLA_2), its conversion by cyclooxygenase (CO) and the subsequent release of PGE_2 are shown. Possible roles for G proteins (G), phospholipase C (PLC), inositol 1,4,5-trisphosphate ([1,4,5]-IP_3), the intracellular Ca^{2+} signal ($[Ca^{2+}]_i$), cAMP-dependent protein kinase (PKA), tyrosine-specific protein kinases (TPK) and inhibition (-X-) by pertussis toxin (PTX), BAPTA, genistein (GEN), tyrphostin 25 (T-25), actinomycin D (ACT D), cyclohexamide (CHX), and indomethacin (INDO) are indicated. (Adapted from Asmis & Dennis.[33])

by the second signal, possibly through phosphorylation by a protein kinase or by interaction with a G protein. Another possibility is that the second signal relates to secretion of an extracellular PLA_2, which also appears to contribute to the formation of PGE_2 in $P388D_1$ cells.[15] The nature of the second signal is currently under investigation.

We have shown that protein kinase C (PKC) most likely does not play a significant role in the regulation of PLA_2 activity in $P388D_1$ cells.[22] Activators of PKC such as oleoyl-acetyl-glycerol[22] and phorbol esters (R. Asmis and E. A. Dennis, unpublished results) neither stimulate prostanoid synthesis nor do they prime for an enhanced release of arachidonic acid and PGE_2 formation in these cells. Furthermore, PAF-stimulated formation of PGE_2 is insensitive to inhibitors of PKC (e.g., H-7).[22] It is, however, inhibited by genistein,[22] a competitive inhibitor of ATP binding to tyrosine-specific protein kinases (TPK),[28] and tyrphostin 25,[26] a competitive inhibitor of substrate binding to TPK.[29] Neither the PAF-induced Ca^{2+} transient nor the release of $(1,4,5)IP_3$ is significantly inhibited by tyrphostin 25.[26] Tyrosine phosphorylation is, therefore, not required for the release of either one of these intracellular second messengers. These results suggest that the PLC involved in the release of $(1,4,5)IP_3$ is not the TPK-activated PLC-γ and remains to be identified. In LPS-primed $P388D_1$

cells, PAF-stimulated PGE_2 synthesis is, therefore, regulated by TPK at a site other than PLC. However, TPK is not a likely candidate for the second signal triggered by PAF, since ionomycin-stimulated PGE_2 formation is also inhibited by tyrphostin 25, although the second signal is not triggered by ionomycin.[26] The important role of TPK in the activation of PLA_2 (arachidonic acid release) and the regulation of eicosanoid synthesis has recently been demonstrated in peritoneal macrophages.[30] However, both phosphorylation by kinases other than TPK[31] and dephosphorylation by phosphatases[32] of PLA_2 appear to play an essential role in the activation and deactivation of PLA_2. These findings suggest the involvement of a complex system of different kinases and phosphatases in the regulation of PLA_2 activity.

In addition to the release of $(1,4,5)IP_3$ and the increase of $[Ca^{2+}]_i$, PAF stimulates the formation of cAMP in $P388D_1$ cells.[33] cAMP synthesis is also stimulated by ionomycin, and LPS priming increases both PAF- and ionomycin-induced cAMP formation. Indomethacin inhibits cAMP formation stimulated by either PAF or ionomycin, but does not affect PGE_2-induced cAMP formation. Of the three major eicosanoids formed by $P388D_1$ cells–PGE_2, PGD_2, and $PGF_{2\alpha}$[21]–only PGE_2 stimulates adenylate cyclase.[34] No significant amount of PGE_2 is retained by the cells. Secreted extracellular PGE_2 is, therefore, the likely mediator responsible for activating cAMP formation in stimulated $P388D_1$ cells. These results demonstrate that cAMP levels in $P388D_1$ cells are regulated by PGE_2 in an autocrine fashion.

cAMP has been shown to block phosphatidyl inositol-bisphosphate turnover and agonist-induced inositol phosphate formation in various cell types and tissues.[32] Furthermore, it has been suggested that cAMP-dependent protein kinase (PKA) directly acts on PLC-γ, inactivating the enzyme through phosphorylation.[35] In resting and stimulated $P388D_1$ cells, $(1,4,5)IP_3$ levels are strongly reduced by exogenous PGE_2 and increased by H-89,[33] an inhibitor of PKA.[36] However, exogenous PGE_2 is unable to down-regulate PAF-stimulated PGE_2 formation in these cells. It is, therefore, unlikely that in $P388D_1$ cells the activation of PKA by PGE_2 results in the inactivation of PLC. The $(1,4,5)IP_3$ mass measurements were carried out in the presence of Li^+, thereby slowing the catabolism of $(1,4,5)IP_3$, which upon stimulation results in a strong but transient accumulation of this second messenger. It is, therefore, more likely that PKA activates the catabolism of $(1,4,5)IP_3$, thereby counteracting the effect(s) of Li^+. A possible target is 5-phosphomonoesterase, which is activated severalfold through multiple phosphorylation.[37,38] PGE_2 has been shown to induce early homologous[39] and late heterologous densensitization,[39,40] probably in a cAMP-dependent process. A possible new role for the activation of PKA by PGE_2 could be to enhance $(1,4,5)IP_3$ metabolism in stimulated cells. This mechanism would avoid an unnecessary accumulation of $(1,4,5)IP_3$, which could lead to a densensitization of the intracellular Ca^{2+} response, rendering the cell unresponsive to this important activation pathway common to many stimuli.

LPS-PRIMING OF $P388D_1$ CELLS

LPS is a potent immunomodulator, and many of its actions in endotoxin shock are mediated by macrophages.[41] Although LPS is not a potent activator of eicosanoid synthesis in $P388D_1$ cells,[21] at low doses it primes these cells for an enhanced arachidonic acid release (2-3-fold) and PGE_2 formation (5-10-fold) upon stimulation.[22] LPS priming is dose dependent, but independent of the type of LPS (i.e., rough or smooth) used. It is maximal at exposure times between 1 to 2 h, before the cells transform into a hyposensitive state where they no longer respond to LPS. The primed state induced by LPS in $P388D_1$ cells deactivates with a half-life of 1.75 h, in contrast to murine peritoneal macrophages, where LPS priming results in a primed state for up to 72 h.[41] LPS priming is inhibited by the transcriptional inhibitor actinomycin D (3 μM), but not by the translational inhibitor cyclohexamide (10 μM).[22] In contrast, cyclohexamide, but not actinomycin D, inhibited PGE_2 formation if present during PAF stimulation of primed $P388D_1$ cells. It appears that LPS involves transcriptional events, whereas PAF stimulates translational and posttranslational events (FIG. 1).

LPS priming does not result in increased $(1,4,5)IP_3$ levels, nor does it affect $[Ca^{2+}]_i$ in $P388D_1$ cells (see above).[26] cAMP formation is increased in cells that have been primed with LPS; however, we have demonstrated that this increase is the result of the increase in basal PGE_2 secretion and not a direct effect of LPS priming.[33] Furthermore, no change in cAMP levels is observed during LPS priming. Interestingly, cholera toxin, but not PGE_2 at high doses, primes $P388D_1$ cells for an enhanced formation of PGE_2. Like LPS, cholera toxin priming is inhibited by the transcriptional inhibitor actinomycin D (3 μM). The involvement of a cholera-sensitive G protein in LPS priming is an intriguing though highly speculative possibility. Protein kinases, particular TPK, and phosphatases are also possible mediators of the priming signal.

PHOSPHOLIPID SPECIFICITY OF ARACHIDONIC ACID RELEASE

Using gas chromatography, we have analyzed the major lipid fractions of $P388D_1$ cells for their arachidonic acid content in order to determine the source of the arachidonic acid that is converted into eicosanoids, in particular into PGE_2. Identifying this particular source will help us to further narrow down the possible candidate(s) for the PLA_2(s) involved in the synthesis of PGE_2 in these cells. Upon LPS priming and subsequent stimulation with PAF, the total arachidonic acid of $P388D_1$ cells is reduced by 34% compared to resting cells (TABLE 1). Of this decrease, 94% is due to the loss of arachidonic acid from phospholipids. There appears to be no significant reduction in the level of arachidonic acid in the neutral lipid fraction, whereas the arachidonic acid found in the free fatty acid fraction is reduced by over 50%. This contribution to the overall decrease in arachidonic acid, however, appears to be minor (<3%).

FIGURE 2 shows the change in arachidonic acid levels in each phospholipid class. To the total reduction of arachidonic acid in the phospholipid fraction (1,987 pmol/10^6 cells), ethanolamine-containing glycerophosphatides (PE), the second largest phospholipid class (24 mol%), contribute over 85%, while most of the rest is lost from the phosphatidylinositol fraction (PI), which constitutes only 7 mol% of the total phospholipids. Interestingly, for the choline-containing glycerophosphatides (PC), the major phospholipid class (47 mol%) in $P388D_1$ cells, we have found no significant change within error in the arachidonic acid content upon stimulation. This is also true for the phosphatidylserine (PS) and phosphatidic acid (PA) fractions. No signifi-

TABLE 1. Arachidonic Acid (AA) Distribution in Resting and LPS-Primed P388D$_1$ Cells Stimulated with PAF

	Resting Cells	Stimulated Cells	Δ(Calculated) (pmol AA/10^6 cells ± SD)
Total (pmol AA/10^6 cells ± SD)	6,280 ± 430	4,120 ± 550	2,160 ± 698
Phospholipids (%)	90.2 ± 1.3	89.3 ± 1.1	−1,987 ± 878
Neutral lipids (%)	8.4 ± 0.7	11.1 ± 0.9	−71 ± 97
Free fatty acids (%)	1.4 ± 0.2	0.6 ± 0.2	−63 ± 30

NOTE: Five × 10^6 cells were either unprimed and unstimulated (resting cells) or primed with LPS (100 ng/ml) and stimulated with PAF (20 nM) as described previously.[26] In a given experiment (either resting or stimulated cells), all lipid fractions of interest were isolated and their arachidonic acid content determined (see FIG. 2). The extraction yields were determined with the use of radiolabeled lipid standards, and the results were corrected accordingly. Cells were extracted according to the method of Bligh and Dyer,[42] except that $CHCl_3$ was replaced by CH_2Cl_2 throughout the experiment. The total lipid extract was separated by ion exchange chromatography into three fractions: neutral lipids, free fatty acids, and phospholipids as described by Kaluzny *et al.*[43] For arachidonic acid determinations, the lipid fractions were transmethylated with Meth-Prep II (Applied Sciences Lab.) and extracted according to the method of van Kuijk *et al.*[44] The fatty acid composition of the lipid fractions was determined by gas chromatography (Hewlett-Packard 5830A equipped with a flame ionization detector and a DB-225 column from J&W Scientific).

cant amount of arachidonic acid was detected in the sphingomyelin fraction in either resting or stimulated cells.

We have also separated PE and PC each into three subclasses according to the bond type in the *sn*-1 position (1-acyl, 1-alkyl, and 1-alkenyl) for the respective phospholipids. The subclasses were analyzed for their arachidonic acid content, and we found that upon stimulation of cells arachidonic acid is lost from all three PE subclasses (1-acyl: 53%; 1-alkyl: 15%; 1-alkenyl: 32%). In contrast, we found only a minor decrease in arachidonic acid in the 1-acyl fraction of PC, which in part was compensated by a slight increase in the 1-alkenyl fraction. No significant change was observed in the 1-alkyl fraction of PC. Thus, the four lipid fractions that significantly contribute to the overall loss of arachidonic acid from LPS-primed and PAF-stimulated P388D$_1$ cells are: 1-acyl-PE, 1-alkenyl-PE, PI, and 1-alkyl-PE, contributing about 42%, 25%, 14%, and 12%, respectively, of the total arachidonic acid released.

Next we estimated the amount of arachidonic acid required for the synthesis of the major prostanoids and leukotrienes. In an average experiment, LPS-primed and PAF-stimulated cells will release approximately 4 ng (11 pmol) of PGE_2/10^6 cells in addition to the basal secretion of PGE_2. PGE_2 constitutes approximately 25% of the total prostanoid synthesis,[21] which appears to be of the same order of magnitude as reported for leukotriene formation in P388D$_1$ cells stimulated with the Ca^{2+} ionophore A23187.[49] The cells, therefore, have to supply approximately 44 pmol of arachidonic acid/10^6 cells for prostanoid synthesis alone, and we estimate that the total could be as high as 100 pmol/10^6 cells for the combined synthesis of prostaglandins and

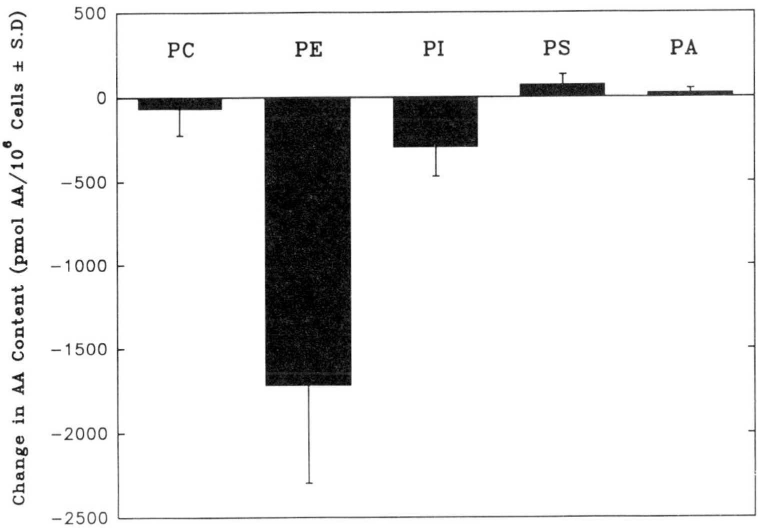

FIGURE 2. Change in arachidonic acid content in the major phospholipid classes after stimulation of LPS-primed P388D$_1$ cells with PAF. Cells were stimulated and extracted as described in TABLE 1. The phospholipid fraction was separated by 2-D thin-layer chromatography using CH_2Cl_2/methanol/concd. NH_4OH (50 : 100 : 5; v : v : v) and CH_2Cl_2/methanol/glacial acetic acid/water (50 : 25 : 8 : 4; v : v : v : v) as solvent systems. Spots were visualized with primuline,[45] scraped, and extracted with CH_2Cl_2/methanol/water (40 : 13 : 2; v : v : v). Phospholipids (phosphatidylcholine, PC; phosphatidylethanolamine, PE; phosphatidylinositol, PI; phosphatidylserine, PS; phosphatidic acid, PA) were quantitated with the phosphate assay described by Mitchell *et al.*[46] The phosphatidylcholine (PC) and phosphatidylethanolamine (PE) fractions were treated with phospholipase C, benzoylated, separated into 1-*O*-acyl-, 1-*O*-alkyl-, and 1-*O*-alkenyl subclasses, and quantitated according to Blank *et al.*[47,48] (not shown; see text). The arachidonic acid (AA) content was determined for each fraction and the result adjusted according to the respective extraction yield. The arachidonic acid content determined in LPS-primed and PAF-stimulated cells was then subtracted from the respective value in untreated cells.

leukotrienes. That amount of arachidonic acid, however, would constitute only 5% of the total amount released from the cells. It is, therefore, impossible to pinpoint the particular source of arachidonic acid utilized by P388D$_1$ cells for the specific synthesis of PGE$_2$ with this approach.

The use of radiolabeled arachidonic acid might help to further elucidate this question. However, we have found that P388D$_1$ cells that were labeled with [^{3}H]-arachidonic acid, LPS-primed, and stimulated with PAF release only 5% of their total incorporated [^{3}H]-arachidonic acid.[22] This result is in stark contrast to the 34% of the total mass of arachidonic acid that is lost under the same conditions as determined by quantitative GC. This discrepancy can be explained by a disparate labeling of the different lipid pools by exogenously added [3H]-arachidonic acid. It would then appear that exogenous arachidonic acid is preferentially incorporated into lipid pools that have a large capacity but do not release substantial amounts of arachidonic acid (i.e., PC). Of course, further rearrangements among lipid classes may occur. The source of the arachidonic acid that serves as the precursor for prostaglandin formation remains to be elucidated. Finally, these experiments have raised a new question: What is the fate of the bulk of arachidonic acid that is released upon stimulation but is not metabolized into prostaglandins?

CONCLUDING REMARKS

While many aspects of the regulation of PLA_2 and eicosanoid formation have been elucidated in $P388D_1$ cells, the true complexity of this particular stimulus-response system is just beginning to emerge. Many questions remain, such as the number and identity of the PLA_2s actually involved as well as their individual contribution to the various eicosanoids synthesized by these cells. Although recent evidence using antisense RNA technology suggests the involvement of the small molecular weight PLA_2,[15] additional PLA_2s[18–20] may also play an important role in this model system. Furthermore, the role of the recently described new mitogen-inducible cyclooxygenase[50] remains to be elucidated. Interestingly, in rat alveolar macrophages the expression of this enzyme is stimulated by LPS.[51] Current studies directed at these questions will help us to better understand the role of PLA_2 in the diverse functions of macrophages.

REFERENCES

1. Davies, P., P. J. Bailey, M. M. Goldberg & A. W. Ford-Hutchinson. 1984. Annu. Rev. Immunol. **2:** 335-357.
2. Smith, W. L. 1989. Biochem. J. **259:** 315-324.
3. Dennis, E. A. 1987. Biotechnology **5:** 1294-1300.
4. Davidson, F. F. & E. A. Dennis. 1990. J. Mol. Evol. **31:** 228-238.
5. Clark, J. D., L.-L. Lin, R. W. Kritz, L. C. Ramesha, L. A. Sultzman, A. Y. Lin, N. Milona & J. L. Knopf. 1991. Cell **65:** 1043-1051.
6. Kramer, R. M., E. F. Roberts, G. Maretta & J. E. Putnam. 1991. J. Biol. Chem. **266:** 5268-5272.
7. Reynolds, L. J., L. L. Hughes, A. I. Louis, R. A. Kramer & E. A. Dennis. 1993. Biochim. Biophys. Acta **1167:** 272-280.
8. Wijkander, J. & R. Sundler. 1991. Eur. J. Biochem. **202:** 873-880.
9. Lin, L.-L., Y. A. Lin & J. L. Knopf. 1992. Proc. Natl. Acad. Sci. USA **89:** 6147-6151.
10. Lin, L.-L., Y. A. Lin & D. L. DeWitt. 1992. J. Biol. Chem. **267:** 23451-23454.
11. Schalkwijk, C. G., E. De Vet, J. Pfeilschifter & H. van den Bosch. 1992. Eur. J. Biochem. **210:** 169-176.
12. Nakamura, T., L.-L. Lin, S. Kharbanda, J. L. Knopf & D. L. Kufe. 1992. EMBO J. **11:** 4917-4922.
13. Svenson, U., M. Houweling, E. Holst & R. Sundler. 1993. Eur. J. Biochem. **213:** 81-86.
14. Nemenoff, R. A., S. Winitz, N. X. Quian, V. Van Putten, G. L. Johnson & L. E. Heasley. 1993. J. Biol. Chem. **268:** 1960-1964.
15. Barbour, S. E. & E. A. Dennis. 1993. J. Biol. Chem. **268:** 21875-21882.

16. PERNAS, P., J. MASLIAH, J. L. OLIVIER, C. SALVAT, T. RYBKINE & G. BÉRÉZIAT. 1991. Biochem. Biophys. Res. Commun. **178:** 1298-1305.
17. MURAKAMI, M., I. KUDO & K. INOUE. 1993. J. Biol. Chem. **268:** 839-844.
18. ROSS, M. I., R. A. DEEMS, A. J. JESAITIS, E. A. DENNIS & R. J. ULEVITCH. 1985. Arch. Biochem. Biophys. **238:** 247-258.
19. ULEVITCH, R. J., M. SANO, Y. WATANABE, M. D. LISTER, R. A. DEEMS & E. A. DENNIS. 1988. J. Biol. Chem. **263:** 3079-3085.
20. LISTER, M. D., R. A. DEEMS, Y. WATANABE, R. J. ULEVITCH & E. A. DENNIS. 1988. J. Biol. Chem. **263:** 7506-7513.
21. LISTER, M. D., K. B. GLASER, R. J. ULEVITCH & E. A. DENNIS. 1989. J. Biol. Chem. **264:** 8520-8528.
22. GLASER, K. B., R. ASMIS & E. A. DENNIS. 1990. J. Biol. Chem. **265:** 8658-8664.
23. AUSSEL, A. & M. FEHLMANN. 1986. Cell. Immunol. **101:** 415-424.
24. VALONE, F. 1988. J. Immunol. **140:** 2389-2394.
25. HONDA, Z., M. NAKAMURA, I. MIKI, M. MINAMI, T. WATANABE, Y. SEYAMA, H. OKADO, H. TOH, K. ITO, T. MIYAMOTO & T. SHIMIZU. 1991. Nature **346:** 342-346.
26. ASMIS, R., C. RANDRIAMAMPITA, R. Y. TSIEN & E. A. DENNIS. 1994. Biochem. J. **298:** 543-551.
27. RANA, R. & L. E. HOKIN. 1990. Physiol. Rev. **70:** 115-164.
28. AKIYAMA, T., J. ISHIDA, S. NAKAGAWA, H. OGAWARA, S. WATANABE, N. ITOH, M. SHIBUYA & Y. FUKAMI. 1987. J. Biol. Chem. **262:** 5592-5595.
29. GAZIT, A., P. YAISH, C. GILON & A. LEVITZKI. 1989. J. Med. Chem. **32:** 2344-2352.
30. GLASER, K. B., A. SUNG, J. BAUER & B. M. WEICHMAN. 1993. Biochem. Pharmacol. **45:** 711-721.
31. ZOR, U., E. HER, T. HARELL, G. FISCHER, Z. NAOR, P. BRAQUET, E. FERBER & N. REISS. 1990. Biochim. Biophys. Acta **1091:** 385-392.
32. DEMARZO, V., S. H. I. GALADARI, J. R. TIPPINS & H. R. MORRIS. 1991. Life Sci. **49:** 247-259.
33. ASMIS, R. & E. A. DENNIS. 1994. Submitted for publication.
34. YAMAMOTO, H. & T. SUZUKI. 1987. J. Immunol. **139:** 3416-3421.
35. KIM, U.-H., W. J. KIM & S. G. RHEE. 1989. J. Biol. Chem. **264:** 20167-20170.
36. CHIJIWA, T., A. MISHIMA, M. HAGIWARA, M. SANO, K. HAYASHI, T. INOUE, K. NAITO, T. TOSHIOKA & H. HIDAKA. 1990. J. Biol. Chem. **265:** 5267-5272.
37. CONNOLLY, T. M., W. J. LAWING, JR. & P. W. MAJERUS. 1986. Cell **46:** 951-958.
38. MOLINA Y VEDIA, L. M. & E. G. LAPENTINA. 1986. J. Biol. Chem. **261:** 10493-10499.
39. SHAMMA, M. S., R. FERNANDEZ-BORTAN & T. SUZUKI. 1988. Prostaglandins **36:** 329-341.
40. COFFEY, R. G., V. A. ALBERTS & L. L. WEAKLAND. 1990. J. Leukocyte Biol. **48:** 557-564.
41. ADEREM, A. A., D. S. COHEN, S. D. WRIGHT & Z. A. COHN. 1986. J. Exp. Med. **164:** 165-179.
42. BLIGH, E. G. & W. J. DYER. 1959. Can. J. Biochem. Physiol **37:** 911-917.
43. KALUZNY, M. A., L. A. DUNCAN, M. V. MERITT & D. E. EPPS. 1985. J. Lipid. Res. **26:** 135-140.
44. VAN KUIJK, F. J. G. M., D. W. THOMAS, R. J. STEPHENS & E. A. DRATZ. 1985. J. Rad. Biol. Med. **1:** 215-225.
45. WRIGHT, R. S. 1971. J. Chromatogr. **59:** 220-221.
46. MITCHELL, K. T., J. E. FERRELL, JR. & W. H. HUESTIS. 1986. Anal. Biochem. **158:** 447-453.
47. BLANK, M. L., M. ROBINSON, V. FITZGERALD & F. SNYDER. 1984. J. Chrom. **298:** 473-482.

48. Blank, M. L., E. A. Cress & F. Snyder. 1987. J. Chrom. **392:** 421-425.
49. Aussel, C. & M. Fehlmann. 1987. Prostaglandins Leukotrienes Med. **28:** 325-336.
50. Xie, W., J. G. Chipman, D. L. Robertson, R. L. Erikson & D. L. Simmons. 1991. Proc. Natl. Acad. Sci. USA **88:** 2692-2696.
51. Lee, S. H., E. Soyoola, P. Chanmugam, S. Hart, W. Sun, H. Zhong, S. Liou, D. Simmons & D. Hwang. 1991. J. Biol. Chem. **267:** 25934-25938.

Cytochrome P450 Arachidonic Acid ω-Hydroxylation in the Proximal Tubule of the Rat Kidney[a]

FANGMING LIN,[b] NADER G. ABRAHAM,[c] AND MICHAL LANIADO SCHWARTZMAN[b]

[b]*Department of Pharmacology*
New York Medical College
Valhalla, New York 10595

[c]*The Rockefeller University*
New York, New York 10021

INTRODUCTION

Cytochrome P450 refers to a family of membrane-bound hemoproteins that are part of the mixed-function monooxygenase system that oxidizes a wide variety of structurally unrelated compounds. The substrates transformed by the cytochrome P450 range from endogenous substrates, such as fatty acids and steroids, to exogenous substrates, including certain drugs, hydrocarbons, and chemical carcinogens. This diversity in substrate specificity has been attributed to the presence of multiple forms of cytochrome P450. This enzyme system in the presence of molecular oxygen and NADPH metabolizes arachidonic acid to several oxygenated metabolites, including: (1) four regioisomeric epoxyeicosatrienoic acids (5,6; 8,9; 11,12; and 14,15 EET), which can be hydrolyzed by epoxide hydrolase to the corresponding diols (DHT); (2) six regioisomeric *cis-trans* conjugated hydroxyeicosatetraenoic acids (HETE); and (3) ω- and ω-1 alcohols (20-HETE and 19-HETE).

In the kidney, the cytochrome P450 system has been localized primarily in the cortex, with the highest concentration in the S_3 segment of the proximal tubules.[1] Although the kidney cortex has been shown to possess the highest concentration of cytochrome P450 content and activity, the outer and inner medulla have this system as well.[2] Cortical and outer medullary microsomes from human, rabbit, and rat kidneys have been shown to metabolize arachidonic acid via the cytochrome P450 monooxygenases to several metabolites, namely EETs, DHTs, and ω/ω-1 alcohols.[2–4] However, the cellular localization of the cytochrome P450-arachidonic acid metabolism within the kidney is not completely elucidated. We have demonstrated cytochrome P450-arachidonate metabolism in isolated cells of the thick ascending limb of the loop of Henle in the rabbit kidney.[5] Koop *et al.*[6] reported that isolated cells from the proximal tubules of the rabbit possess a high activity of arachidonate ω/ω-1 hydroxylases, while Lapuerta *et al.*[7] showed the formation of EETs in the pars

[a]This work was supported in part by National Institutes of Health Grant P0134300.

recta of the rabbit kidney. Madhun *et al.*[8] recently reported that cultured proximal tubule cells from the rabbit kidney metabolized arachidonic acid to EETs. The vascular bed of the dog kidney also has been shown to possess significant activity of arachidonic acid ω-hydroxylase.[9] We have demonstrated a preferred localization of arachidonic acid ω/ω-1 hydroxylases to the proximal tubule of the rat, whereas epoxygenase activity is equally distributed along the entire nephron.[10]

Whereas many of the renal cytochrome P450-derived arachidonic acid metabolites are biologically active and demonstrate either vasoreactivity and/or inhibition/stimulation of tubular transport mechanisms in micromolar concentrations,[11–18] 20-HETE affects vascular tone, Ca^{2+} flux, and cell growth in picomolar to nanomolar concentrations.[19–21] 20-HETE has also been shown to cause natriuresis when injected *in situ* to the euvolemic rat[13] and to inhibit ^{86}Rb uptake in thick ascending limb of the loop of Henle *in vitro.*[14] 20-HETE is the product of a cytochrome P450 ω-hydroxylase activity, and it constitutes the major arachidonate metabolite in the kidney cortex. The ω-hydroxylation of arachidonic acid, as well as that of other fatty acids, was demonstrated as an activity associated with several forms of the cytochrome P450 4A gene family, mainly 4A1, 4A2 and 4A3 in the rat kidney.[22,23] Sharma *et al.*[24] showed that the rat liver and kidney P450 4A1, although very active as lauric acid ω-hydroxylase, carries out relatively well the ω-hydroxylation of arachidonic acid and is susceptible to induction by hypolipidemic drugs, such as clofibrate. Funae and co-workers[25] isolated several P450 isozymes from rat liver and kidney that exhibit a high specificity towards arachidonic acid ω-hydroxylation.

More than 85% of total 20-HETE formation along the entire rat nephron was found in the proximal tubules, with highest production in the S3 segment, suggesting a specific localization of arachidonic acid ω-hydroxylase activity in this segment of the nephron.[10] Through the use of GC/MS analysis, 20-HETE has been shown to be present in rat urine,[26] suggesting it is an endogenous constituent of the kidney. In the present study we further characterized arachidonic acid metabolism to 20-HETE in the proximal tubules, demonstrated the endogenous production of 20-HETE, and provided a correlation of 20-HETE synthesis with the expression of cytochrome P450 4A1.

METHODS

Isolation of Proximal Tubules

Five- to seven-week-old male Sprague Dawley rats (150-200 g, Charles River) were anesthetized by an intraperitoneal injection of pentobarbital (50 mg/kg body weight). Kidneys were perfused *in situ* via the abdominal aorta with 10 ml ice-cold Tyrode's solution containing 5 mM $NaHCO_3$ followed by 10 ml Tyrode's solution containing 1 mg/ml of paramagnetic iron oxide, which becomes entrapped in glomeruli and other renal vasculature. Kidneys were removed and decapsulated. The cortex was separated from the medulla and minced (1 mm^3) with a single-edged razor blade. Cortical slices were incubated in 5 ml of Tyrode's digestion solution containing collagenase (200 U/ml), hyaluronidase (200 U/ml), soybean trypsin inhibitor (0.5 mg/ml), albumin (10 mg/ml), glucose (10 mM), sodium succinate (1 mM) and L-alanine (5 mM). The digestion procedure was performed in a shaking water bath at 37 °C, and the incubation mixture was continuously bubbled with 95% O_2/5% CO_2. Every 5 min the incubation solution was allowed to settle for 1 min, the supernatant was drawn off, and 5 ml of fresh digestion solution was added to the undigested tissue. This procedure was repeated 3-4 times. Tubules in the pooled supernatant were separated from vascular tissue and glomeruli using a side pull magnet to trap the iron-containing vascular tissue while the tubules were decanted off. The magnetic separation was repeated twice. Tubular fractions were gently layered onto a 35% Percoll solution over a 100% Percoll solution and centrifuged for 10 min at 13000 *g*. The bottom layer consisted mainly of proximal tubules as measured by alkaline phosphatase staining (Sigma diagnostics alkaline phosphatase kit).

^{14}C-Arachidonic Acid Metabolism

Proximal tubules were sonicated to homogeneity using an ultrasonic cell disrupter (Heat System, Farmingdale, NY). Protein concentration was determined using the method of Bradford (Bio-Rad). The homogenates (~100 μg protein) were incubated with ^{14}C-arachidonic acid (56 mCi/mmol, 0.2 μCi) with or without NADPH (1 mM) in a final volume of 0.5 ml of 0.1 M potassium phosphate buffer, pH 7.4, for 30 min at 37°C. In some experiments the homogenates were preincubated with indomethacin (10 μM), 17-ODYA (5 μM) or 7-ER (5 μM), and with anti-rat P450 4A1 antibody (a gift from Dr. R. T. Okita[27]) or anti-epoxygenases antibody[28] at a protein ratio of 1 : 1 for 10 minutes at 37 °C. Incubation with boiled tubules was carried out as control. The reaction was terminated by acidification to pH 3.5-4.0 with formic acid and metabolites extracted with ethyl acetate. The final extract was evaporated under nitrogen, resuspended in 50 μl of methanol and injected onto the HPLC column. Reverse-phase HPLC was performed on a C_{18} Microsorb column (Rainin, Emeryville, CA) using a linear gradient from acetonitrile : water : acetic acid (50 : 50 : 0.1) to acetonitrile : acetic acid (100 : 0.1) at a flow rate of 1 ml/min for 30 min. This solvent system separates 20-HETE from EETs, DHTs, and 20-COOH-AA, but not from 19-HETE. However, previous studies showed that conversion of arachidonic acid to 19-HETE by cortical and proximal tubular preparation is very low and does not amount to more than 5% of 20-HETE production. The elution profile of the radioactive products was monitored by a flow detector (Radiomatic Instruments and Chemical Co.) The identity of each metabolite was confirmed by its comigration with authentic standard.

Endogenous Formation of 20-HETE

Freshly isolated proximal tubules (20-100 μg protein) were placed in an Eppendorf microcentrifuge tube, resuspended in 100 μl of PBS, pH 7.4, containing $CaCl_2$ (1.2 mM), NADPH (1 mM), glucose (11 mM), and phenanthroline (1.2 mM) and incubated for 0-30 min at 37 °C. The reaction was terminated by addition of 0.5 ml of ethanol containing 5 ng of deuterated [14,15-$^2H_2$]-20-HETE as an internal standard and [5,6,8,9,11,12,14,15-^{3}H]-20-HETE (5000 cpm) as a tracer for recovery efficiency. The incubation tubes were centrifuged for 1 min, and the supernatant was collected.

The pooled supernatant was evaporated, dissolved in 1 ml distilled water, and extracted with ethyl acetate after acidification to pH 3.5 with formic acid. The pellet was resuspended in 0.2 ml of 0.5 N NaOH containing 0.1% SDS and incubated for 30 min at 37 °C to solubilize the membrane. Protein content was determined by the method of Bradford (Bio-Rad).

The amount of 20-HETE released into the medium was determined by selected ion monitoring-negative chemical ionization-gas chromatography/mass spectrometry (SIM-NCI-GC/MS) analysis as described previously.[26] The incubation extract was evaporated to dryness, resuspended in 100 μl acetonitrile, and converted to the pentafluorobenzyl (PFB) ester by adding 10 μl of pentafluorobenzyl bromide and 10 μl of *N,N*-diisopropylethylamine and incubating it at room temperature for 30 min. The sample was evaporated under nitrogen and further incubated with 100 μl of *N*-methyl-*N*-(*tert*-butyldimethylsilyl)-trifluoroacetamide for 30 min at 65 °C to obtain the PFB ester, *tert*-butyldimethylsilyl (TBDMS) ether derivative of 20-HETE. The sample was concentrated to dryness and dissolved in 20 μl of octane for GC/MS analysis. NCI-GC/MS was performed on a HP 5989A (Hewlett Packard, Palo Alto, CA) mass spectrometer interfaced with a SP-2330 capillary column (15 m × 0.25 mm inside diameter, 0.2 μm film thickness; Supelco, Bellefonte, PA) programmed from 100-250 °C at 25 °C/min using helium as the carrier gas. The mass spectrometer was scanned from 200-600 mass units in one second or single ions, *m/z* 433 for the endogenous derivatized 20-HETE (PFB ester TBDMS ether) and *m/z* 435 for the derivatized internal standard. An estimation of the total 20-HETE content in the purified biological sample was made by comparison of the ion intensities at *m/z* 433 : 435 vs. a standard curve of 20-HETE-PFB-TBDMS/$^2H_2$-20-HETE-PFB-TBDMS molar ratio constructed by NCI-GC/MS analysis.

RNA Extraction and Northern Blot Analysis

Total RNA was isolated from proximal tubules of control and rats treated with dexamethasone (300 μg/kg/day) or clofibrate (10 mg/kg/day) for 2 days by the guanidium isothiocynate method and quantitated by spectrophotometry.[29] Total RNA (10 μg) was denatured and size separated by electrophoresis on 1.2% agarose gels containing 6% formaldehyde. To verify the integrity of RNA, gels were stained for 5 min with 0.5 μg/ml ethidium bromide in diethylpyrocarbonate-treated water, destained, and photographed. RNA was transferred to nylon membranes (Amersham) under vacuum and cross-linked to the support by UV irradiation. Hybridization was performed using the full-length cDNA of the rat liver P450 4A1 (a gift from Dr. Hardwick[30]). After hybridization the filters were washed at 45 °C in 0.1 × SSC (1 × SCC-150 mM NaCl/15 mM Na citrate) and 0.1% SDS and exposed to Kodak XAR film at −70 °C. As a control, all filters were reprobed with cDNA encoding GAPDH.

RESULTS

We have previously demonstrated using a microdissection technique that the major site of 20-HETE synthesis along the rat nephron is the proximal tubule with the highest specific activity present in the S_3 segment.[10] In the present study, we used

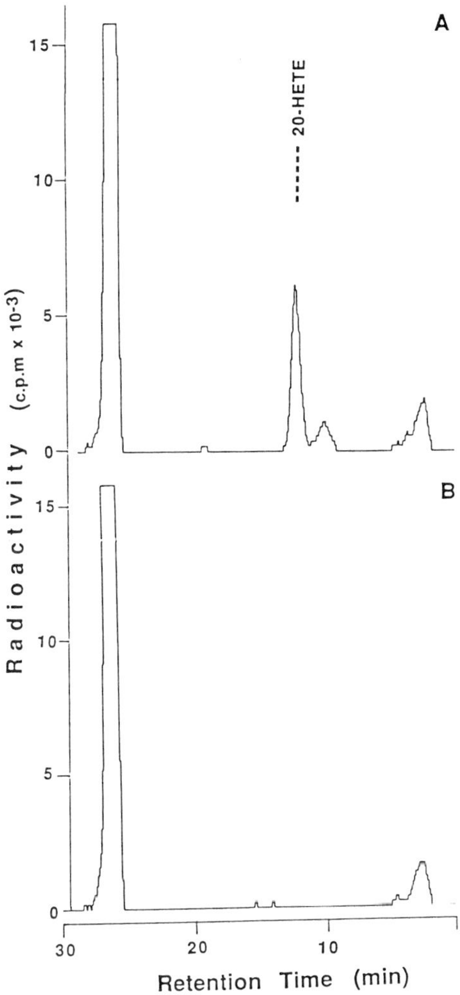

FIGURE 1. Representative HPLC chromatograms of arachidonic acid metabolites formed by proximal tubular homogenates in the (**A**) presence and (**B**) absence of NADPH. Homogenates (100 μg) were incubated with ^{14}C-arachidonic acid (6.9 μM) with or without NADPH (1 mM) for 30 min at 37 °C. Metabolites were extracted and separated by reverse phase HPLC as described in METHODS.

a different method, which allows the isolation of relatively large amounts of purified proximal tubules with little contamination of other nephron segments, microvessels, and glomeruli. The homogeneity of this preparation was determined by histochemical staining of alkaline phosphatase, an enzyme marker for the brush border of the proximal tubule.[31] More than 95% of the tubules were strongly stained positive, providing a homogenous proximal tubular preparation for the characterization of AA ω-hydroxylation activity.

Incubation of homogenates of proximal tubules with ^{14}C-arachidonic acid in the presence of NADPH resulted in the formation of one major metabolite, 20-HETE, which was separated by HPLC and identified by its comigration on HPLC with authentic standard (FIG. 1). In the absence of NADPH or in incubations with boiled

TABLE 1. Effect of NADPH and Inhibitors of Arachidonic Acid Metabolism on 20-HETE Formation in the Proximal Tubules

Incubation	20-HETE (pmol/mg/min)
None	0
NADPH (1 mM)	65.5 ± 1.1
NADPH + indomethacin (10 μM)	92.2 ± 15.2
NADPH + 17-ODYA (5 μM)	11.5 ± 2.6[a]
NADPH + 7-ER (5 μM)	54.0 ± 2.4[b]
NADPH + anti-rat P450 4A1 IgG (1 : 1)	29.9 ± 1.2[a]
NADPH + anti-human P450-AA IgG (1 : 1)	70.9 ± 5.5

NOTE: Proximal tubule homogenates were preincubated with the indicated inhibitors/antibodies or their vehicle for 10 min prior to the addition of ^{14}C-arachidonic acid and NADPH. Incubations were carried out as described in METHODS. 20-HETE was separated following extraction of the incubates by reverse phase HPLC, and the amount formed was quantitated from the percent conversion of the added substrate. Anti-human P450-AA IgG was raised against a P450 form that was isolated from the human liver and shown to oxidize arachidonic acid to EETs.[28] The results are mean ± SE, $n = 3$.

[a] $p < 0.05$.

[b] $p < 0.001$.

homogenates, no formation of 20-HETE was observed. Proximal tubular homogenates demonstrated a high specific activity of NADPH-dependent arachidonic acid conversion to 20-HETE (65.6 ± 1.1 and 92.2 ± 15.2 pmol/mg/min in the absence and presence of indomethacin, respectively). This activity accounts for more than 50% of that in the whole cortex microsomes from the rat (164.7 ± 6.6 pmol/mg/min).

Preincubation of proximal tubules with indomethacin (10 μM) did not affect the production of 20-HETE (TABLE 1). Moreover, there was an increase in the production of 20-HETE suggesting that either more substrate, arachidonic acid, is available to the hydroxylase enzyme and/or the increased 20-HETE production is due to a diminished 20-HETE metabolism by cyclooxygenase activity. On the other hand, preincubation with 17-ODYA or 7-ER, inhibitors of cytochrome P450-dependent reactions including metabolism of AA to 20-HETE,[32,33] resulted in 83% and 20% inhibition of 20-HETE formation from arachidonic acid, respectively (TABLE 1). The ability of these inhibitors to decrease 20-HETE formation provides strong evidence for cytochrome P450-dependent monooxygenase activity in this reaction.

The presence of AA ω-hydroxylase activity and the formation of 20-HETE *in vitro* by homogenized tubules do not necessarily indicate that this metabolite is an endogenous constituent of the proximal tubule and is then formed *in vivo*. We developed a quantitative method, selected ion monitoring by negative ion chemical ionization-gas chromatography/mass spectrometry (NCI-GC/MS), for the measurement of the endogenous level of 20-HETE. A representative trace of such measurements is depicted in FIGURE 2. Using this method we demonstrated that freshly isolated proximal tubules possess a basal arachidonic acid ω-hydroxylase activity as

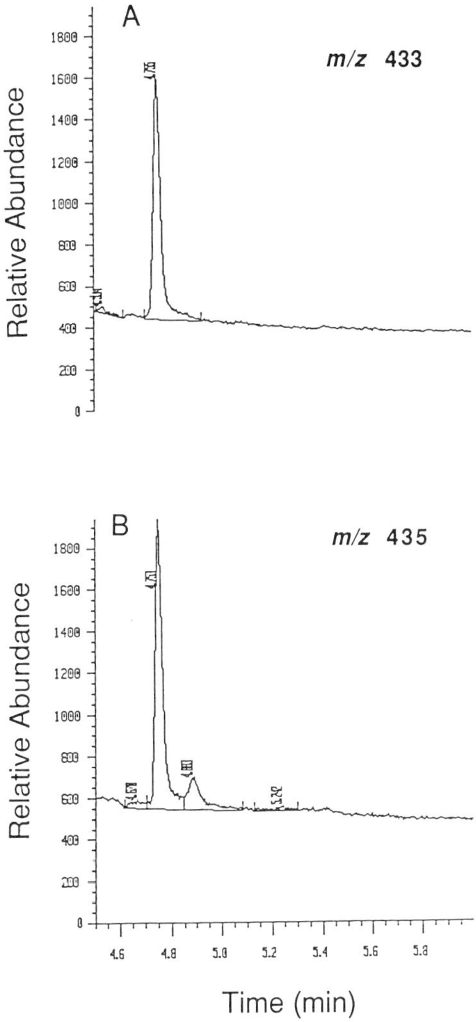

FIGURE 2. Selected ion monitoring-NCI/GC-MS chromatogram of selected PFB-TBDMS-derived ions present in extract of proximal tubule incubation. Derivatized sample was injected onto the capillary GC column, and column eluent was monitored by mass fragmentography at (**A**)*m/z* 433 for the endogenous 20-HETE and (**B**)*m/z* 435 for the internal standard, $^{2}H_2$-20-HETE. The identity of the peak as 20-HETE was further verified by its comigration with authentic 20-HETE standard (including deuterated) derivatized under the same conditions.

determined by the gradual increase in 20-HETE released into the incubation medium, which amounts to 8.7 ± 0.9 and 22.5 ± 2.4 pmol/mg at 10 and 30 min, respectively.

As mentioned before, the ω-hydroxylation of arachidonic acid is carried out by a member of the P450 4 gene family. To date, however, no specific isozyme has been purified and cloned that selectively uses arachidonic acid as the substrate. Since P450 4A1 is a major form found in the rat, we used polyclonal antibodies raised against P450 4A1 to perform immunoprecipitation. As seen in TABLE 1, preincubation of this antibody with the proximal tubule in a ratio of 1 : 1 resulted in 54% inhibition of arachidonic acid conversion to 20-HETE, suggesting that P450 4A1 or a P450 4A1-like protein is involved in 20-HETE production in the proximal tubule. In

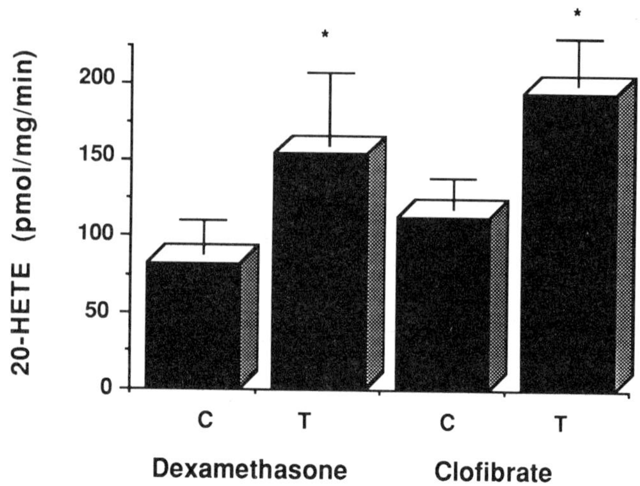

FIGURE 3. Effect of dexamethasone and clofibrate on 20-HETE production in the proximal tubule. Rats were treated with dexamethasone (300 μg/kg/day) and clofibrate (10 mg/kg/day) for 2 days. Proximal tubules were isolated from control (C) and treated (T) rats, and ^{14}C-arachidonic acid conversion to 20-HETE was measured as described in METHODS. Results are the mean ± SE, $n = 3$; $*p < 0.05$ from untreated control group.

contrast, antibodies raised against cytochrome P450-AA epoxygenase[28] did not affect 20-HETE formation (TABLE 1).

The ω-hydroxylation of fatty acids, including arachidonic acid, is readily induced in different tissues by hypolipidimic drugs such as clofibrate.[24,34] This induction process includes increased levels of P450 4A mRNAs, specifically that of the P450 4A1.[34,35] We therefore used this tool to determine whether induction of 4A1 in the proximal tubules will also correlate with formation of 20-HETE. Rats were also treated with dexamethasone, previously shown to increase 20-HETE synthesis in the rabbit kidney.[36] As seen in FIGURE 3, proximal tubules from rats treated with clofibrate or dexamethasone demonstrated a twofold increase over the controls (vehicle-treated rats) in converting arachidonic acid to 20-HETE. Total RNA was extracted from control and treated kidneys, and Northern blot analysis was performed using the full-length cDNA probe to the rat liver P450 4A1. The blot was reprobed with cDNA encoding GAPDH. As seen in FIGURE 4, P450 4A1 mRNA levels were greatly increased in both dexamethasone- and clofibrate-treated proximal tubules, implicating a P450 4A1 or P450 4A1-like protein in the increased enzymatic activity of 20-HETE formation. No change in GAPDH mRNA was seen under these conditions (data not shown).

DISCUSSION

The present study further characterizes arachidonic acid ω-hydroxylation in the proximal tubule, the nephron segment that has been previously shown to possess the

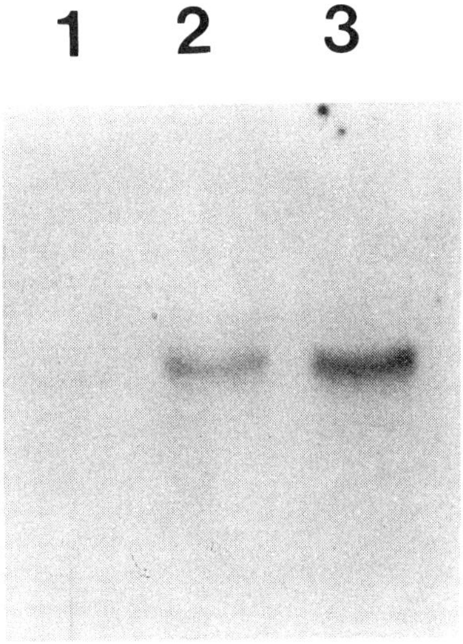

FIGURE 4. Northern blot of proximal tubular RNA using the rat P450 4A1 cDNA as a probe. Total RNA (10 μg) was extracted from proximal tubules of control (**lane 1**) and dexamethasone- (**lane 2**) or clofibrate- (**lane 3**) treated rats. Hybridization was performed as described in METHODS.

highest level of this activity along the nephron.[10] Using a different isolation technique, we demonstrated again that 20-HETE, the ω-hydroxylation metabolite of arachidonic acid, is the major metabolite in the proximal tubule. Furthermore, we provided evidence for its endogenous production in the proximal tubule using a selected ion monitoring NCI-GC/MS method. In fact, the capacity of the tubules to generate 20-HETE under basal conditions (22.5 pmol/mg in 30 min) may be sufficient to reach physiological concentrations. The specific activity of arachidonic acid ω-hydroxylation in homogenates from freshly isolated proximal tubules was more than 50% of that found in microsomes from cortex, a region where the proximal tubule makes up roughly 70% of structural components. This comparison is not quite adequate since it is made between homogenates and microsomes, the latter being a further purification step for measuring P450 activity. However, one would assume that if proximal tubules are the sole source of 20-HETE in the kidney cortex, then specific activity should be higher when the enzyme source is enriched. Preliminary experiments and reports by others[9,37,38] demonstrated that renal cortical microvasculature possess a significant arachidonic acid ω-hydroxylase activity, suggesting an additional source for 20-HETE in the kidney cortex.

That the conversion of arachidonic acid to 20-HETE in the proximal tubule homogenates was undetected in the absence of NADPH, unaffected by a cyclooxgenase inhibitor such as indomethacin, but inhibited by 17-ODYA and 7-ER and by antibodies against P450 4A1 strongly argues for a cytochrome P450-dependent reaction in the formation of 20-HETE. 17-ODYA is a mechanism-based inhibitor of the arachidonic acid ω-hydroxylation by cytochrome P450,[33] and as such it acts as a strong inhibitor of 20-HETE synthesis in the proximal tubule. 7-ER, on the other hand, is a substrate for several cytochrome P450 enzymes and is also an inhibitor of NADPH cytochrome P450 (c) reductase,[38] the essential component of the cytochrome P450-dependent oxidation reaction. Therefore, its effect would be dependent on the concentration of NADPH as well as its affinity for the hemoprotein involved in a particular reaction. Our results indicated a significant inhibition of 20-HETE synthesis in the proximal tubules in the presence of 7-ER. The level of cyclooxygenase in the proximal tubule is very low as evidenced by histochemical analysis[39] and enzymatic activity.[10] However, an increase in 20-HETE synthesis was observed when proximal tubules were incubated with indomethacin, suggesting an increase in substrate availability to the ω-hydroxylase and/or diminished metabolism of the formed 20-HETE. We have previously demonstrated that 20-HETE is a substrate for cyclooxygenase and that its endoperoxide metabolites are biologically active while its prostaglandin analogues are lacking measurable biological activity.[40] Thus, the presence of even low levels of cyclooxygenase in the proximal tubule may be important in either increasing or decreasing the biological activity of 20-HETE.

Of interest was the ability of anti-rat P450 4A1 antibody to inhibit 20-HETE synthesis in the proximal tubule. A simple interpretation would be that the P450 4A1 form is responsible for the arachidonic acid ω-hydroxylation to 20-HETE. However, the multiplicity of the P450 gene family and the close homology (75–90%) between the three major P450 4A fatty acid ω-hydroxylating hemoproteins (4A1, 4A2, and 4A3) in the rat kidney confound the interpretation of these results. The fact that anti-rat P450 4A1 antibody can detect 4A2 and 4A3[27] would argue that the latter may be equally important to 4A1 in generating 20-HETE. The same applies for the use of the full-length P450 4A1 cDNA probe to measure the level of 4A1 mRNA. It may also detect 4A2 and 4A3 mRNAs as well. Nevertheless, the experiments conducted in this study show that 4A1 or a 4A1-like protein is responsible for 20-HETE synthesis in the proximal tubules and that increased production of 20-HETE in this segment by dexamethasone and clofibrate is associated with increased level of 4A1 or 4A1-like mRNA. Since the molecular masses of these forms (51.0, 51.5, and 52.5 kDa for 4A1, 4A2, and 4A3, respectively) as well as their cDNA sequences differ slightly, additional studies using SDS-gel electrophoresis, specific antibodies, and specific oligonucleotide probes are needed to distinguish between the three forms to indicate which one or more may be important for the synthesis of 20-HETE.

We have previously shown that the production of 20-HETE in rat cortical microsomes is also age dependent. It shows a remarkable 60-fold increase between fetus and 5 weeks of age.[41,42] The sharp increase in production rate of 20-HETE may be accounted for by either an increase in enzyme activity brought about by cellular and/or extracellular factors/hormones or by an induction of *de novo* enzyme synthesis by affecting transcriptional and/or translational steps. This unique increase of 20-HETE production suggests the involvement of a specific and distinct P450 isozyme that is under developmental control and that specifically metabolizes AA to 20-HETE, possibly implicating this eicosanoid in the developmental growth of the nephron.

Thus, our findings and those of others concerning 20-HETE as the major endogenous AA metabolite in cortical structures, where cyclooxygenase activity is low, suggest a possible role for 20-HETE in the regulation of renal function. 20-HETE is an excellent substrate of cyclooxygenase.[40] In fact, its vasoreactivity in isolated rat aorta and rabbit mesenteric, carotid, and renal arteries is attributed to its conversion by endothelial cyclooxygenase activity to the vasoconstrictor metabolites, the 20-OH-endoperoxides.[20,40] It may affect renal function directly or via its conversion to biologically active prostanoids; or it can compete and interfere with AA metabolism to prostanoids that are important in the modulation of renal physiology and pathophysiology. It is, therefore, important to understand the mechanisms underlying its production in the kidney.

SUMMARY

20-Hydroxy-5,8,11,14-eicosatetraenoic acid (20-HETE) is a major cytochrome P450-dependent arachidonate metabolite in the rat kidney. In the present study we characterized the formation of 20-HETE in the proximal tubule, the nephron segment with the highest concentration of cytochrome P450 activities, including P450 arachidonic acid metabolism. Freshly isolated tubules showed a basal formation of 20-HETE, implying that it is an endogenous constituent of the proximal tubule. Conversion of exogenous arachidonic acid to 20-HETE in proximal tubule homogenates was enzymatic and NADPH-dependent (i.e., 0 and 65.5 ± 1.1 pmol/mg/min in the absence and presence of NADPH, respectively). That its formation was not affected by indomethacin but inhibited following preincubation with 17-ODYA (17-octadecynoic acid) and 7-ER (7-ethoxyresorufin) suggested that a P450 monooxygenase activity was involved in its synthesis. This was further strengthened by the demonstration that antibody raised against the rat cytochrome P450 4A1, a major fatty acid ω-hydroxylase isozyme, inhibited 20-HETE formation, suggesting the involvement of a P450 4A1 or P450 4A1-like activity in this reaction. Pretreatment of rats with clofibrate and dexamethasone, inducers of the P450 4A gene family, yielded a twofold increase in the proximal tubular synthesis of 20-HETE as well as an increase in P450 4A1 mRNA. These results, together with previous demonstrations that 20-HETE vasoconstricts isolated blood vessels, namely, renal microvessels, and affects tubular ion transport, suggest a role for 20-HETE in the regulation of renal vascular tone and transport functions and further stress the importance of understanding the regulation of 20-HETE synthesis in the kidney.

REFERENCES

1. Endou, H. 1983. Distribution and some characteristics of cytochrome P450 in the kidney. J. Toxicol. Sci. **8:** 165-179.
2. Schwartzman, M. L., N. G. Abraham, M. A. Carroll, R. Levere & J. C. McGiff. 1986. Regulation of arachidonic acid metabolism by cytochrome P450 in rabbit kidney. Biochem. J. **238:** 283-290.
3. Sacerdoti, D., N. G. Abraham, J. C. McGiff & M. L. Schwartzman. 1987. Renal cytochrome P450-dependent metabolism of arachidonic acid in spontaneously hypertensive rats. Biochem. Pharmacol. **37:** 521-527.
4. Schwartzman, M. L., P. Martasek, A. R. Rios, R. D. Levere, K. Solangi, A. I. Goodman & N. G. Abraham. 1990. Cytochrome P450-dependent arachidonic acid metabolism in human kidney. Kidney Int. **37:** 94-99.
5. Schwartzman, M. L., N. R. Ferreri, M. A. Carroll, E. Songu-Mize & J. C. McGiff. 1985. Renal cytochrome P-450-related arachidonate metabolites inhibit (Na^+-K^+)-ATPase. Nature **314:** 620.

6. Koop, D., A. Morrison & J. Douglas. 1988. Co-expression of cytochrome P450 and metabolism of arachidonic acid to novel products by early proximal tubular epithelial cells. Kidney Int. **33:** 62.
7. Lapuerta, L., N. S. Chacos, J. R. Falck, H. R. Jacobson & J. H. Capdevila. 1988. Renal microsomal cytochrome P-450 and the oxidative metabolism of arachidonic acid. Am. J. Med. Sci. **295:** 275-279.
8. Madhun, Z. T., D. A. Goldthwait, D. McKay, U. Hopfer & J. G. Douglas. 1991. An epoxygenase metabolite of arachidonic acid mediates angiotensin II-induced rises in cytosolic calcium in rabbit proximal tubule epithelial cells. J. Clin. Invest. **88:** 456-461.
9. Clark, J. E., R. J. Roman, Y.-H. Ma, K. Muerhoff, D. R. Dauser, D. R. Harder & B. S. Masters. 1990. Cytochrome P450-dependent metabolism of arachidonic acid by microsomes from renal vessels. FASEB J. **4:** A836.
10. Omata, K., N. G. Abraham & M. Laniado-Schwartzman. 1992. Arachidonic acid ω/ω-1 hydroxylase along the nephron of the spontaneously hypertensive rat. Am. J. Physiol. **262** (Renal Fluid Electrolyte Physiol. **31**): F592-599.
11. Carroll, M. A., M. Schwartzman, J. Capdevila, J. R. Falck & J. C. McGiff. 1987. Vasoactivity of arachidonic acid epoxides. Eur. J. Pharmacol. **138:** 218-283.
12. Jacobson, H. R., S. Corona, J. Capdevila, N. Chacos, S. Manna, A. Womack & J. R. Falck. 1984. Effects of epoxyeicosatrienoic acid on ion transport in the rabbit cortical collecting tubule. *In* Prostaglandin and Membrane Ion Transport. P. Braquet, R. P. Garay, J. C. Frolich & S. Nicosia, Eds.: 311-318. Raven Press. New York, NY.
13. Takahashi, K., J. Capdevila, A. Karara, J. R. Falck, H. R. Jacobson & K. F. Badr. 1990. Cytochrome P450 arachidonate metabolites in rat kidney: Characterization and hemodynamic responses. Am. J. Physiol. **258:** F781-F789.
14. Escalante, B., D. Erlij & J. C. McGiff. 1991. Effect of cytochrome P450 arachidonate metabolites on ion transport in rabbit kidney loop of Henle. Science **251:** 799-801.
15. Katoh, T., K. Takahashi, J. Capdevila, A. Karara, J. R. Falck, H. R. Jacobson & K. F. Badr. 1991. Glomerular stereospecific synthesis and hemodynamic actions of 8,9-epoxyeicosatrienoic acid in rat kidney. Am. J. Physiol. **261:** F578-F586.
16. Henrich, W. L., J. R. Falck & W. B. Campbell. 1990. Inhibition of renin release by 14,15-epoxyeicosatrienoic acid in renal cortical slices. Am. J. Physiol. **258:** E269-E274.
17. Harris, R. C., T. Homma, H. R. Jacobson & J. H. Capdevila. 1990. Epoxyeicosatrienoic acids activate Na^+/H^+ exchange and are mitogenic in cultured rat glomerular mesangial cells. Cell Physiol. **144:** 429-437.
18. Escalante, B., J. R. Falck, P. Yadagiri, L. Sun & M. L. Schwartzman. 1988. 19(S)-hydroxyeicosatetraneoic acid is a potent stimulator of renal Na^+/K^+-ATPase. Biochem. Biophys. Res. Commun. **152:** 1269-1274.
19. Escalante, B., W. C. Sessa, J. R. Falck, P. Yadagiri & M. L. Schwartzman. 1989. Vasoactivity of 20-hydroxyeicosatetraenoic acid is dependent on metabolism by cyclooxygenase. J. Pharmacol. Exp. Ther. **248:** 229-232.
20. Escalante, B., K. Omata, W. Sessa, S.-G. Lee, J. R. Falck & M. L. Schwartzman. 1993. 20-Hydroxyeicosatetraenoic acid is an endothelium-dependent vasoconstrictor in rabbit arteries. Eur. J. Pharmacol. **235:** 1-7.
21. Ma, Y.-H., D. Gebremedihin, M. L. Schwartzman, J. R. Falck, J. E. Clark, B. S. S. Masters, D. R. Harder & R. J. Roman. 1993. 20-Hydroxyeicosatetraenoic acid is an endogenous vasoconstrictor of canine renal arcuate arteries. Circ. Res. **72:** 126-136.
22. Gibson, G. G. 1989. Comparative aspects of the mammalian cytochrome P450 IV gene family. Xenobiotics **19:** 1123-1148.

23. JOHNSON, E. F., D. L. WALKER, K. J. GRIFFIN, J. E. CLARK, R. T. OKITA, A. S. MUERHOFF & B. S. MASTERS. 1990. Cloning and expression of three rabbit kidney cDNAs encoding lauric acid ω-hydroxylase. Biochemistry **29:** 873-879.
24. SHARMA, R. K., M. V. DOIG, D. F. V. LEWIS & G. G. GIBSON. 1989. Role of hepatic and renal cytochrome P450 IV A1 in the metabolism of lipid substrates. Biochem. Pharmacol. **38:** 3621-3629.
25. TANAKA, S., S. IMAOKA, E. KUSUNOSE, M. KUSUNOSE, M. MAEKAWA & Y. FUNAE. 1990. ω and (ω-1) hydroxylation of arachidonic acid, lauric acid and prostaglandin A1 by multiple forms of cytochrome P450 purified from rat hepatic microsomes. Biochim. Biophys. Acta **1043:** 171-181.
26. LANIADO-SCHWARTZMAN, M., K. OMATA, F. LIN, R. K. BHATT, J. R. FALCK & N. G. ABRAHAM. 1991. Detection of 20-hydroxyeicosatetraenoic acid in rat urine. Biochem. Biophys. Res. Commun. **180:** 445-449.
27. OKITA, J. R., P. CASTLE & R. T. OKITA. 1993. Characterization of cytochrome P450 in liver and kidney of rats treated with DI-(2-ethylhexyl)phthalate (DEHP). J. Biochem. Toxicol. **8:** 135-144.
28. SCHWARTZMAN, M. L., K. L. DAVIS, J. C. MCGIFF, R. D. LEVERE & N. G. ABRAHAM. 1988. Purification and characterization of cytochrome P-450-dependent arachidonic acid epoxygenase from human liver. J. Biol. Chem. **263** (5): 2536-2542.
29. CHMCZYNSKI, P. & N. SACCHI. 1987. Single-step method of RNA isolation by acid guanidinum thiocyanate-phenol-chloroform extraction. Anal. Biochem. **162** (1): 156-159.
30. HARDWICK, J. P., B.-J. SONG, E. HUBERMAN & F. J. GONZALEZ. 1987. Isolation of complementary DNA sequence and regulation of rat hepatic lauric acid ω-hydroxylase (cytochrome $P450_{L\omega}$). Identification of a new cytochrome P450 gene family. J. Biol. Chem. **261:** 801-810.
31. GESEK, F. A., D. W. WOLFF & J. W. STRANDHOY. 1987. Improved separation method for rat proximal and distal renal tubules. Am. J. Physiol. **253** (Renal Fluid Electrolite Physiol. **22**): F358-F365.
32. LIN, F., M. L. SCHWARTZMAN & N. G. ABRAHAM. 1991. Characterization of inhibitors of cytochrome P450-dependent arachidonic acid metabolism. FASEB J. **5:** A1518.
33. MA, Y.-H., D. R. HARDER & R. J. ROMAN. 1191. Inhibitors of cytochrome P-450 attenuate the myogenic response of dog renal arcuate arteries. Circ. Res. **68:** 1154-1163.
34. SUNDSETH, S. S. & D. J. WAXMAN. 1993. Sex-dependent expression and clofibrate inducibility of cytochrome P450 4A fatty acid ω-hydroxylases. J. Biol. Chem. **267:** 3915-3921.
35. SHARMA, R. K., B. G. LAKE, R. MAKOWSKI, T. BRADSHAW, D. EARNSHAW, J. W. DALE & G. G. GIBSON. 1989. Differential induction of peroxisomal and microsomal fatty-acid-oxidizing enzymes by peroxisome proliferators in rat liver and kidney. Eur. J. Biochem. **184:** 69-78.
36. DECARLO, L., C. TEMU, W. SESSA, M. L. SCHWARTZMAN & A. NASJLETTI. 1991. Dexamethsone (DEX) increases renal arachidonic acid metabolism by a non-cyclooxygenase pathway. FASEB J. **5** (1): A608.
37. MA, Y.-H., D. GEBREMEDIHIN, M. L. SCHWARTZMAN, J. R. FALCK, J. E. CLARK, B. S. S. MASTERS, D. R. HARDER & R. J. ROMAN. 1993. 20-Hydroxyeicosatetraenoic acid is an endogenous vasoconstrictor of canine renal arcuate arteries. Circ. Res. **72:** 126-136.
38. CLARK, J. E., R. J. ROMAN, Y.-H. MA, D. R. HARDER & B. S. S. MASTERS. 1992. Cytochrome P450 ω-hydroxylation of fatty acid and eicosanoid by microsomes from renal arteries, medulla and cortex. FASEB J. **6** (2): A3661.
39. SMITH, W. L. & T. G. BELL. 1978. Immunohistochemical localization of the prostaglandin-forming cyclooxygenase in renal cortex. Am. J. Physiol. (Renal Fluid Electrolyte Physiol.) **4:** (5): F451-F457.

40. Laniado-Schwartzman, M., J. R. Falck, P. Yadagiri & B. Excalante. 1989. Metabolism of 20-hydroxyeicosatetraenoic acid by cyclooxygenase: Formation and identification of novel endothelium-dependent vasoconstrictor metabolites. J. Biol. Chem. **264:** 1165-1162.
41. Omata, K., N. G. Abraham, B. Escalante & M. Laniado-Schwartzman. 1992. Age-related changes in renal cytochrome P450 arachidonic acid metabolism in spontaneously hypertensive rats. Am. J. Physiol. **262** (Renal Fluid Electrolyte Physiol. **31**): F8-F16.
42. Tiefenthaler, M., F. Lin, N. G. Abraham & M. L. Schwartzman. 1992. Genetic expression and activity of cytochrome P450 arachidonate ω-hydroxylation during development. FASEB J. **6** (2): A3631.

Novel Transformations of HPETEs by Cytochrome P450s[a]

ALAN R. BRASH, MIN S. CHANG, COLIN D. FUNK, AND WENCHAO SONG

Division of Clinical Pharmacology
Vanderbilt University Medical Center
Nashville, Tennessee 37232-6602

The conversion of prostaglandin endoperoxides to the mediators thromboxane A_2 and prostaglandin I_2 is catalyzed by distinct cytochrome P450 enzymes.[1–3] In essence, these specialized P450s react with and rearrange the peroxy oxygens of an endoperoxide to produce the unstable mediators. By analogy, in the lipoxygenase pathways of arachidonic acid metabolism there is the potential for P450-dependent transformation of hydroperoxy fatty acids (HPETEs, hydroperoxyeicosatetraenoic acids). The conversion of fatty acid hydroperoxides to allene oxides is the first example of this type of metabolism to be established[4,5] (FIG. 1). Allene oxides are unstable epoxides that have been detected in plants and lower animals.[6] We purified and subsequently cloned one of the plant allene oxide synthases and identified the enzyme as a novel type of cytochrome P450.[5,6] We are interested in the relationship of this enzyme to other P450s and the potential for analogous transformations of HPETE substrates by the cytochromes P450 of higher animals.

PURIFICATION OF AN ALLENE OXIDE SYNTHASE

We isolated the enzyme from flaxseed,[4] the source in which the activity was originally reported in 1966.[7] The UV-visible spectrum of the purified enzyme is characteristic of the ferric form of a cytochrome P450 in the high-spin state. The carbon monoxide difference spectrum of the dithionite-reduced enzyme gave the 450-nm chromophore typical of a cytochrome P450. However, it was noticed that the absorbance did not develop fully, and subsequent studies by others have shown that the allene oxide synthase has a weak affinity for carbon monoxide compared to other P450s.[8]

MOLECULAR CLONING OF THE ALLENE OXIDE SYNTHASE OF FLAXSEED

To establish the relationship of the flaxseed allene oxide synthase to other P450s we obtained the sequence of the cDNA by a combination of RT-PCR and library

[a]This work was supported by National Institutes of Health Grant GM-49502.

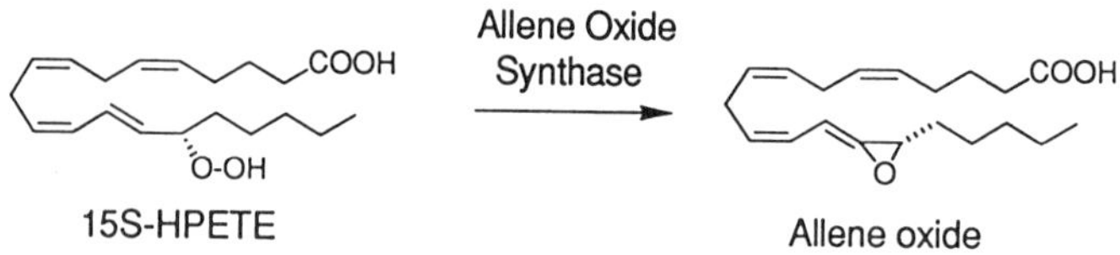

FIGURE 1. Allene oxide biosynthesis from 15S-HPETE.

screening approaches.[5] The deduced amino acid sequence revealed that the enzyme is related to the cytochrome P450 family of hemoproteins. The relationship is seen mainly in the C-terminal half of the protein, where a few sequences characteristic of P450s are located in the correct positions relative to one another in the allene oxide synthase. The heme-binding cysteine can be recognized near the carboxyl terminus. Overall, the enzyme has no more than 25% amino acid identity to any other cytochrome P450s. One unusual element of the structure of the allene oxide synthase is the absence of a conserved threonine residue in the region of the protein that normally helps comprise the oxygen binding pocket of a typical monooxygenase P450. Notably, the allene oxide synthase is designed to interact with a hydroperoxide substrate, and molecular oxygen is not involved in the reaction. This unusual property of the allene oxide synthase may correlate also with the relatively weak binding of carbon monoxide.

EPOXYALCOHOL SYNTHESIS BY A PURIFIED ALLENE OXIDE SYNTHASE

Some insight into the mechanism of the allene oxide synthase reaction derived from our observation that the metabolism of certain "unnatural" hydroperoxide substrates is associated with formation of epoxyalcohol by-products. When the 8R-hydroxy-15S-hydroperoxy derivative of eicosapentaenoic acid is metabolized by the flaxseed allene oxide synthase, in addition to the predicted allene oxide product (14,15-epoxide), the enzyme also forms a mixture of 13-hydroxy-14,15-epoxy derivatives of the substrate.[9] Significantly, both oxygens of the original 15-hydroperoxy group are retained in these epoxyalcohol products (^{18}O experiment). The synthesis of the epoxyalcohol by-products reveals that the allene oxide synthase has the potential to use one of the original hydroperoxy oxygens to hydroxylate an intermediate of the enzymatic reaction. This suggests a straightforward relationship of the catalytic properties of the allene oxide synthase to those of other P450s.

MECHANISTIC ANALOGIES OF ALLENE OXIDE SYNTHASE AND MONOOXYGENASE P450s

The prototypical P450 is a monooxygenase that hydroxylates its substrate. The initial steps of enzyme activation involve the splitting of an oxygen molecule and elimination of one half as H_2O using the reducing equivalents from NADPH. At this point the activated enzyme controls an iron-oxygen complex that is a very powerful

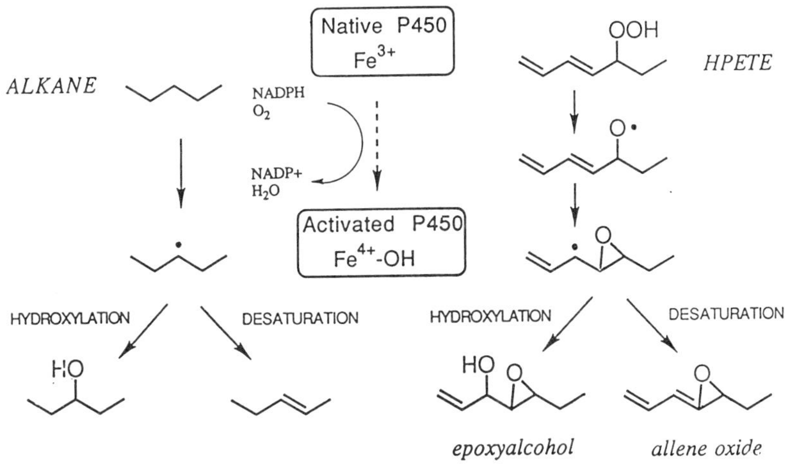

FIGURE 2. Mechanistic analogies in monooxygenase and allene oxide synthase P450s.

oxidant. With an appropriate substrate held in the active site, hydroxylation is nearly always the outcome of the enzyme-substrate interaction. Occasionally, however, desaturation of the substrate has been observed to compete with hydroxylation. This was reported in the metabolism of testosterone,[10] valproic acid,[11] and prostaglandin E_2.[12]

It is at the point when the enzyme has the potential to hydroxylate or desaturate the intermediate that there are close mechanistic parallels to the reactions of the allene oxide synthase (FIG. 2). The allene oxide synthase is activated by interaction with its hydroperoxide substrate. Usually reaction is completed by desaturation of the intermediate giving the allene oxide product. However, there does remain the potential for hydroxylation at this stage, and detection of the epoxyalcohol by-product is evidence of this type of interaction.

EPOXYALCOHOL FORMATION FROM 15-HPETE IN RAT LIVER MICROSOMES

As an initial approach to study of the P450-dependent transformations of HPETEs in higher animals, we reexamined the metabolism of 15-HPETE in rat liver microsomes. It was already known from the work of Estabrook and colleagues[13] that liver microsomes from phenobarbital-treated rats convert 15S-HPETE mainly to epoxyalcohols and their trihydroxy hydrolysis products. Our aim was to establish the stereochemistry of the transformation. We were looking for evidence of stereoselective formation of epoxyalcohols, as we had observed earlier in the reactions of the allene oxide synthase. In a nonenzymatic heme-catalyzed reaction with HPETEs, epoxyalcohols are formed as equimolar mixtures of diastereomers.

After incubation of 15S-HPETE with liver microsomes from phenobarbital-treated rats, the products were extracted and analyzed by reversed-phase HPLC. The profile

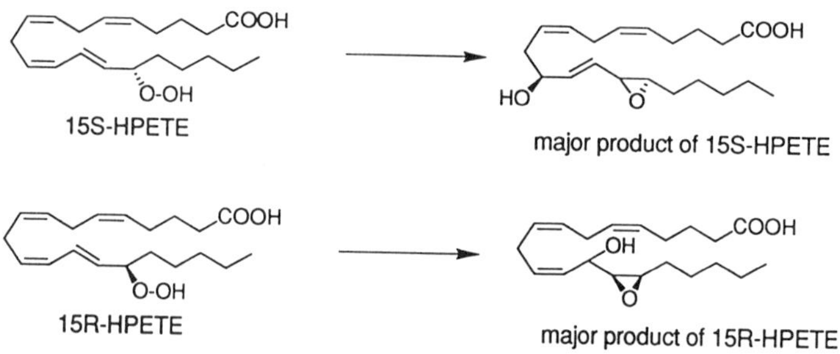

FIGURE 3. Main products of 15-HPETE metabolism in rat liver microsomes.

of epoxyalcohol products differed markedly from that obtained using free hematin. From the microsomal incubations, the major 15S-HPETE metabolite was identified as a diastereomer of the allylic epoxyalcohol 11S-hydroxy-14,15*trans*-epoxyeicosa-5Z,8Z,12E-trienoic acid (Fig. 3). A further indication of the stereoselectivity of the transformation stemmed from the observation that the 15R enantiomer of 15-HPETE is converted by liver microsomes to a distinctly different pattern of metabolites; the main product from 15R-HPETE was identified as 13-hydroxy-14,15*cis*-epoxyeicosa-5Z,8Z,11Z-trienoic acid (Fig. 3). The very likely involvement of P450s in catalyzing these reactions was indicated by the observation that very similar patterns of epoxyalcohol products were obtained upon incubation of 15S-HPETE or 15R-HPETE with purified cytochrome P450 2B1.

We examined the reversed-phase HPLC chromatograms carefully for evidence of allene oxide synthesis from the 15-HPETE substrates. The predicted stable end-product of allene oxide hydrolysis is an α-ketol (14-keto-15-hydroxyeicosatrienoic acid, a standard of which was prepared using the flaxseed allene oxide synthase). However, there was no trace of this product in extracts of the incubations of 15-HPETE with rat liver microsomes or purified P450 2B1.

CONCLUDING REMARKS

In mammalian tissues, there is no evidence as yet for the natural occurrence of P450-dependent transformations of lipoxygenase-derived hydroperoxides. The biosynthesis of allene oxides is confined to lower animals as far as we know, but it is clear that the formation of epoxyalcohols is a related reaction. While there are numerous reports of the synthesis of epoxyalcohols from arachidonic acid in mammalian tissues,[14] it is quite likely that many of these reactions involve heme-catalyzed breakdown of hydroperoxides; thus they will be nonenzymatic and nonstereoselective. Our studies in liver microsomes confirm that there is the potential to observe P450-dependent stereoselective transformations to epoxyalcohols. This is important, for if any of these reactions are to be of physiological significance, it might be expected that the product will be formed with some defined stereochemistry, and not with the lack of stereochemical control characteristic of the nonenzymatic reactions.

For the "natural occurrence" of mediator formation via these types of reactions, we assume that the tissue will need to contain a lipoxygenase to provide a stereospecific hydroperoxide substrate in addition to the P450 as catalyst of the second stereoselective step. This eliminates the liver as a likely source, as hepatocytes do not contain a lipoxygenase. There may be a better chance of uncovering an enzyme-controlled synthesis by concentrating on tissues reported to contain both a lipoxygenase and a P450; this includes, for example, tissues of the skin and the eye.[15–18] Reactions in these tissues should be examined for the stereoselectivity of the transformations and for the potential involvement of cytochrome P450s.

ACKNOWLEDGMENT

We thank Dr. F. P. Guengerich for the sample of pure cytochrome P450 2B1.

REFERENCES

1. Ullrich, V. & H. Graf. 1984. Prostacyclin and thromboxane synthase as P-450 enzymes. Trends Pharmacol. Sci. **5:** 352-355.
2. Yokoyama, C., A. Miyata, H. Ihara, V. Ullrich & T. Tanabe. 1991. Molecular cloning of human platelet thromboxane synthase. Biochem. Biophys. Res. Commun. **178:** 1479-1484.
3. Miyata, A., S. Hara, C. Yokoyama, H. Inoue, V. Ullrich & T. Tanabe. 1994. Molecular cloning and expression of human prostacyclin synthase. Biochem. Biophys. Res. Commun. **200:** 1728-1734.
4. Song, W-C. & A. R. Brash. 1991. Purification of an allene oxide synthase and identification of the enzyme as a cytochrome P450. Science **253:** 781-784.
5. Song, W-C., C. D. Funk & A. R. Brash. 1993. Molecular cloning of an allene oxide synthase: A cytochrome P450 specialized for the metabolism of fatty acid hydroperoxides. Proc. Natl. Acad. Sci. U.S.A. **90:** 8519-8523.
6. Song, W-C. & A. R. Brash. 1991. Allene oxide biosynthesis and metabolism. *In* Prostaglandins, Leukotrienes, Lipoxins, and PAF. J. M. Bailey, Ed. Plenum Press. New York, NY.
7. Zimmerman, D. C. 1966. A new product of linoleic acid oxidation by a flaxseed enzyme. Biochem. Biophys. Res. Commun. **23:** 398-402.
8. Lau, S-M. C., P. A. Harder & D. P. O'Keefe. 1993. Low carbon monoxide affinity allene oxide synthase is the predominant cytochrome P450 in many plant tissues. Biochemistry **32:** 1945-1950.
9. Song, W-C., S. W. Baertschi, T. M. Harris & A. R. Brash. 1993. Formation of epoxyalcohols by a purified allene oxide synthase. Implications for the mechanism of allene oxide synthesis. J. Biol. Chem. **268:** 6293-6298.
10. Nagata, K., J. Liberato, J. R. Gillette & H. A. Sasame. 1986. An unusual metabolite of testosterone. 17β-Hydroxy-4,6-androstadiene-3-one. Drug Metab. Dispos. **14:** 559-565.
11. Rettie, A. E., A. W. Rettenmeier, W. N. Howald & T. A. Baillie. 1987. Cytochrome P-450-catalyzed formation of delta-4-VPA, a toxic metabolite of valproic acid. Science **235:** 890-893.
12. Oliw, E. H. & Ø. Johnsen. 1988. Rapid and slow hydroxylators of seminal E prostaglandins. Biochim. Biophys. Acta **963:** 295-301.

13. WEISS, R. H., J. L. ARNOLD & R. W. ESTABROOK. 1986. Transformation of an arachidonic acid hydroperoxide into epoxyhydroxy and trihydroxy fatty acids by liver microsomal cytochrome P-450. Arch. Biochem. Biophys. **252:** 334-338.
14. PACE-ASCIAK, C. R. & S. ASOTRA. 1989. Biosynthesis, catabolism, and biological properties of HPETEs, hydroperoxide derivatives of arachidonic acid. Free Radical Biol. Med. **7:** 409-433.
15. NUGTEREN, D. H. & G. A. A. KIVITS. 1987. Conversion of linoleic acid and arachidonic acid by skin epidermal lipoxygenases. Biochim. Biophys. Acta **921:** 135-141.
16. BICKERS, D. R., T. DUTTA-CHOUDHURY & H. MUKHTAR. 1982. Epidermis: A site of drug metabolism in neonatal rat skin. Studies on cytochrome P-450 content and mixed-function oxidase and epoxide hydrolase activity. Mol. Pharmacol. **21:** 239-247.
17. WILLIAMS, R. N., P. BHATTACHERJEE & K. E. EAKINS. 1983. Biosynthesis of lipoxygenase products by ocular tissues. Exp. Eye Res. **36:** 397-402.
18. SCHWARTZMAN, M. L. & N. G. ABRAHAM. 1990. Ocular cytochrome P450 metabolism of arachidonate: Synthesis and bioassay. Methods Enzymol. **187:** 372-384.

Suicide Inactivation of Leukotriene A_4 Hydrolase/Aminopeptidase[a]

F. A. FITZPATRICK,[b] ROBERT LEPLEY,[b]
LARS ORNING,[c] AND KEVIN DUFFIN[d]

[b]*University of Colorado Health Sciences Center*
Department of Pharmacology C236
4200 East Ninth Avenue
Denver, Colorado 80262

[d]*Monsanto Company*
Physical Science Center
Chesterfield Village Parkway
St. Louis, Missouri 63110

LEUKOTRIENE MEDIATORS OF HYPERSENSITIVITY AND INFLAMMATION

Leukotrienes (LT) are lipid mediators of inflammation.[1-3] LTB_4 stimulates adhesion of circulating neutrophils to vascular endothelium, directs their migration to sites of inflammation,[4] catalyzes the release of their granule constituents, and produces hyperalgesia.[5] The sulfidopeptide LTs (LTC_4, LTD_4, and LTE_4) cause vascular permeability and constrict bronchial smooth muscle.[1-3] Leukotriene biosynthesis originates from sequential enzymatic reactions involving: (i) phospholipases, which liberate arachidonic acid from membrane phospholipids; (ii) 5-lipoxygenase (5-LO), which produces 5(S)-hydroperoxyeicosatetraenoic acid (5-HPETE) and converts it into LTA_4, the pivotal intermediate in this biosynthetic pathway; (iii) LTC_4 synthase, which conjugates reduced glutathione and LTA_4, yielding LTC_4; and (iv) LTA_4 hydrolase, which hydrates LTA_4 stereospecifically, producing LTB_4. Nonenzymatic hydration yields biologically inactive 5,12- or 5,6-dihydroxyeicosatetraenoic acids (diHETE). Pathological actions of leukotrienes are best understood in terms of their roles in immediate hypersensitivity and asthma.[4-9]

REGULATION OF CELLULAR LEUKOTRIENE BIOSYNTHESIS: CRITICAL ROLE OF LTA_4 HYDROLASE IN THE ACCUMULATION AND REDISTRIBUTION OF LTA_4

LTA_4 hydrolase is a pivotal element in cellular leukotriene biosynthesis. Inactivation of LTA_4 hydrolase has two important consequences. First, its biologically active

[a]This work was supported by a component of United States Public Health Service Grant HLB PO1 34303 awarded to F.A.F.

[c]Current address: Nycomed Bioreg, Gaustadalleen 21, N-0371 Oslo, Norway.

product, LTB_4, declines. Second, its substrate, LTA_4, can accumulate.[10,11] In the latter case, neutrophils can secrete LTA_4, thereby enhancing transcellular biosynthesis of LTB_4 by erythrocytes[12] and of LTC_4 by platelets and endothelial cells.[13–15] An unusual enzymological process termed *suicide inactivation* regulates LTA_4 hydrolase activity.[17–21] Suicide inactivation of LTA_4 hydrolase can deplete the level of active enzyme in cells.[20] The blood cells, which are rich sources of LTA_4 hydrolase, erythrocytes,[22] and neutrophils,[23] do not synthesize proteins *de novo;* thus, depletion of LTA_4 hydrolase by suicide inactivation might last for the lifetime of the cell. Derangements in this suicide process would alter the accumulation and redistribution of LTA_4 in neutrophil-mediated inflammatory disorders.

SUICIDE INACTIVATION OF LTA_4 HYDROLASE

To clarify the biological significance of suicide inactivation of LTA_4 hydrolase we determined if it was an irreversible process accompanying catalysis. The kinetic and stoichiometric data satisfy several criteria defining a mechanism-based process (SCHEME 1).

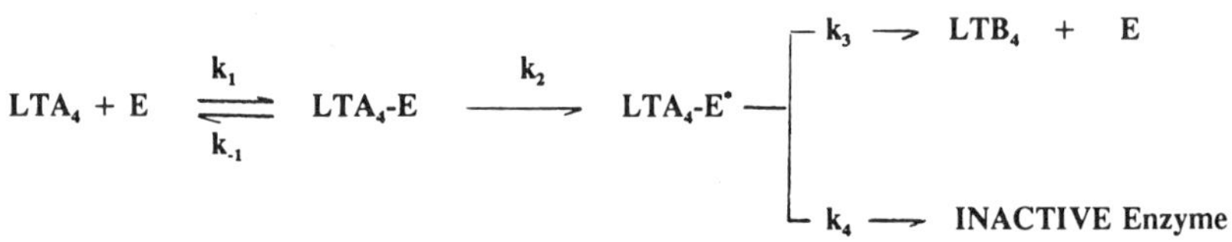

SCHEME 1. Mechanism-based inactivation of LTA_4 hydrolase/aminopeptidase.

Criteria for mechanism-based inactivation are: (1) time-dependent loss of LTA_4 hydrolase activity by pseudo-first-order kinetics; (2) a direct relationship between turnover and inactivation; (3) irreversible, covalent addition of LTA_4 with a 1 : 1 stoichiometry of inactivation; and (4) protection by competitive inhibitors demonstrating active-site specificity.

Enzymatic hydration of LTA_4 starts instantaneously and ends abruptly, with maximal LTB_4 formation by 1 min. Consistent with criterion 1, there is a corresponding exponential decline in LTA_4 hydrolase activity, which is inversely related to product formation: half-maximal LTB_4 formation and half-maximal loss of LTA_4 hydrolase activity each occur at 21 sec (FIG. 1). Inactivation was saturable with apparent inactivation rate constants, $k_{inact} = 0.8\ min^{-1}$ for LTA_4 and 1.1 min^{-1} for LTA_4 methyl ester. The apparent binding constant, at 2 °C, was $K_I = 3.7\ \mu M$ for LTA_4 and 3.6 μM for LTA_4 methyl ester. The apparent $K_m = 3.6\ \mu M$ was indistinguishable from K_I, consistent with the participation of equivalent enzyme-substrate complexes in both turnover and mechanism-based inactivation. The loss of catalytic activity accompanying LTB_4 formation and the pseudo-first-order kinetics typify a mechanism-based process.

To examine criterion 2, we determined the partition ratio from experiments involving multiple, consecutive additions of 5 μM LTA_4 to LTA_4 hydrolase (1.0 ml, 6 μ/ml) at 5-min intervals. The partition ratio (k_3/k_4) was calculated by plotting residual enzymatic activity as a function of LTB_4 formation per μM enzyme. Extrapolation of

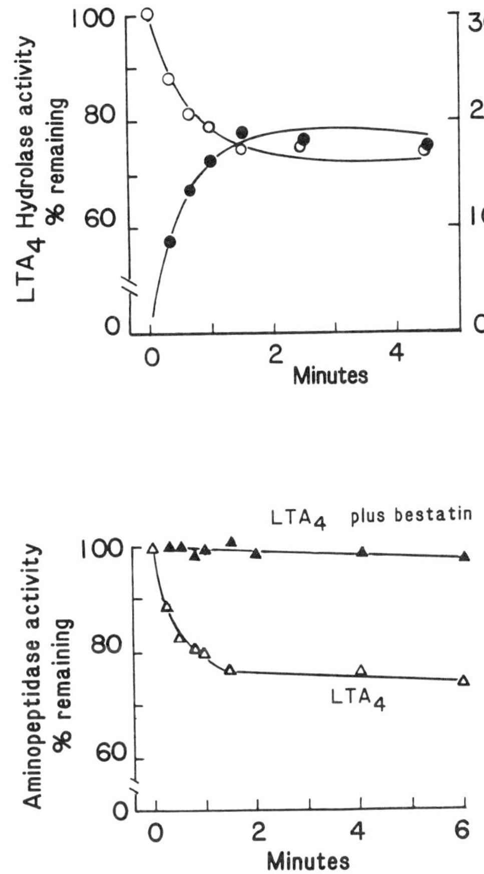

FIGURE 1. Upper panel: Inactivation of LTA_4 hydrolase activity by LTB_4 formation (*filled circles*); enzyme activity remaining after incubation with LTA_4 (*open circles*). Conditions: Enzyme (20 μg/ml) in 50 mM Tris, pH 8, LTA_4 (20 μM), 2 °C. **Lower panel:** Inactivation of aminopeptidase by LTA_4 (*open triangles*); "protection" of aminopeptidase by bestatin (*filled triangles*). Conditions: Enzyme (165 μg/ml) in 50 mM Tris, pH 8.0; LTA_4 (220 μM), bestatin (4 μM). Aliquots were diluted 100-fold before aminopeptidase assay. Data shown are from experiments with enzyme purified from human leukocytes. Similar data were obtained with recombinant enzyme.

this line to the abscissa (total inactivation) gives an estimate of the average number of turnover events per inactivation event. LTA_4 hydrolase activity declined exponentially. Plots of activity remaining versus turnover (μmol LTB_4 formed/μmol enzyme) show that inactivation was linearly related to catalysis.[20] The value for the average number of substrate turnovers before inactivation, that is, the partition ratio (k_3/k_4) = 129 ± 16 (mean ± SE, n = 13 experiments). Under conditions where V_{max} reflects k_{cat} the partition ratio is equivalent to k_{cat}/k_{inact}. As reported above k_{inact} = 0.8 min^{-1} for LTA_4.

From experiments using 10-s incubations at 2 °C, k_{cat} = 117 min^{-1}. The ratio k_{cat}/k_{inact} = 146 agrees with the value 129 ± 16 determined by measuring inactivation as a function of product formation.

We examined recombinant human LTA_4 hydrolase by electrospray ionization mass spectrometry to establish criterion 3. Enzyme alone (1 μM in 0.01 M Tris, pH 8) or enzyme incubated with 100 μM LTA_4 for 10 min was washed to remove all traces of Na^+ and then concentrated to approximately 10 μM enzyme. A portion of the incubation mixture (25 μl) was diluted with 25 μl of methanol : formic acid (50 : 1, v : v) and introduced at 4 μl/min with a syringe pump to a Sciex API III triple quadrupole electrospray ionization mass spectrometer equipped with an atmospheric pressure ion source to sample positive ions produced from a pneumatically assisted electrospray interface.[21] The average molecular mass of the pure enzyme and enzyme modified covalently during catalysis of LTA_4 were calculated from the multiply protonated ions in the mass spectrum, as described.[21] Electrospray ionization mass spectrometry of LTA_4 hydrolase showed a distribution of multiply protonated ions from m/z = 1543.2 to m/z = 2313.3, originating from a single species with a molecular weight = 69,399 ± 4, consistent with the amino acid content of the enzyme. After incubation of the enzyme with 100 μM LTA_4, ions from two separate molecular species were evident: the intact enzyme with a molecular weight of 69,399 and an adjacent peak representing a species with a molecular weight of 69,717 ± 4. The latter corresponds to the addition of one mole LTA_4 (MW = 318) per mole enzyme.[21]

To examine criterion 4 we first had to discover a competitive inhibitor of LTA_4 hydrolase. Malfroy *et al.*[24] and Vallee and Auld[25] facilitated our search by drawing attention to a homologous sequence of amino acids that occurs in LTA_4 hydrolase and the active site of certain peptidase enzymes. The disclosure that the enzymes share the unique "signature" sequence common among Zn^{++}-metallohydrolases prompted us to examine their functional resemblance. Among several peptidase inhibitors tested, bestatin was the most potent.[26,27] The reversible, competitive inhibitor bestatin protected the enzyme from irreversible inhibition by LTA_4. Enzyme alone or enzyme-inhibitor mixtures were incubated with LTA_4, then gel-filtered through PD-10 columns to separate enzyme from inhibitor, and, finally, analyzed for their remaining LTA_4 hydrolase activity. Enzyme incubated three times with 5 μM LTA_4 had a specific activity of 117 ± 3 nmol LTB_4/mg enzyme (mean ± SD, n = 3) after gel filtration. This was a 66% reduction ($p < 0.001$) compared to the control value of 340 ± 29 (n = 5) nmol LTB_4/mg enzyme, consistent with irreversible inactivation. Enzyme incubated with LTA_4 in the presence of bestatin had a specific activity of 280 ± 10 nmol LTB_4/mg enzyme after gel filtration, a value statistically indistinguishable from a control consisting of an untreated enzyme-bestatin mixture that had a specific activity of 275 ± 13 nmol LTB_4/mg enzyme after gel filtration. Bestatin also protected the aminopeptidase activity from inactivation by LTA_4 (FIG. 1).

CHARACTERIZATION OF LEUKOTRIENE A_4 HYDROLASE AS A BIFUNCTIONAL PEPTIDASE

LTA_4 hydrolase is also an aminopeptidase[26–29] belonging to a metallohydrolase superfamily.[22–33] This catalytic versatility suggests that LTA_4 hydrolase may have a

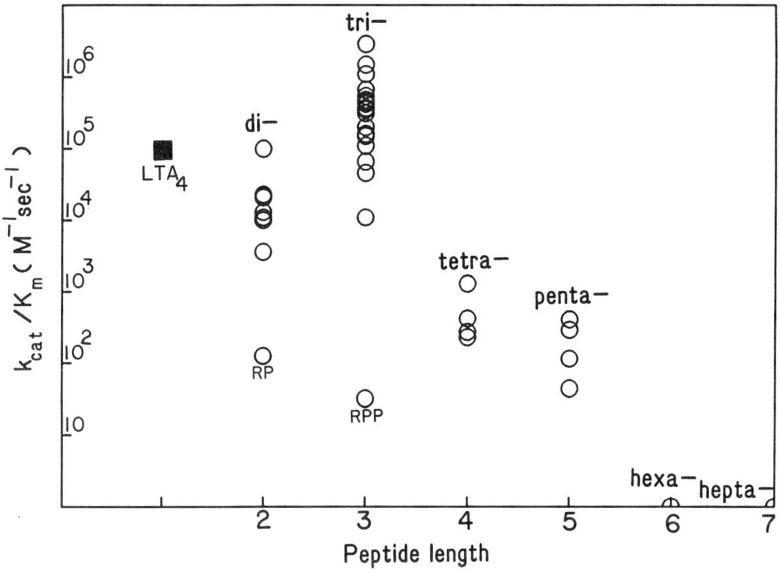

FIGURE 2. Catalytic specificity [k_{cat}/K_m] as a function of peptide length. *Circles* depict the k_{cat}/K_m for a battery of different di-, tri-, tetra-, penta-, hexa-, and hepta peptides hydrolyzed by LTA_4 hydrolase/aminopeptidase. Peptides with proline in the second position (RP and RPP) and the lipid substrate LTA_4 (*filled square*) are designated individually.

biological role, distinct from LTB_4 formation, involving its proteolytic trait. This implies that hydrolysis of peptide substrates might occur with efficiency and selectivity approaching that of the lipid substrate, LTA_4. We examined 50 representative peptides to clarify the substrate specificity. LTA_4 hydrolase (E.C. 3.3.2.6) removes the N-terminal amino acid from several tripeptides, typified by arginyl-glycyl-aspartic acid, arginyl-glycyl-glycine, and arginyl-histidyl-phenylalanine, with catalytic efficiencies (k_{cat}/K_m) $\geq 1 \times 10^6$ M^{-1} s^{-1}. This exceeds by 10-fold the k_{cat}/K_m for its lipid substrate, LTA_4. Catalytic efficiency declined for dipeptides that had k_{cat}/K_m ratios 10- to 100-fold lower than tripeptides. Tetrapeptides and pentapeptides were even poorer substrates, with catalytic efficiencies below 10^3 M^{-1} s^{-1} (FIG. 2). The enzyme preferentially hydrolyzed tripeptide substrates and single amino acid p-nitroanilides (pNA) with L-arginine at the N-terminus. Two known inhibitors of arginine peptidases, arphamenine A and B, inhibited hydrolysis of L-arginine p-nitroanilide with dissociation constants = 2.0 and 2.5 μM, respectively. The four peptides, RG, RGG, RGD, and RGDS, illustrate the relationship between peptide length and substrate specificity. The dipeptide RG has a K_m = 2.2 mM and k_{cat} = 23 s^{-1}. Addition of a third residue, either D- or G-, to the carboxyl terminus enhances substrate binding affinity: K_m = 0.43 mM for RGD and 0.21 mM for RGG. There is a corresponding increase in activity; k_{cat} = 184 s^{-1} for RGD and 205 s^{-1} for RGG. Conversely, addition of a fourth residue, S, to the carboxyl terminus reduces substrate binding affinity and activity: K_m = 8.6 mM and k_{cat} = 2 s^{-1} for RGDS. Data for 11 tripeptides with the formula

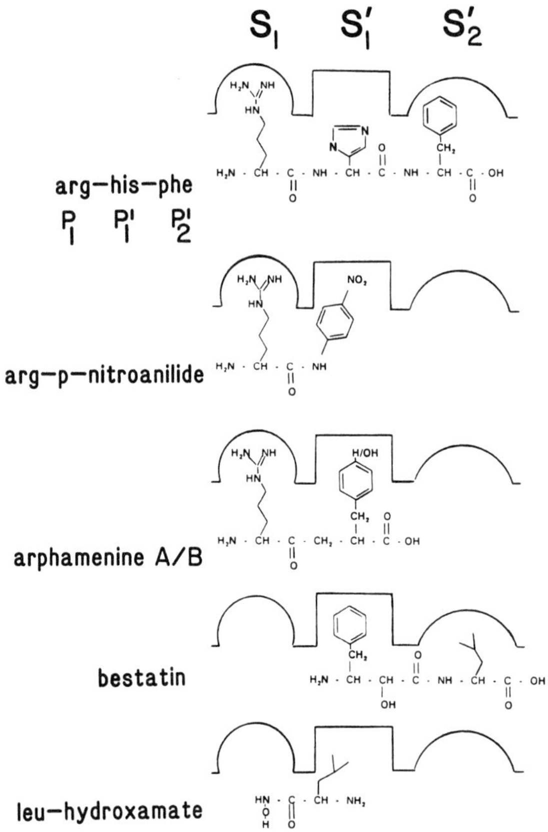

FIGURE 3. Proposed model for interpreting the interaction of substrates and inhibitors with three subsites (S_1, S_1' and S_2') of aminotripeptidase (E.C. 4.4.2.6).

xGG indicated that the effect of their N-terminal amino acid on catalytic selectivity was maximal for arginine, minimal for valine and glycine, and intermediate for leucine, lysine, methionine, alanine, and proline.

FIGURE 3 depicts our model to interpret the interaction of substrates and inhibitors with LTA_4 hydrolase/aminopeptidase. S_1, S_1', and S_2' represent three separate subsites on the enzyme that bind residues, designated P_1, P_1', and P_2', of any corresponding ligand. The active site is asymmetric. Aminopeptidase activity implies that subsites begin at S_1. The decline in k_{cat}/K_m for peptides longer than three residues and the lack of inhibitor effect of tetrapeptide analogs like amastatin support the conclusion that there is no S_3' subsite. Interpretation of our data (k_{cat}/K_m) according to this model shows that the enzyme prefers dipeptides and tripeptides with arginine at the N-terminal residue, P_1, consistent with the data for p-nitroanilide substrates. The low k_{cat}/K_m ratios for DpNA, EpNA and glutathione (γECG) indicate that S_1 does not accommodate acidic residues.

Although L-proline is acceptable at S_1 (PGG, PpNA), it is unacceptable at S_1' (RP, RPP), suggesting a requirement for hydrogen at the amide bond connecting P_1 and P_1'. Arphamenine A and B are analogs of RF and RY, respectively. Precedents show that they can bind to the S_1 and S_1' subsites of arginine peptidases.[34] Amino acid hydroxamates also bind to the S_1' substrate of peptidases. Another potent inhibitor of LTA_4 hydrolase/aminopeptidase, bestatin, is an FL analog. Inhibitor binding to arginine aminopeptidase, typified by bestatin, is not necessarily dominant at S_1 and S_1', where peptide and amino acid p-nitroanilide substrates bind. Consistent with a precedent from Harbeson and Rich,[34] it may involve S_1' and S_2'. Our data for bestatin (K_i = 174 nM) are consistent with the apparent preference for aromatic residues at P_1' interacting at the S_1' subsite. Like the S_1 subsite, the S_1' subsite does not accommodate acidic residues. The high-affinity binding of bestatin and nitrobestatin and their proposed interaction at S_1' and S_2' are consistent with this interpretation. This model may be useful for design of inhibitors of the enzyme.

The investigation of LTA_4 hydrolase/aminopeptidase from two distinct catalytic perspectives has facilitated the discovery of inhibitors and the characterization of their mechanism of action. It is interesting to note that our model can explain the fact that the active sites for peptides and LTA_4 substrates overlap, but are not identical.[35] In our model LTA_4 would be situated at the S_1' and S_2' subsites because they best accommodate its hydrophobic features and carboxyl group. The positioning of bestatin at the S_1' and S_2' sites is consistent with its potent inhibition of both catalytic processes of the enzyme, while the positioning of arphamenines at S_1 and S_1' is consistent with its potent inhibition of peptide hydrolysis, but not LTA_4 hydrolysis.

REFERENCES

1. Holtzman, M. J. 1991. Am. Rev. Respir. Dis. **143:** 188-203.
2. Lewis, R. J., K. F. Austen & R. Soberman. 1990. N. Engl. J. Med. **323:** 645-655.
3. Goetzl, E. J., D. Payan & D. Goldman. 1984. J. Clin. Immunol. **4:** 79-84.
4. Snyderman, R. & E. Goetzl. 1981. Science **213:** 837-839.
5. Levine, J. D., W. Lau, G. Kwiat, & E. Goetzl. 1984. Science **225:** 743-745.
6. Dahlen, S.-E., G. Hansson & P. Hedquist. 1983. Proc. Natl. Acad. Sci. USA **89:** 1712-1716.
7. Sautebin, L., T. Vigano, E. Grassi, M. Crivellari, G. Galli, F. Berti, M. Mezzetti & G. Folco. 1985. J. Pharmacol. Exp. Ther. **234:** 217-221.
8. Freeland, H. S., R. P. Schleimer, E. S. Schulman, L. M. Lichtenstein & S. P. Peters. 1988. Am. Rev. Respir. Dis. **138:** 389-394.
9. Voelkel, N. F., K. R. Stenmark, J. T. Reeves, M. M. Mathias & R. C. Murphy. 1988. J. Appl. Physiol Respir. Environ. Exercise Physiol. **57**(3): 860-867.
10. Jakschik, B. A. & C. G. Kuo. 1983. Prostaglandins **25:** 767-782.
11. Sun, F. F. & J. C. McGuire. 1984. Biochim. Biophys. Acta **794:** 56-64.
12. McGee, J. E. & F. A. Fitzpatrick. 1986. Proc. Natl. Acad. Sci. USA **83:** 1349-1353.
13. Marcus, A., M. J. Broekman, L. B. Safier, H. L. Ullman, N. Islam, C. Serhan, L. Rutherford, H. Korchak & G. Weismann. 1982. Biochem. Biophys. Res. Commun. **109:** 130-137.
14. Maclouf, J. & R. C. Murphy. 1988. J. Biol. Chem. **263:** 174-181.
15. Feinmark, S. & P. Cannon. 1986. J. Biol. Chem. **261:** 16466-16472.
16. McGee, J. E. & F. A. Fitzpatrick. 1985. J. Biol. Chem. **260:** 12832-12837.

17. Evans, J. F., D. Nathaniel, R. J. Zamboni & A. W. Ford-Hutchinson. 1985. J. Biol. Chem. **260:** 10966-10970.
18. Nathaniel, D. J., J. F. Evans, Y. Leblanc, C. Leveille, B. Fitzsimmons & A. W. Ford-Hutchinson. 1985. Biochem. Biophys. Res. Commun. **131:** 827-835.
19. Ohishi, N., T. Izumi, M. Minami, S. Kitamura, Y. Seyama, S. Ohkawa, S. Terao, H. Yotsumoto, F. Takaku & T. Shimizu. 1987. J. Biol. Chem. **262:** 10200-10205.
20. Orning, L., D. A. Jones & F. A. Fitzpatrick. 1990. J. Biol. Chem. **265:** 14911-14916.
21. Orning, L., J. Gierse, K. Duffin, G. Bild, G. Krivi & F. A. Fitzpatrick. 1992. J. Biol. Chem. **267:** 22733-22739.
22. Fitzpatrick, F., W. Liggett, J. McGee, S. Bunting, D. Morton & B. Samuelsson. 1984. J. Biol. Chem. **259:** 11403-11407.
23. Borgeat, P. & B. Samuelsson. 1979. Proc. Natl. Acad. Sci. USA **76:** 3213-3217.
24. Malfroy, B., H. Kado-Fong, C. Gros, B. Giros, J. Schwartz & R. Helmiss. 1989. Biochem. Biophys. Res. Comm. **161:** 236-241.
25. Vallee, B. & D. Auld. 1990. Biochemistry **29:** 5647-5659.
26. Orning, L., G. Krivi & F. A. Fitzpatrick. 1991. J. Biol. Chem. **266:** 1375-1378.
27. Orning, L., G. Krivi, G. Bild, J. Gierse, S. Aykent & F. A. Fitzpatrick. 1991. J. Biol. Chem. **266:** 16507-16511.
28. Minami, H., N. Ohishi, H. Mutoh, T. Izumi, H. Bito, H. Wada, Y. Seyama, H. Toh & T. Shimizu. 1990. Biochem. Biophys. Res. Comm. **173:** 620-626.
29. Haeggstrom, J., A. Wetterholm, B. Vallee & B. Samuelsson. 1990. Biochem. Biophys. Res. Comm. **173:** 431-437.
30. Wetterholm, A. & J. Z. Haeggstrom. 1992. Biochim. Biophsy. Acta **1123:** 275-281.
31. Orning, L. & F. A. Fitzpatrick. 1992. Biochemistry **31:** 4218-4223.
32. Haeggstrom, J., A. Wetterholm, R. Shapiro, B. Vallee & B. Samuelsson. 1990. Biochem. Biophys. Res. Comm. **172:** 965-970.
33. Toh, H., M. Minami & T. Shimizu. 1990. Biochem. Biophys. Res. Comm. **171:** 216-221.
34. Harbeson, S. & D. H. Rich. 1988. Biochemistry **27:** 7301-7310.
35. Orning, L., J. Gierse & F. A. Fitzpatrick. 1994. J. Biol. Chem. **269:** 11269-11273.

Metabolism of Leukotriene B_4 (LTB_4) and 12-Hydroxy-5,8,10,14-Eicosatetraenoic Acid (12-HETE) in Human Keratinocytes[a]

PAT WHEELAN,[b] JEFFREY B. TRAVERS,[c] JOSEPH G. MORELLI,[c] AND ROBERT C. MURPHY[b,d]

[b]*Department of Pediatrics*
National Jewish Center for Immunology and Respiratory Medicine
Denver, Colorado 80206

[c]*Department of Dermatology*
University of Colorado
School of Medicine
Denver, Colorado 80262

INTRODUCTION

Metabolism of leukotriene B_4 (LTB_4) is thought to be the major mechanism by which the biological activity of this important lipid mediator is terminated, and particular interest has centered around specific metabolic events that take place in human cells. Shortly after the discovery that human polymorphonuclear leukocytes synthesize LTB_4, it was found that this cell catalyzed degradation of LTB_4, initially by ω-oxidation utilizing a specific cytochrome P-450 followed by further ω-oxidation ultimately leading to 20-carboxy-LTB_4.[1] More recently, another metabolic pathway has been identified in certain human cells, including human lung macrophages,[2] monocytes,[3] and kidney mesangial cells[4] that oxidizes the 12-hydroxy group of LTD_4 with formation of the intermediate 12-oxo-LTB_4.

Powell and co-workers[5] found that an enzyme in the porcine neutrophil carried out not only the oxidation of LTB_4 into 12-oxo-LTB_4, but also catalyzed the metabolism of 12(S)-hydroxy-5,8,10,14-eicosatetraenoic acid [12(S)-HETE] as well as its enantiomer 12(R)-HETE into 12-oxo-ETE. They termed this enzymatic activity 12-hydroxy-eicosanoid dehydrogenase and found that it was associated with the microsomal fraction of the porcine neutrophil and used NAD^+ as oxidizing cofactor. A second, more specific enzyme was later isolated and purified from porcine kidney cells (PK-15 cells) with similar enzymatic properties, but Shimizu and co-workers[6] found that this enzyme was of cytosolic origin, required $NADP^+$ as oxidizing factor, and did not recognize 12(S)-HETE or 12(R)-HETE as substrates.

[a]This work was supported, in part, by a grant from the National Institutes of Health (HL 25785).

[d]Address for correspondence: Robert C. Murphy, Ph.D., Department of Pediatrics, National Jewish Center for Immunology and Respiratory Medicine, 1400 Jackson Street, Denver, Colorado 80206.

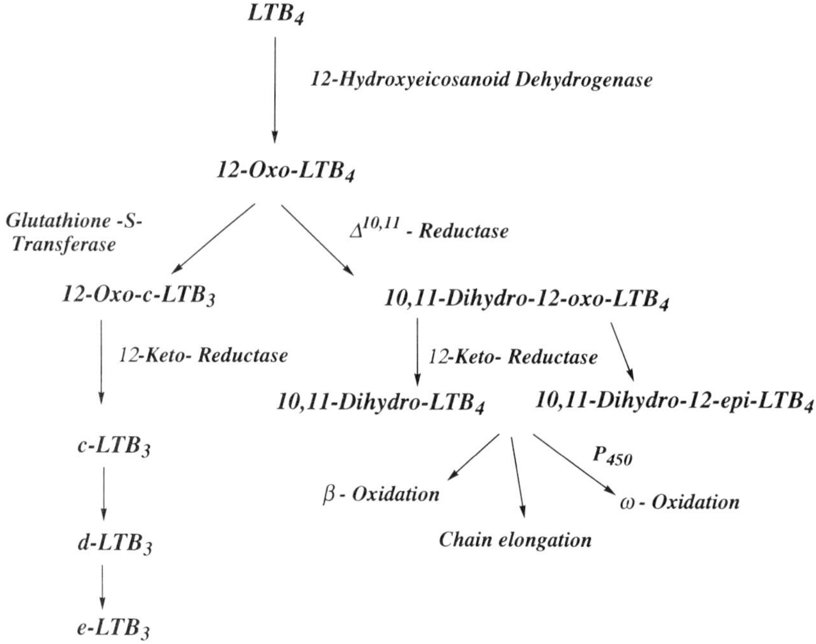

FIGURE 1. Metabolism of LTB_4 in human keratinocytes. LTB_4 metabolism occurs by two distinct pathways, the 10,11-reductase pathway and the glutathione conjugation pathway. The two pathways share the common intermediate 12-oxo-LTB_4.

This pathway of LTB_4 metabolism to 12-oxo-LTB_4 is typically characterized not by 12-oxo-LTB_4 as a major metabolite, but rather by metabolites that have a conjugated diene following reduction of the 10,11-double bond in LTB_4. Therefore, other enzymes utilize 12-oxo-LTB_4 as substrate, including a $\Delta^{10,11}$-reductase that converts a conjugated triene to a conjugated diene and a keto reductase that reduces the 12-oxo group to a 12-hydroxy moiety. Recently we found that the human keratinocyte is another cell that metabolizes LTB_4 into numerous products following initial oxidation of LTB_4 into 12-oxo-LTB_4.[7] Interestingly, two separate classes of metabolites were identified. One class of metabolites was characterized by having the 10,11-double bond of LTB_4 reduced, and a second novel pathway corresponded to glutathione conjugates of LTB_4 (FIG. 1).

$\Delta^{10,11}$-Reductase Pathway

The 10,11-reductase pathway of LTB_4 metabolism in the human keratinocyte results in the formation of numerous conjugated diene metabolites. The formation of these metabolites was consistent with metabolism of 12-oxo-LTB_4 by $\Delta^{10,11}$-reductase to form 10,11-dihydro-12-oxo-LTB_4 followed by a second reduction step by 12-keto reductase to form the two epimers 10,11-dihydro-LTB_4 and 10,11-dihydro-12-epi-LTB_4. These two compounds differ by the configuration of the hydroxyl group at the C12 position. It was further found that both isomers undergo epimerization when incubated with human keratinocytes. The 10,11-dihydro-LTB_4 epimers were proposed as intermediates in the formation of additional diene metabolites with further metabolism by ω-oxidation to generate 20-OH-10,11-dihydro-LTB_4, by carboxy terminus β-oxidation leading to 10-hydroxy-4,6,12-octadecatrienoic acid (10-HOTrE), and by chain elongation to produce 3,7,14-trihydroxy-8,10,16-docosatrienoic acid (3,7,14-triHDoTrE).

Glutathione Conjugation

The second pathway of LTB_4 metabolism in human keratinocytes following initial oxidation to 12-oxo-LTB_4 results in the formation of glutathione conjugates of LTB_4.[7] This metabolic pathway has not been identified in any other cell type and may be due to a uniquely high concentration of glutathione in human keratinocytes. Conjugation of glutathione at carbon 6 of 12-oxo-LTB_4 followed by reduction of the 12-oxo group by 12-keto-reductase results in the formation of the glutathione conjugate that has been termed c-LTB_3.[7] This is further metabolized by γ-glutamyltranspeptidase to the cysteinylglycine adduct termed d-LTB_3 and finally by a dipeptidase to form the cysteinyl adduct termed e-LTB_3.

Since all of the metabolites of LTB_4 generated by the human keratinocytes involve an intermediate formation of 12-oxo-LTB_4, it became of interest to investigate the formation of metabolites as a function of incubation parameters. Results revealed that LTB_4 metabolism occurred by both the $\Delta^{10,11}$-reductase pathway and the glutathione conjugation pathway to an equal extent at various time points, suggesting that formation of the intermediate 12-oxo-LTB_4 was the rate-determining intermediate. In addition, the substrate specificity of the initial dehydrogenase that would result in the formation of 12-oxo-LTB_4 has been investigated using 12(S)-HETE as substrate. Metabolism of 12(S)-HETE by the 12-hydroxyeicosanoid dehydrogenase described by Powell and co-workers[5] would be expected to result in the formation of 10,11-dihydro-12(R,S)-HETE, but if the enzyme was more closely related to that described by Shimizu and co-workers, 12(S)-HETE would not be reductively metabolized. The results reported here suggest that the human keratinocyte 12-hydroxyeicosanoid dehydrogenase does not recognize 12(S)-HETE as substrate, suggesting the more specific enzyme directed towards LTB_4 metabolism, likely the leukotriene B_4 12-hydroxydehydrogenase purified from cultured porcine kidney cells.[6]

METHODS

Leukotriene B_4 (LTB_4) and 12(S)-hydroxy-5,8,10,14-eicosatetraenoic acid [12(S)-HETE] were purchased from Cayman Chemical (Ann Arbor, MI) [5,6,8,9,11,12,14,15-$^3H_8$] LTB_4 (195 Ci/mmol) and [5,6,8,9,11,12,14,15-$^3H_8$] 12(S)-HETE (178 Ci/mmol) were purchased from Dupont-New England Nuclear (Boston, MA). Human keratinocytes were isolated from neonatal foreskin as described previously.[8] Keratinocytes were grown in 25-cm^2 tissue culture flasks at 37°C, 5% CO_2 in keratinocyte serum-free medium (KSFM, Gibco Laboratories, Grand Island, NY) until fully confluent (7.5-10.0 × 10^5 cells/flask). For LTB_4 time course studies, culture medium from 5 flasks was replaced with medium (4 mL/flask) containing LTB_4 (3.5 μM) and [$^3H_8$] LTB_4 (0.20 μCi/flask). Following incubation of each flask at 37°C for 4, 8, 12, 16, or 24 h, cell supernatant was removed and purified by reverse phase HPLC as described previously.[7] Metabolite identification was based on reverse phase HPLC retention time of radioactive metabolites and UV absorbance as determined earlier.[7] The sum of integration values for each peak in the radiochromatogram was used to determine total radioactivity at each time point, and the amounts of remaining LTB_4 and of each metabolite were calculated as the integrated value for those peaks as a percent of the total supernatant radioactivity. For 12(S)-HETE metabolism studies, culture medium was replaced with medium (5 mL/flask) containing 12(S)-HETE (3.1 μM) and [$^3H_8$] 12(S)-HETE (0.25 μCi/flask). Following incubation at 37°C for 24 h, cell supernatant was removed and purified by reverse phase HPLC and metabolites or hydrogenated metabolites derivatized to the pentafluorobenzyl ester/trimethyl silyl ether derivatives as described previously.[7] GC/MS analysis was performed using a Finnigan SSQ 70 (San Jose, CA); electron capture ionization (ECI) mass spectra (negative ions) were obtained using methane as the moderating gas, and electron ionization (EI) mass spectra (positive ions) were obtained using an electron energy of 70 eV. A 5 m × 0.25 mm DB-1 (J&W, Folsom, CA) GC capillary column was used with 0.25 μm film thickness. The injector temperature was maintained at 275°C and the transfer line at 300°C. The initial column temperature of 150°C was followed by a 15°C/min linear program to 300°C. Equivalent carbon values (EC values) were determined by comparison to PFB derivatives of straight-chain fatty acids.

RESULTS

The disappearance of LTB_4 in culture medium of human keratinocytes over a 24-h incubation period is shown in FIGURE 2A. At 24 h, less than 5% of LTB_4 remained unmetabolized. The disappearance of LTB_4 followed first-order kinetics with a half-life of approximately 5 h. The time course for the formation of metabolites resulting from glutathione conjugation is shown in FIGURE 2B. The c-LTB_3 metabolite (5,12-dihydroxy-6-glutathionyl-7,9,14-eicosatrienoic acid) was the most abundant metabolite formed following a 4-h incubation. Further accumulation of this metabolite occurred more slowly throughout 24 h as it was further metabolized by γ-glutamyl-transpeptidase to produce d-LTB_3 (5,12-dihydroxy-6-cysteinylglycyl-7,9,14-eicosatrienoic acid). Metabolism of d-LTB_3 by dipeptidase to form e-LTB_3 (5,12-dihydroxy-6-cysteinyl-7,9,14-eicosatrienoic acid) was evident only at 24 h, when the formation of e-LTB_3 accounted for 2% of the total radioactivity. The time course for the formation of the four major LTB_4-derived metabolites resulting from the 10,11-reductase pathway is shown in FIGURE 2C. The formation of 10,11-dihydro-LTB_4 as an intermediate in the formation of additional reduced metabolites was evidenced by the increased accumulation of this metabolite up to 16 h, followed by a decrease at 24 h. The chain-elongated metabolite 3,7,14-trihydroxy-8,10,16-docosatrienoic acid (3,7,14-triHDoTrE) initially accumulated as rapidly as 10,11-dihydro-LTB_4 and

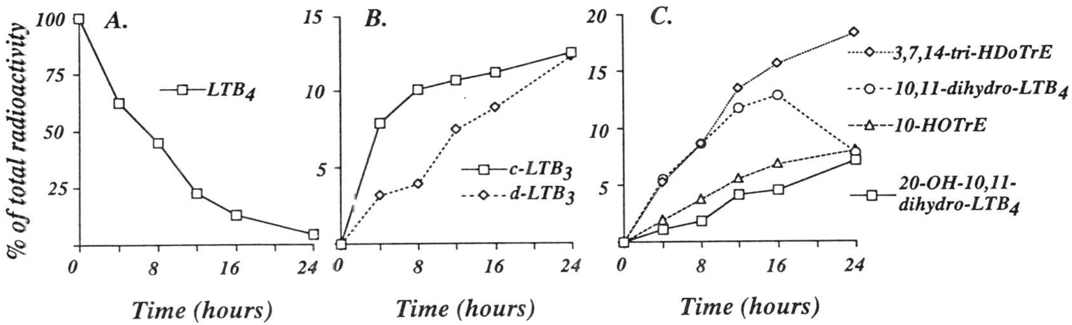

FIGURE 2. Time courses for the disappearance of LTB_4 and the accumulation of metabolites following LTB_4 incubation with human keratinocytes. **A.** Time course for the disappearance of LTB_4. **B.** Time course for the formation of products formed by the glutathione conjugation pathway: c-LTB_3 (5,12-dihydroxy-6-glutathionyl-7,9,14-eicosatrienoic acid) and c-LTB_3 (5,12-dihydroxy-6-cysteinylglycyl-7,9,14-eicosatrienoic acid). **C.** Time course for the formation of products resulting from the 10,11-reductase pathway: 3,7,14-triHDoTrE (3,7,14-trihydroxy-8,10,16-docosatrienoic acid), 10,11-dihydro-LTB_4, 20-OH-10,11-dihydro-LTB_4, and 10-HOTrE (10-hydroxy-4,6,12-octadecatrienoic acid).

continued to accumulate over 24 h to become the most abundant LTB_4-derived metabolite. Metabolism of 10,11-dihydro-LTB_4 by β-oxidation to form 10-hydroxy-4,6,12-octadecatrienoic acid (10-HOTrE) and by ω-oxidation to form 20-OH-10,11-dihydro-LTB_4 occurred at comparable rates. The accumulation of these products was, at all time points, less than that of the chain-elongated metabolite.

An additional minor metabolite observed only following 24-h incubation has been tentatively identified as 20-OH-10,11,14,15-tetrahydro-LTB_4. This metabolite eluted on reverse phase HPLC slightly after 20-OH-10,11-dihydro-LTB_4 and c-LTB_3 and contained a UV chromophore indicative of a conjugated diene. Negative-ion FAB mass spectral analysis of this metabolite resulted in an observed ion at *m/z* 355. The ECI GC/MS spectrum of the pentafluorobenzyl ester/trimethyl silyl ether (PFB/TMS) derivative (EC value = 25.9) revealed abundant ions at *m/z* 571 and *m/z* 481, and analysis of the derivatized, hydrogenated metabolite revealed an abundant ion at *m/z* 575. This data was consistent with an eicosadienoic acid metabolite containing three hydroxy substituents. EI GS/MS analysis of the PFB/TMS derivatized metabolite resulted in an observed ion at *m/z* 303, consistent with α-OTMS fragmentation at the C-11,C-12 bond and confirmed reduction of the C-14,C-15 double bond. This metabolite most likely results from reduction of the ω-6 double bond of 20-OH-10,11-dihydro-LTB_4.

Incubation of 12(S)-HETE with human keratinocytes resulted in the production of two major metabolites (retention times 35.0 min and 35.9 min) and several minor metabolites as identified by radioactivity monitoring following reverse phase HPLC separation (FIG. 3A). Both of the major metabolites contained a conjugated diene chromophore with λ_{max} = 234 nm (FIG. 3B). No evidence was found indicating formation of 12-oxo-ETE or a dihydro-12-HETE metabolite.

Electron capture ionization (ECI, negative ions) GC/MS analysis of the pentafluorobenzyl ester/trimethyl silyl ether (PFB/TMS) derivative of metabolite I (EC value = 21.1) resulted in an observed carboxylate anion, [A^-], at *m/z* 453 formed by loss of PFB from the molecular ion (FIG. 4A). Additional ions at *m/z* 363 [A^--TMSOH] and at *m/z* 291 [A^--TMSOTMS] indicated the presence of two derivatized hydroxy groups. ECI GC/MS analysis of the derivatized, hydrogenated metabolite (EC value = 21.6) resulted in observed ions at *m/z* 459, *m/z* 369, and *m/z* 279 (data not shown). These results were consistent with a dihydroxy C-18 carboxylic acid containing three double bonds. The electron ionization (EI, positive ions) GC/MS spectrum of derivatized metabolite I (FIG. 4B) contained abundant fragment ions at *m/z* 523 and *m/z* 433 (*m/z* 523-TMSOH). This was consistent with a C-10 derivatized hydroxy substituent, and the fragment ion contained an additional derivatized hydroxy substituent and two double bonds between C-10 and the PFB carboxylate terminus. EI GC/MS analysis of the PFB/TMS derivatized, hydrogenated metabolite located the second derivatized hydroxy substituent at C-3 by the observed ion at *m/z* 341 (FIG. 4C) and confirmed the C-10 position of the derivatized hydroxy substituent by the observed ions at *m/z* 527 and *m/z* 215. The above data is consistent with the assignment of metabolite I as 3,10-dihydroxyoctadecatrienoic acid.

The ECI GC/MS spectrum of PFB/TMS derivatized metabolite II (EC value = 17.6) showed a carboxylate anion, [A^-], at *m/z* 337 and additional ions at *m/z* 247, [A^--TMSOH], and at *m/z* 203, [A^--TMSOH-CO_2] (FIG. 5A). This was consistent with a C-16 carboxylic acid containing one derivatized hydroxy group and three double

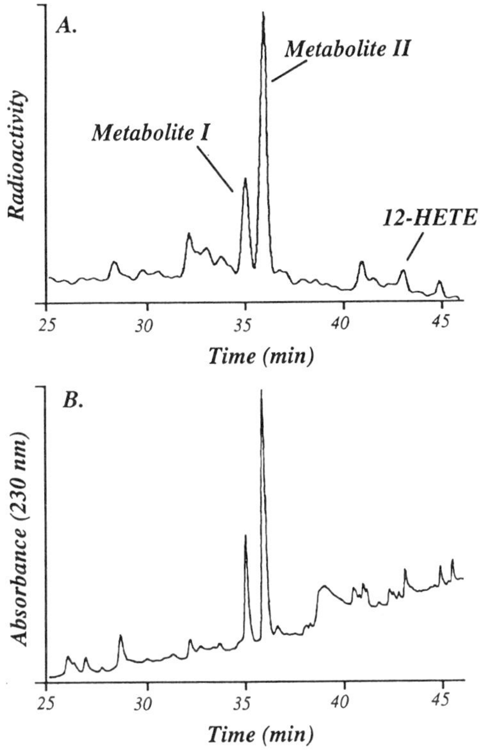

FIGURE 3. Reverse phase HPLC separation of 12(S)-HETE metabolites formed by human keratinocytes. **A.** Radioactivity monitor profile showing two major metabolites at retention times 35.0 min and 35.9 min. **B.** UV monitor profile at 230 nm.

bonds. The ECI GC/MS spectrum of the derivatized, hydrogenated metabolite II (EC value – 18.0) confirmed the presence of three double bonds by the observed ions at *m/z* 343 and *m/z* 253 (data not shown). The EI GC/MS spectrum located the derivatized hydroxy substituent at C-8 by the observed ion at *m/z* 407 (FIG. 5B), which was confirmed in the EIGC/MS spectrum of derivatized, hydrogenated metabolite II by the observed ions at *m/z* 411 and *m/z* 215 (FIG. 5C). The data was consistent with the assigned structure of metabolite II of 8-hydroxyhexadecatrienoic acid.

DISCUSSION

Incubation of human keratinocytes with LTB_4 results in an almost equal production of metabolites from both the glutathione conjugation pathway and the 10,11-reductase pathway at all time points over a 24-h period. The biological activity of these metabolites toward keratinocyte function is presently unknown; however, 10,11-dihydro-LTB_4 has been shown to be much less potent than LTB_4 as a neutrophil chemoattractant.[6,7] Possible biological activities of the novel metabolites formed by the glutathione conjugation pathway, c-LTB_3 and d-LTB_3, are presently under investigation due to the structural similarity between these metabolites and the potent lipid

Metabolite I

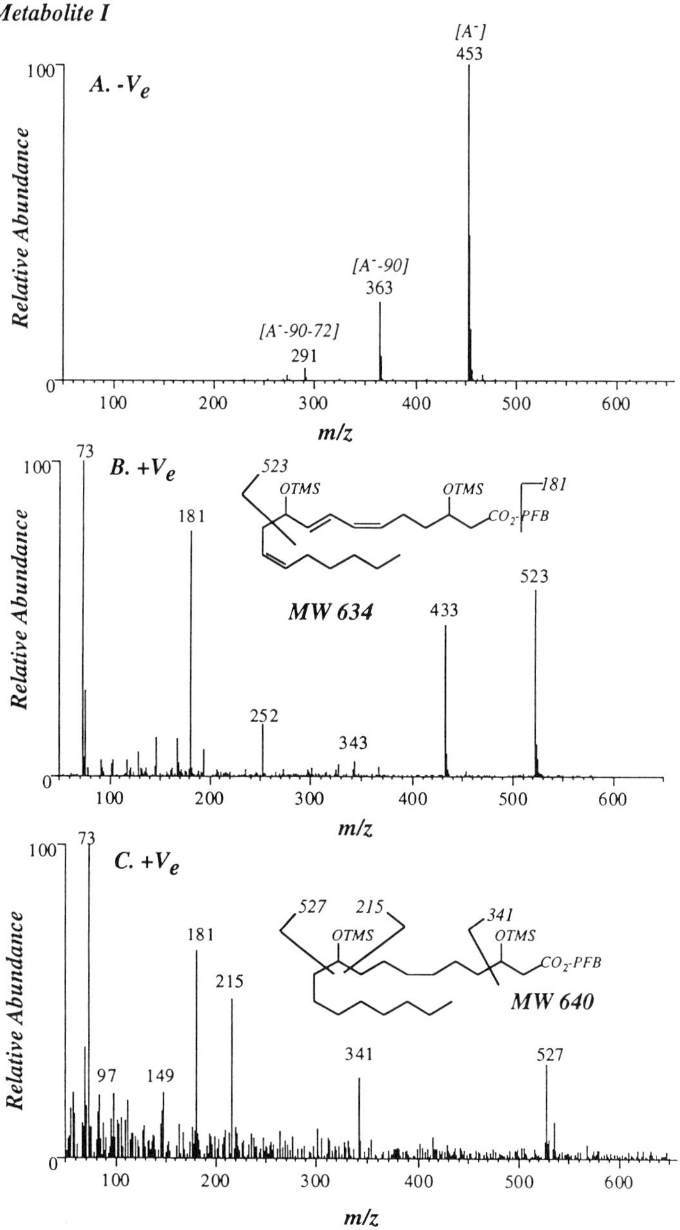

FIGURE 4. GC/MS analysis of PFB/TMS derivatized metabolite I from 12(S)-HETE incubation with human keratinocytes. **A.** Electron capture ionization spectrum showing the carboxylate anion, [A⁻], at *m/z* 453. **B.** Electron ionization mass spectrum showing the observed C-1 to C-10 fragment ion at *m/z* 523. **C.** Electron ionization mass spectrum of the derivatized, hydrogenated metabolite.

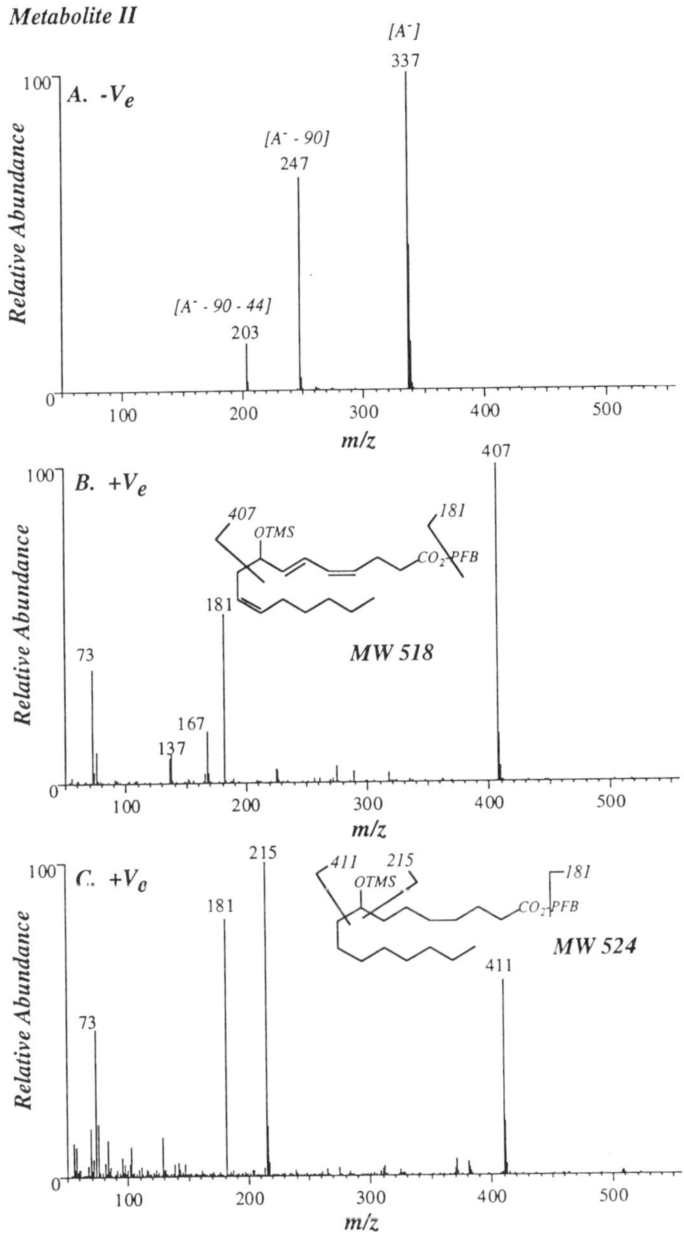

FIGURE 5. GC/MS analysis of PFB/TMS derivatized metabolite II from 12(S)-HETE incubation with human keratinocytes. **A.** Electron capture ionization spectrum showing the carboxylate anion, [A⁻], at *m/z* 337. **B.** Electron ionization mass spectrum showing the observed C-1 to C-8 fragment ion at *m/z* 407. **C.** Electron ionization mass spectrum of the derivatized, hydrogenated metabolite.

mediators LTC_4 and LTD_4. Both LTC_4 and LTD_4 are mitogenic for keratinocytes and also for melanocytes.[8,9] Interestingly, metabolism of LTC_4 to LTD_4 by γ-glutamyltrans-peptidase has been suggested to involve an ectoenzyme.[10] Therefore, observation of the metabolites d-LTB_3 and e-LTB_3 suggests release of c-LTB_3 from the keratinocyte prior to further metabolism to the peptide cleavage products.

Identification of an additional metabolite in the 10,11-reductase pathway, 20-OH-10,11,14,15-tetrahydro-LTB_4, suggests reduction of the C-14 double bond of 20-OH-10,11-dihydro LTB_4 and an additional enzymatic activity involved in LTB_4 metabolism. Analogous reduction of the ω-6 double bond has also been reported in porcine kidney metabolism of LTB_4 by identification of 12-oxo-10,11,14,15-tetrahy-dro-LTB_4.[3]

Incubation of human keratinocytes with 12(S)-HETE resulted in the formation of two metabolites, 3,10-dihydroxyoctadecatrienoic acid and 8-hydroxyhexadecatrienoic acid. The 8-hydroxyhexadecatrienoic acid metabolite has been identified previously as a 12-HETE metabolite produced by smooth muscle cells[11] and by MOLT-4 lymphocytes.[12] The 3,10-dihydroxyoctadecatrienoic acid metabolite was also identified during 12(S)-HETE metabolism by MOLT-4 lymphocytes. Location of double bonds has not been unambiguously assigned; however, based on the EI GC/MS fragmentation and the likelihood that these metabolites arise from β-oxidation at the carboxy terminus, it is assumed the remaining three double bonds retain the position and stereochemistry of the original double bonds in 12(S)-HETE. No evidence was found for metabolites produced from 12-hydroxyeicosanoid dehydrogenase metabolism of 12(S)-HETE. In particular, formation of 10,11-dihydro-12-HETE and 10,11-dihydro-12-oxo-ETE was not observed, although these were the major metabolites produced during 12-HETE incubation with porcine PMNL.[16] Although both porcine PMNL and human keratinocytes metabolize LTB_4 by the 12-hydroxyeicosanoid dehydrogenase pathway, the finding that human keratinocytes do not metabolize 12(S)-HETE by this pathway suggests human keratinocyte 12-hydroxyeicosanoid dehydrogenase is a more substrate-specific enzyme and more closely resembles the substrate specificity of the enzyme recently purified from porcine kidney that has been termed leukotriene B_4 12-hydroxydehydrogenase.[6]

ACKNOWLEGMENT

J.B.T. would like to thank Ms. Tamara Zekman for technical assistance in the cell culture work.

REFERENCES

1. SHAK, S. & I. M. GOLDSTEIN. 1984. ω-Oxidation is the major pathway for the catabolism of leukotriene B_4 in human polymorphonuclear leukocytes. J. Biol. Chem. **259:** 10181-10187.
2. SCHÖNFELD, W., B. SCHLÜTER, R. HILGER & W. KÖNIG. 1988. Leukotriene generation and metabolism in isolated human lung macrophages. Immunology **65:** 529-536.
3. FAULER, J., K.-H. MARX, V. KAEVER & J. C. FRÖLICH. 1989. Human monocytes convert leukotriene B_4 to two dihydro-leukotriene B_4 metabolites. Prostaglandins Leukotrienes Essent. Fatty Acids **37:** 193-196.

4. KAEVER, V., M. MARTIN, J. FAULER, K.-H. MARX & K. RESCH. 1987. A novel metabolic pathway for leukotriene B_4 in different cell types: Primary reduction of a double bond. Biochim. Biophys. Acta **922:** 337-344.
5. WAINWRIGHT, S. L. & W. S. POWELL. 1991. Mechanism for the formation of dihydro metabolites of 12-hydroxyeicosanoids. J. Biol. Chem. **266:** 20899-20906.
6. YOKOMIZO, T., T. IZUMI, T. TAKAHASHI, T. KASAMA, Y. KOBAYASHI, F. SATO, Y. TAKETANI & T. SHIMIZU. 1993. Enzymatic inactivation of leukotriene B_4 by a novel enzyme found in porcine kidney. J. Biol. Chem. **268:** 18128-18135.
7. WHEELAN, P., J. A. ZIRROLLI, J. G. MORELLI & R. C. MURPHY. 1993. Metabolism of leukotriene B_4 by cultured human keratinocytes. Formation of glutathione conjugates and dihydro metabolites. J. Biol. Chem. **268:** 25439-25448.
8. YOHN, J. J., J. G. MORELLI, S. J. WALCHAK, K. B. RUNDELL, D. A. NORRIS & M. R. ZAMORA. 1993. Cultured human keratinocytes synthesize and secrete endothelin-I. J. Invest. Dermatol. **100:** 23-26.
9. KAEVER, V., J. BRUUNS, J. WUNDER, B. DAMERAU, G. ZIMMER, J. FAULER, K. WESSEL, J. FLOEGE, N. TOPLEY, H. RADEKE & K. RESCH. 1990. Human glomerular mesangial cells inactivate leukotriene B_4 by reduction into dihydro-leukotriene B_4 metabolites. Life Sci. **46:** 1465-1470.
10. KUMLIN, M., J. R. FALCK, J. RAUD, Y. HARADA, S.-E. DAHLÉN & E. GRANSTRÖM. 1990. Identification and biological activity of dihydro-leukotriene B_4: A prominent metabolite of leukotriene B_4 in the human lung. Biochem. Biophys. Res. Commun. **170:** 23-29.
11. KRAGBALLE, K., L. DESJARLAIS & J. J. VOORHEES. 1985. Leukotriene B_4, C_4 and D_4 stimulate DNA synthesis in cultured human epidermal keratinocytes. Br. J. Dermatol. **113:** 43-52.
12. MORELLI, J. G., J. J. YOHN, M. B. LYONS, R. C. MURPHY & D. A. NORRIS. 1989. Leukotrienes C_4 and D_4 as potent mitogens for cultured human neonatal melanocytes. J. Invest. Dermatol. **93:** 719-722.
13. ISHIKAWA, T., M. MÜLLER, C. KLÜNEMANN, T. SCHAUB & D. KEPPLER. 1990. ATP-dependent primary active transport of cysteinyl leukotrienes across liver canalicular membrane. J. Biol. Chem. **265:** 19279-19286.
14. HADJIAGAPIOU, C., H. SPRECHER, T. L. KADUCE, P. H. FIGARD & A. A. SPECTOR. 1987. Formation of 8-hydroxyhexadecatrienoic acid by vascular smooth muscle cells. Prostaglandins **34:** 579-589.
15. HADJIAGAPIOU, C., J. TRAVERS, R. FERTEL & H. SPRECHER. 1992. β-oxidation of 12(S)-hydroxy-5,8,10,14-eicosatetraenoic acid by MOLT-4 lymphocytes. Arch. Biochem. Biophys. **292:** 112-120.
16. WAINWRIGHT, S., J. R. FALCK, P. YADAGIRI & W. S. POWELL. 1990. Metabolism of 12(S)-hydroxy-5,8,10,14-eicosatetraenoic acid and other hydroxylated fatty acids by the reductase pathway in porcine polymorphonuclear leukocytes. Biochemistry **29:** 10126-10135.

Interactions of PGH Synthase Isozymes-1 and -2 with NSAIDs[a]

WILLIAM L. SMITH,[b] ELIZABETH A. MEADE, AND DAVID L. DEWITT

Department of Biochemistry
Michigan State University
East Lansing, Michigan 48824

INTRODUCTION

Prostaglandin endoperoxide (PGH) synthase catalyzes the committed step in prostaglandin biosynthesis. During the last two years, it has become apparent that there are two different isoforms of PGH synthase, referred to as PGH synthase-1 (PGHS-1) and PGH synthase-2 (PGHS-2) or COX I and COX II (TABLE 1) encoded by separate genes.[1-3] The amino acid sequences of PGH synthase isozymes from a single species are about 60% identical; moreover, all of the residues required for catalysis by PGHS-1 are conserved in PGHS-2.[4] Thus, both enzymes probably catalyze reactions that are mechanistically identical. The two enzymes even exhibit the same K_m values for arachidonic acid.[4,5]

The biological basis for the existence of two distinct PGH synthases is not understood. The most dramatic difference between the two isozymes relates to their expression and its regulation. PGHS-1 is expressed "constitutively" in most tissues,[7] although at different levels in different cell types. Vascular endothelial cells[8] and platelets[9] express relatively large amounts of PGHS-1. PGHS-1 appears to be maintained at constant levels, although two- to four-fold changes can occur following stimulation with growth factors.[10] PGHS-2 is undetectable in most tissues, but its expression can be dramatically increased during inflammation or in fibroblasts and endothelial cells following exposure to mitogenic stimuli. Growth factors, phorbol esters, and IL-1 induce PGHS-2 in fibroblasts;[7,11-13] LPS stimulates PGHS-2 expression in monocytes and macrophages;[14-16] and tumor necrosis factor, phorbol esters, LPS, and IL-1 can induce PGHS-2 in vascular endothelial cells.[17,18] PGHS-2 is also increased in synoviocytes during inflammation,[19] and there is a dramatic transient increase in PGHS-2 in rat follicles immediately preceding ovulation.[20,21] In fact, it seems likely that PGHS-2 can be expressed in almost any cell or tissue during different stages of the cell cycle. Another feature that distinguishes PGHS-1 and PGHS-2 is the fact that induction of PGHS-2 can be inhibited by dexamethasone.[13,14,22,23] PGHS-2 appears to belong to a family of glucocorticoid-sensitive inflammatory-response genes.[24]

[a] This work was supported by National Institutes of Health Grants DK22042, DK42509, and GM40713, and National Institutes of Health Training Grant HL07404.

[b] Address for correspondence: William L. Smith, 510 Biochemistry Building, Department of Biochemistry, Michigan State University, East Lansing, Michigan 48824.

TABLE 1. Comparison of PGH Synthase-1 and PGH Synthase-2

	PGHS-1	PGHS-2
Homology	Approximately 60% identical, 75% homologous[9,11,28,32] Protein regions believed to be important for enzyme function that are conserved include: EGF homology domain,[33] heme ligand sites, aspirin acetylation ''active'' site, and glycosylation sites.[11,28,34–36] Enzymatic activities (V_{max}) and affinity for arachidonate (K_m) are essentially identical.[4,5]	
Regulation	Constitutive[11,12]	Inducible[11,12]
Range of expression	Can increase 2-4 fold[10,37–39]	Can increase 10-80 fold[11,12]
Tissue expression	Platelets,[9] endothelial cells,[17,18] stomach,[25] kidney,[25] smooth muscles,[25] most tissues[7]	Prostate,[7] brain,[7] activated monocytes,[14–16] or fibroblasts,[11,12] synoviocytes during inflammation,[19] follicles preceding ovulation;[20,21] can be expressed in most tissues but requires stimulation by growth factors, cytokines, phorbol esters, or hormones.
Effect of glucocorticoids	Little or none[14,22]	Inhibits expression[13,14,22,23]
Proposed role of enzyme	''Housekeeping gene,'' to produce prostaglandins involved in the autocrine/paracrine response of cells to circulating hormones, and that help regulate normal kidney and stomach function and vascular homeostasis.	1. ''Inflammatory response gene,'' member of family of genes induced during inflammation,[24] may produce prostaglandins involved in inflammation. 2. ''Immediate early gene'' member of gene family thought to control mitogenesis; may produce prostaglandins involved in cell growth.[26,27] 3. Produce prostaglandins essential for signal transduction regulating ovulation.[20,21]

PGHS-1 is hypothesized to be the enzyme involved in the synthesis of prostanoids involved in ''housekeeping'' functions such as attenuating responses to circulating hormones.[4] Using immunofluorescent staining, we have found that PGHS-1 is the PGH synthase isozyme expressed in kidney, stomach, and vascular smooth muscle;[25] Funk *et al.*[9] have shown that PGHS-1 is the isozyme expressed in platelets. PGHS-1 is most likely the enzyme involved in producing prostaglandins that regulate gastrointestinal and kidney functions and vascular homeostasis.

PGHS-2 is probably involved in the production of prostanoids involved in inflammation, differentiation, and mitogenesis. The induction of PGHS-2 by inflammatory agents and during inflammatory responses, in cells that mediate inflammation, and the reduction of expression of this enzyme by antiinflammatory agents such as dexamethasone provide circumstantial evidence to support this role. Because PGHS-2 is induced rapidly in fibroblasts in response to growth factors and because the PGHS-2 gene is superinduced in the presence of cycloheximide, the PGHS-2 gene qualifies as a member of the "immediate early" gene family.[26,27] Thus, prostaglandins produced by PGHS-2 may also be involved in regulating cellular proliferation.

Physiological roles for the two PGH synthase isozymes will be determined definitively only when specific inhibitors for each isozyme have been developed or when gene "knockout" animal models have been constructed that express only one of the two isozymes. PGHS-1 is inhibited by aspirin and most other nonsteroidal antiinflammatory drugs (NSAIDs). The discovery of PGHS-2 prompted us to define the pharmacology of the two enzymes. Our goals were to determine (a) if PGHS-2 is inhibited by NSAIDs and, if so, (b) whether there are any differences in the ability of common NSAIDs to inhibit PGHS-1 versus PGHS-2. Here, we review studies in which we have expressed cDNAs encoding murine PGHS-1 and PGHS-2 *in vitro* and examined their relative sensitivities to *instantaneous* inhibition by a group of common NSAIDs.

RESULTS AND DISCUSSION

The differences in the amino acid sequences between PGHS-1 and PGHS-2 suggested that there may be differences in their abilities to interact with fatty acid substrates and pharmacological inhibitors. To examine this possibility, we developed a system for expressing mouse PGHS-1 and PGHS-2 *in vitro*. cDNAs encoding each of the two murine PGH synthase isozymes were used to construct plasmid expression vectors that could be used for the expression of each of the isozymes in cos-1 cells.[6] Cos-1 cells were transfected with plasmid constructs containing the coding regions of either PGHS-1 or PGHS-2, and microsomal membranes were prepared from the cells. Cyclooxygenase activity was determined using an O_2 electrode. *Instantaneous* inhibition of cyclooxygenase activity was measured for a series of different NSAIDs, and an IC_{50} value was determined for each drug examined. Assays were performed using 10 μM arachidonate and 200 μM O_2; all other conditions were as previously described.[28] Oxygen electrode cuvettes were prepared with buffer, substrate, and varying concentrations of inhibitors, and the reactions were initiated by adding microsomal preparations of cos-1 cells expressing either murine PGHS-1 or PGHS-2. This ordered addition permitted us to eliminate the complication of time-dependent effects that can result from preincubation of inhibitors with the enzyme.[29,30]

Seven common NSAIDs from various chemical families of NSAIDs[5] were tested. All of these agents inhibited PGHS-2 as well as PGHS-1, indicating that PGHS-2 can be a target of NSAIDs. In addition, we found that some NSAIDs discriminated between PGHS-1 and PGHS-2 (FIG. 1). Indomethacin, piroxicam, and sulindac sulfide (the active metabolite of sulindac) were much better inhibitors of PGHS-1 than of PGHS-2. Ibuprofen, flurbiprofen, and meclofenamate each inhibited murine PGHS-1 and PGHS-2 with comparable potencies. 6-Methoxy-naphthyl-acetic acid (6-MNA), the active metabolite of nabumetone (Relafen®),[31] showed some preference for inhibiting PGHS-2.

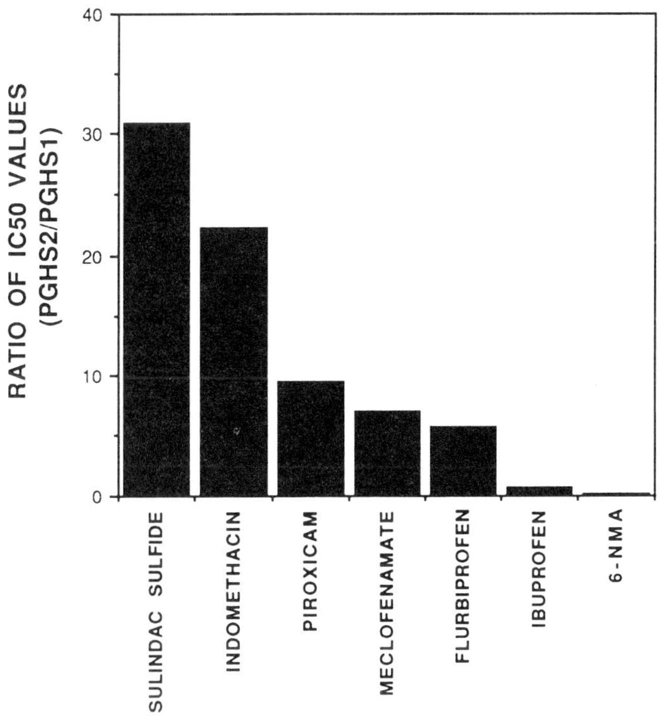

FIGURE 1. Relative selectivity of seven common NSAIDs for the murine PGHS-1 and PGHS-2 isozymes. Mouse PGHS-1 and PGHS-2 were expressed in cos-1 cells.[6] The inhibitor dose that caused 50% instantaneous inhibition of cyclooxygenase activity (IC_{50}) was determined by oxygen electrode assay of oxygen consumption.[6,28] The ratio of the IC_{50} for inhibition of PGHS-2 relative to the IC_{50} for PGHS-1 inhibition are plotted. All NSAIDs were better inhibitors of PGHS-1, except for 6-methoxy-naphthylacetic acid (6-MNA), the active metabolite of Nabumetone, which was about a sevenfold better inhibitor of murine PGHS-2.

We have recently extended our findings with the mouse PGHS isozymes to studies with the human homologs. Our preliminary results indicate that the patterns of inhibition observed with the murine enzymes also occur with the human PGHS isozymes. Although our results do establish that PGHS-1 and PGHS-2 interact differently with several NSAIDs, our measurements of instantaneous inhibition do not correlate with the antiinflammatory activities of various NSAIDs. Obviously, issues of pharmacokinetics need to be considered. At least one additional point is that some NSAIDs, including indomethacin, meclofenamate, and flurbiprofen, cause time-dependent inhibition of PGHS-1.[29,30] We have not investigated this aspect of the activities of these NSAIDs with PGHS-1 and PGHS-2, but the time dependence of inhibition may be important clinically.

In summary, our results establish that murine PGHS-1 and PGHS-2 are pharmacologically distinct; moreover, preliminary work indicates that the human isozymes exhibit the same patterns of differences observed with the murine isozymes.[25] These studies suggest that agents can be developed that will inhibit PGHS-1 and/or PGHS-2 specifically. Thus, it may be possible to identify NSAIDs that will block proinflammatory prostaglandin synthesis mediated via PGHS-2 but spare prostaglandin synthesis in stomach, presumably mediated via PGHS-1; that is, development of an NSAID lacking the ulcerogenic activity of the current generation of NSAIDs may now be possible.

SUMMARY

There are two isozymes of prostaglandin endoperoxide (PGH) synthase (cyclooxygenase) called PGH synthase-1 and -2 or COX I and II. Both isozymes catalyze the same two reactions: oxygenation of arachidonate to yield PGG_2 and reduction of PGG_2 to PGH_2. PGH synthase-1 is expressed constitutively and is found in most tissues. PGH synthase-2 is undetectable in most cells but can be induced in fibroblasts, endothelial cells, ovarian follicles, and macrophages by various mitogens, cytokines, and tumor promoters. PGH synthase-1 (PGHS-1) has been presumed to be the site of action of nonsteroidal antiinflammatory drugs (NSAIDs). However, the discovery of the second isozyme, PGH synthase-2 (PGHS-2), and its association with inflammation has suggested that this latter enzyme may be the therapeutic target of NSAIDs functioning in their antiinflammatory capacities. We have cloned cDNAs for murine PGHS-1 and PGHS-2, expressed these enzymes in cos-1 cells, and compared the relative sensitivities of the two isozymes to some common NSAIDs. Indomethacin, piroxicam, and sulindac sulfide were found to preferentially inhibit PGHS-1. Ibuprofen and meclofenamate inhibit both enzymes with comparable potencies. 6-Methoxy-2-naphthylacetic acid, the active metabolite of Relafen®, inhibits murine PGHS-2 preferentially. Aspirin irreversibly inhibits PGHS-1, preventing this isozyme from forming PGH_2 or any other oxygenated product; in contrast, aspirin treatment of PGHS-2 causes this enzyme to form 15-hydroxy-5c,8c,11c,13t-eicosatetraenoic acid (15-HETE) instead of PGH_2. Our results indicate mouse PGHS-1 and PGHS-2 are pharmacologically distinct. Thus, it should be possible to develop agents highly selective for each PGHS isozyme. PGHS-2 is not expressed in stomach but is increased by inflammatory cytokines in cells such as macrophages. Thus, a selective inhibitor of PGHS-2 could be an antiinflammatory agent but without being ulcerogenic.

REFERENCES

1. YOKOYAMA, C. & T. TANABE. 1989. Cloning of human gene encoding prostaglandin endoperoxide synthase and primary structure of the enzyme. Biochem. Biophys. Res. Commun. **165:** 888-894.
2. KRAEMER, S. A., E. A. MEADE & D. L. DEWITT. 1992. Prostaglandin endoperoxide synthase gene structure: Identification of the transcriptional start site and 5′-flanking regulatory sequences. Arch. Biochem. Biophys. **293:** 391-400.
3. FLETCHER, B. S., D. A. KUJUBU, D. M. PERRIN & H. R. HERSCHMAN. 1992. Structure of the mitogen-inducible TIS10 gene and demonstration that the TIS10-encoded protein is a functional prostaglandin G/H synthase. J. Biol. Chem. **267:** 4338-4344.

4. SMITH, W. L. 1992. Prostanoid biosynthesis and mechanisms of action. Am. J. Physiol. **263:** F181-F191.
5. MEADE, E. A., W. L. SMITH & D. L. DEWITT. 1993. Differential inhibition of prostaglandin endoperoxide synthase (cyclooxygenase) isozymes by aspirin and other non-steroidal anti-inflammatory drugs. J. Biol. Chem. **268:** 6610-6614.
6. MEADE, E. A., W. L. SMITH & E. W. DEWITT. 1992. Expression of the murine prostaglandin (PGH) synthase-1 and PGH synthase-2 isozymes in cos-1 cells. *In* Prostaglandins, Thromboxanes, Leukotrienes and Related Compounds: Update 1992. L. S. Wolfe & A. W. Ford-Hutchinson, Eds. Elsevier Science Publishers. Amsterdam. In press.
7. SIMMONS, D. L., W. XIE, J. G. CHIPMAN & G. E. EVETT. 1991. Multiple cyclooxygenases: Cloning of a mitogen-inducible form. *In* Prostaglandins, Leukotrienes, Lipoxins, and PAF. J. M. Bailey, Ed. 67-78. Plenum Press. New York, NY.
8. DEWITT, D. L., J. S. DAY, W. K. SONNENBURG & W. L. SMITH. 1983. Concentrations of prostaglandin endoperoxide synthase and prostaglandin I_2 synthase in the endothelium and smooth muscle of bovine aorta. J. Clin. Invest. **72:** 1882-1888.
9. FUNK, C. D., L. B. FUNK, M. E. KENNEDY, A. S. PONG & G. A. FITZGERALD. 1991. Human platelet/erythroleukemia cell prostaglandin G/H synthase: cDNA cloning, expression, and gene chromosomal assignment. FASEB J. **5:** 2304-2312.
10. DEWITT, D. L. 1991. Prostaglandin endoperoxide synthase: Regulation of enzyme expression. Biochim. Biophys. Acta **1083:** 121-134.
11. KUJUBU, D. A., B. S. FLETCHER, B. C. VARNUM, R. W. LIM & H. R. HERSCHMAN. 1991. TIS10, a phorbol ester tumor promoter inducible mRNA from Swiss 3T3 cells, encodes a novel prostaglandin synthase/cyclooxygenase homologue. J. Biol. Chem. **266:** 12866-12872.
12. KUJUBU, D. A., S. T. REDDY, B. S. FLETCHER & H. R. HERSCHMAN. 1993. Expression of the protein product of the prostaglandin synthase-2/TIS10 gene in mitogen-stimulated Swiss 3T3 cells. J. Biol. Chem. **268:** 5425-5430.
13. O'BANION, M. K., V. D. WINN & D. A. YOUNG. 1992. cDNA cloning and functional activity of a glucocorticoid-regulated inflammatory cyclooxygenase. Proc. Natl. Acad. Sci. USA **89:** 4888-4892.
14. LEE, S. H., E. SOYOOLA, P. CHANMUGAM, S. HART, W. SUN, H. ZHONG, S. LIOU, D. SIMMONS & D. HWANG. 1992. Selective expression of mitogen-inducible cyclooxygenase in macrophages stimulated with lipopolysaccharide. J. Biol. Chem. **267:** 25934-25938.
15. O'SULLIVAN, G. M., F. H. CHILTON, E. M. HUGGINS, JR. & C. E. MCCALL. 1992. Lipopolysaccharide priming of alveolar macrophages for enhanced synthesis of prostanoids involves induction of a novel prostaglandin H synthase. J. Biol. Chem. **267:** 14547-14550.
16. O'SULLIVAN, M. G., E. M. HUGGINS, E. A. MEADE, D. L. DEWITT & C. E. MCCALL. 1992. Lipopolysaccharide induces prostaglandin H synthase-2 in aveolar macrophages. Biochem. Biophys. Res. Commun. **187:** 1123-1127.
17. MAIER, J. A. M., T. HLA & T. MACIAG. 1990. Cyclooxygenase is an immediate-early gene induced by interleukin-1 in human endothelial cells. J. Biol. Chem. **265:** 10805-10808.
18. JONES, D. A., D. P. CARLTON, T. M. MCINTYRE, G. A. ZIMMERMAN & S. M. PRESCOTT. 1993. Molecular cloning of human prostaglandin endoperoxide synthase type II and demonstration of expression in response to cytokines. J. Biol. Chem. **268:** 9049-9054.
19. SANO, H., T. HLA, J. A. M. MAIER, L. J. CROFFORD, J. P. CASE, T. MACIAG & R. L. WILDER. 1992. *In vivo* cyclooxygenase expression in synovial tissue of patients with rheumatoid arthritis and osteoarthritis and rats with adjuvant and streptococcal cell wall arthritis. J. Clin. Invest. **89:** 97-108.

20. SIROIS, J. & J. S. RICHARDS. 1992. Purification and characterization of a novel, distinct isoform of prostaglandin endoperoxide synthase induced by human chorionic gonadotropin in granulosa cells of rat preovulatory follicles. J. Biol. Chem. **267:** 6382-6388.
21. SIROIS, J., D. L. SIMMONS & J. S. RICHARDS. 1992. Hormonal regulation of messenger ribonucleic acid encoding a novel isoform of prostaglandin endoperoxide H synthase in rat preovulatory follicles. J. Biol. Chem. **267:** 11586-11592.
22. KUJUBU, D. A. & H. R. HERSCHMAN. 1992. Dexamethasone inhibits mitogen induction of the TIS10 prostaglandin synthase/cyclooxygenase gene. J. Biol. Chem. **267:** 7991-7994.
23. MASFERRER, J. L., K. SEIBERT, B. ZWEIFEL & P. NEEDLEMAN. 1992. Endogenous glucocorticoids regulate an inducible cyclooxygenase enzyme. Proc. Natl. Acad. Sci. USA **89:** 3917-3921.
24. JONAT, C., H. J. RAHMSDORF, K. PARK, A. C. CATO, S. GEBEL, H. PONTA & P. HERRLICH. 1990. Antitumor promotion and antiinflammation: Down-modulation of AP-1 (fos/jun) activity by glucocorticoid hormone. Cell **62:** 1189-1204.
25. SMITH, W. L. & D. L. DEWITT. Unpublished observations.
26. HERSCHMAN, H. R. 1991. Primary response genes induced by growth factors and tumor promoters. Ann. Rev. Biochem. **60:** 281-319.
27. RYSECK, R., C. RAYNOSCHEK, H. MACDONALD-BRAVO, K. DORFMAN & R. BRAVO. 1992. Identification of an immediate early gene, PGHS-b, whose protein product has prostaglandin synthase/cyclooxygenase activity. Cell Growth Differ. **3:** 443-450.
28. DEWITT, D. L., E. A. EL-HARITH, S. A. KRAEMER, M. J. ANDREWS, E. YAO, R. L. ARMSTRONG & W. L. SMITH. 1990. The aspirin and heme-binding site for the ovine and murine prostaglandin endoperoxide synthases. J. Biol. Chem. **265:** 5192-5198.
29. KULMACZ, R. J. & W. E. M. LANDS. 1985. Stoichiometry and kinetics of the interaction of prostaglandin H synthase with anti-inflammatory agents. J. Biol. Chem. **260:** 12572-12578.
30. WALENGA, R. W., S. F. WALL, B. N. Y. SETTY & M. J. STUART. 1986. Time-dependent inhibition of platelet cyclooxygenase by indomethacin is slowly reversible. Prostaglandins **31:** 625-637.
31. HADDOCK, R. E., D. J. JEFFERY, J. A. LLOYD & A. R. THAWLEY. 1984. Metabolism of nabumetone (BRL 14777) by various species including man. Xenobiotics **14:** 327-337.
32. HLA, T. & K. NEILSON. 1992. Human cyclooxygenase-2 cDNA. Proc. Natl. Acad. Sci. USA **89:** 7384-7388.
33. TOH, H. 1989. Prostaglandin endoperoxide synthase contains an EGF-like domain. FEBS Lett. **258:** 317-319.
34. SHIMOKAWA, T. & W. L. SMITH. 1991. Essential histidines of prostaglandin endoperoxide synthase. J. Biol. Chem. **266:** 6168-6173.
35. SHIMOKAWA, T., R. J. KULMACZ, D. L. DEWITT & W. L. SMITH. 1990. Tyrosine 385 of prostaglandin endoperoxide synthase is required for cyclooxygenase catalysis. J. Biol. Chem. **265:** 20073-20076.
36. SHIMOKAWA, T. & W. L. SMITH. 1992. Prostaglandin endoperoxide synthase: The aspirin acetylation region. J. Biol. Chem. **267:** 12387-12392.
37. SIMONSON, M. S., J. A. WOLFE, M. KONIECZKOWSKI, J. R. SEDOR & M. J. DUNN. 1991. Regulation of prostaglandin endoperoxide synthase gene expression in cultured rat mesangial cells: Induction by serum via a protein kinase C-dependent mechanism. Mol. Endocrinol. **5:** 441-451.

38. DeWitt, D. L., S. A. Kraemer & E. A. Meade. 1991. Serum induction and superinduction of PGH synthase mRNA levels in 3T3 fibroblasts. *In* Prostaglandins and Related Compounds II. Vol. 21. B. Samuelsson, Ed. 75-90. Raven Press. New York, NY.
39. Oshima, T., T. Yoshimoto, S. Yamamoto, M. Kumegawa, C. Yokoyama & T. Tanabe. 1991. cAMP-dependent induction of fatty acid cyclooxygenase mRNA in mouse osteoblastic cells (MC3T3-E1). J. Biol. Chem. **266:** 13621-13626.

Prostaglandin H Synthase and Lipoxygenase Gene Families in the Epithelial Cell Barrier[a]

MICHAEL J. HOLTZMAN,[b] VIVIAN ZHANG, HAMEDA HUSSAIN, WILLIAM T. ROSWIT, AND JOHN D. WILSON

Washington University School of Medicine
St. Louis, Missouri 63110

Epithelial tissues (in skin, gut, and airway) bear the responsibility for maintaining and regulating the boundary between host and environment. Because this task may be regulated at a molecular level by products of phospholipid-arachidonic acid metabolism, we have defined some of the characteristics of gene expression and enzyme regulation for critical members of this metabolic pathway in normal and inflamed epithelial tissues and in epithelial cells isolated from mucosal and epidermal tissue. In particular, our studies have concentrated on two enzymatic systems for fatty acid oxygenation that are prominently expressed in epithelial barrier cells: (i) the arachidonate 12- and 15-lipoxygenases that (at least in some cases) are capable of peroxidatively damaging cellular membranes; and (ii) the prostaglandin H (PGH) synthases that catalyze the rate-limiting step in the formation of biologically active prostaglandins. This chapter will review information obtained by us (generally using epithelial cells as a model system) and by others (often using hematopoietic cells as a model) for each of these enzyme families. The studies are aimed at uncovering novel mechanisms governing the expression and regulation of these enzyme systems and for providing a rational basis for physiologic versus inflammatory roles of phospholipid-arachidonic acid metabolism in the epithelial cell barrier.

THE LIPOXYGENASE GENE FAMILY

Lipoxygenases catalyze the incorporation of one oxygen molecule into polyunsaturated fatty acids containing a 1,4-*cis,cis*-pentadiene structure to yield a 1-hydroperoxy-2,4-*trans,cis*-pentadiene product. The initial catalytic step is stereospecific removal of a hydrogen atom followed by antarafacial addition of molecular oxygen. This step therefore gives rise to the lipoxygenase signature for generating arachidonate products containing a single S-stereoisomer. Mammalian lipoxygenases also possess

[a]This research was supported by Grants HL-40078, HL-07317, and DK-38111 from the National Institutes of Health.

[b]Address for correspondence: Michael J. Holtzman, Washington University School of Medicine, 660 South Euclid Avenue, Box 8052, St. Louis, Missouri 63110.

regiospecificity during interaction with substrate and on that basis have been designated as arachidonate 5-, 12-, or 15-lipoxygenase. The three distinct enzyme types insert oxygen at carbon 5, 12, or 15 of arachidonate acid, and the primary product is a 5*S*-, 12*S*-, or 15*S*-hydroperoxyeicosatetraenoic acid (5-, 12-, or 15-HPETE). The enzymes may also catalyze the formation of unstable epoxide intermediates that can be further metabolized to a variety of compounds, including leukotrienes (LTs), dihydroxy acids (diHETEs), epoxyhydroxy acids (hepoxilins), trihydroxy acids (lipoxins), aldehydes, and sulfidopeptide adducts (e.g., LTC_4). As noted below, each of these characteristics can be exploited to permit definition of an eicosanoid-forming activity as due to a lipoxygenase versus another type of enzyme (e.g., cytochrome P450 monooxygenase).

Human Airway Epithelial Cells Contain a Leukocyte-Type 15-Lipoxygenase

Following the discovery of arachidonate 5-, 12-, and 15-lipoxygenases in leukocytes, platelets, and reticulocytes, a natural question was whether these enzymes were also expressed in nonhematopoietic cell types. In this regard, perhaps the first definitive identification of high-level lipoxygenase activity came with the discovery that mucosal epithelial cells freshly isolated from postmortem human trachea contained a 15-lipoxygenase activity.[1] Subsequent work provided evidence that tracheal epithelial cells isolated from other species of animals (especially cows) contained a high level of a 12-lipoxygenase activity[2,3] (which, as noted below, turned out to be a homologous enzyme). The prominent expression of a 15-lipoxygenase in human tracheal epithelial cells (and 12-lipoxygenase in several animal species of tracheal epithelial cells) initially appeared to be distinctly unusual. For example, among other cell types in humans, only eosinophils and reticulocytes exhibit comparable levels of activity.[4] However, comparing the levels of 12- and 15-lipoxygenase activities in isolated airway epithelial cells to levels in other isolated (often cultured) cells may be misleading, because the activities appear to be lost under some culture conditions[5] and may be present in a cryptic form in other conditions (see below). Several investigators (including our group) have also found evidence of a 5-lipoxygenase activity in canine, ovine, and equine airway epithelial cell preparations,[6–9] but the relatively low levels of activity have raised questions of the biologic significance of this finding, especially for humans.

In an attempt to reconcile the differences in lipoxygenase expression between species, we took advantage of reagents developed in studies of the hematopoietic cell system. In that system, it had been established that arachidonate 12-lipoxygenases existed in two distinct forms—a so-called leukocyte-type that was prominently expressed in peripheral blood granulocytes of different animal species as well as a platelet-type that was contained in platelets of several species (including humans). These two distinct "types" of 12-lipoxygenases were distinguished initially on the basis of substrate specificity (the leukocyte-type enzyme exhibiting a much broader range of acceptable fatty acid substrates versus being restricted to arachidonic acid), subcellular localization (the leukocyte-type enzyme being only cytosolic versus membrane-bound), and by antigenic reactivity (with both polyclonal and monoclonal antibodies capable of distinguishing the two isoforms).[10]

By extending these findings to studies of bovine and then human tracheal epithelial cells, it was soon learned that these cells contained leukocyte-type 12- (or homologous 15-) lipoxygenase: in the case of bovine tracheal epithelial cells, we found a cytosolic, stereospecific 12-lipoxygenase that cross-reacted with anti-leukocyte-type but not anti-platelet-type 12-lipoxygenase antibodies.[2,11] Interestingly, the epithelial form of the enzyme exhibited subtle differences in substrate specificity even from the leukocyte form, but these small differences in catalytic activity may not be due to differences in enzyme structure. In the case of human tracheal epithelial cells, the same profile for antibody reactivities was observed for the 15-lipoxygenase.[12,13] This finding linked the two enzymes as members of an extended family of leukocyte-type 12/15-lipoxygenases and also provided a basis for studies of *in situ* expression of these enzymes using cross-reactive but selective antibodies. Interestingly, the first immunohistochemical studies of the distribution of 12-lipoxygenase (in bovine lung) and 15-lipoxygenase (in humans) indicated that the enzyme may be expressed (at the highest levels) in the uppermost airways (nose and trachea)[12,13]—a finding that explains the fortuitous discovery of enzymatic activity by investigators who had access only to those cells.[1,7,14]

The possibility that the 12- and 15-lipoxygenase heterogeneity among species represented molecular divergence within a family of closely related enzymes was soon confirmed by molecular cloning studies: the leukocyte- and platelet-type enzymes were only ~65% homologous (as opposed to the ~85% homology observed within the family of leukocyte-type lipoxygenases) and so were apparently derived from two distinct genes.[15–17] Using the same approach, it was soon learned that the mucosal epithelial cells contained a leukocyte-type 12-lipoxygenase (in the case of the bovine cells) or a homologous 15-lipoxygenase (in the case of human cells).[18]

Airway Epithelial Cell Lipoxygenases Are Induced by Inflammation

Biology of the 15-lipoxygenase pathway takes on added interest because it is the predominant pathway for arachidonic acid metabolism in whole lung tissue,[19,20] and its activity may be increased in asthmatic bronchial tissue.[21] In addition, the major 15-lipoxygenase end product derived from arachidonic acid (15-HETE) is found in lavage fluid from asthmatic airways[22] and is released selectively by airway epithelial cells and eosinophils,[4] two cell types that are implicated in the development of asthmatic airway inflammation.

Accordingly, we utilized selective anti-lipoxygenase antibodies to examine the expression of 15-lipoxygenase in bronchial biopsy tissue from normal subjects versus those with airway disease.[23] Immunohistochemistry of bronchial tissue using two antibodies against distinct epitopes of the 15-lipoxygenase and indirect biotin-avidin-peroxidase detection demonstrated that, in contrast to airway tissue from normal subjects in which 15-lipoxygenase antigen was confined to the uppermost airways (nose and trachea) and was almost undetectable in bronchi, the bronchial tissue obtained from subjects with asthma or chronic bronchitis exhibited markedly positive immunostaining of mucosal epithelial cells with both anti-12/15-lipoxygenase antibodies. Specificity of 15-lipoxygenase immunostaining was verified by antigen competition experiments and by the lack of immunostaining with preimmune serum or control anti-5-lipoxygenase antibodies. The increased levels of 15-lipoxygenase antigen in the bronchial epithelial cells of asthmatic and bronchitic subjects compared to the same cell population in normal subjects coupled with the previous findings of increased 15-lipoxygenase activity in asthmatic airways suggest that epithelial 15-lipoxygenase is induced by airway inflammatory disease. Determining the factors that cause induction of epithelial lipoxygenase and the consequence of overexpression are important questions (see below).

Colonic Mucosal Epithelial Cells Overexpress a Related 12-Lipoxygenase at Active Sites of Inflammatory Bowel Disease

To further determine whether epithelial lipoxygenases are regulated by mucosal inflammation, we examined the distribution of lipoxygenase in healthy colonic tissue and in involved and uninvolved sections of colon with inflammatory bowel disease.[24] Immunohistochemistry of formaldehyde-fixed, paraffin-embedded intestinal tissue using two anti-leukocyte-type 12/15-lipoxygenase antibodies and indirect biotin-avidin-peroxidase detection demonstrated that, in contrast to tissue from normal colon, in which 12/15-lipoxygenase antigen was undetectable in mucosal epithelial cells, the mucosal and glandular epithelium of inflamed colon in ulcerative colitis or Crohn's disease exhibited markedly positive immunostaining with anti-12/15-lipoxygenase antibodies. Immunoblotting of whole-cell extracts with anti-12/15-lipoxygenase antibodies showed immunoperoxidase staining of a single protein band, which comigrated with purified 12-lipoxygenase (M_r = 72,000) in mucosal samples from inflamed colon from subjects with ulcerative colitis or Crohn's disease but no comparable band in samples from uninvolved sections of the same colons or from colons of subjects without inflammatory bowel disease. Assays of enzymatic activity indicated a corresponding increase in 12-lipoxygenase activity in the same mucosal samples. The increased levels of lipoxygenase antigen in the mucosal and glandular epithelial cells in regions of colon affected by inflammatory bowel disease, the corresponding increases in 12-lipoxygenase activity, and the absence of detectable lipoxygenase antigen or enzymatic activity is the same cell types in noninflamed colonic tissue all suggest that epithelial cell 12-lipoxygenase is induced by local mediators of colonic inflammation. Structural evidence that the human colonic epithelial enzyme (predominant 12-lipoxygenation) may be distinct from the airway epithelial enzyme (predominant 15-lipoxygenation) is under study.

Keratinocytes Contain a Platelet-Type 12-Lipoxygenase That Is Overexpressed in Psoriatic Inflammation

The involvement of arachidonate oxygenation products in the development of inflammatory skin diseases, particularly psoriasis, has been the subject of a variety of studies. Approaches to understanding the pathophysiologic basis of altered arachidonate metabolism in inflammatory skin diseases have included determining the effects of intradermal or topical administration of arachidonate metabolites and assaying the levels of such compounds in biologic specimens, including biopsies from the epidermis of patients with psoriasis. Results of such studies indicate that:

(a) concentrations of arachidonic acid, 12-hydroxyeicosatetraenoic acid (12-HETE), prostaglandins, and leukotrienes are increased in psoriatic lesional epidermis;[25–28] (b) whole human epidermis from healthy or diseased skin has the enzymatic capacity to convert arachidonic acid to some of these same compounds;[25–30] and (c) topical or intradermal administration of 12-HETE or leukotrienes may cause epidermal inflammation and proliferation.[31,32]

The precise enzymatic mechanisms responsible for the generation of the major arachidonate product 12-HETE by epidermis and the localization of the responsible enzyme(s) to specific epidermal cell types were not precisely established when we began our studies. Dermal elements such as fibroblasts as well as inflammatory cells are often included in dermatome preparations of human skin, and the enzymatic pathways leading to HETE formation in the skin have only been assumed to represent the activity of an arachidonate lipoxygenase. In fact, the observation that the 12-HETE in psoriatic scale is not the expected lipoxygenase-derived 12*S*-hydroxy enantiomer but instead may consist predominantly of the 12*R*-isomer[33] suggested that an enzyme other than the 12-lipoxygenase may be activated in inflamed epidermis. For example, cytochrome P450 monooxygenases have been reported to produce predominantly 12*R*-HETE rather than 12*S*-HETE from arachidonate in hepatic tissue.[34]

In order to better define the type(s) and the site(s) of specific oxygenases in human skin, we set out to define arachidonate oxygenation pathways in human epidermal cells purified from healthy subjects. In our initial study, we found evidence of a microsomal NADPH-dependent monooxygenase that selectively converted arachidonic acid to 12-HETE (12*S*- and 12*R*-enantiomers) and that exhibited the properties of a cytochrome P450.[35] The absence of cytosolic 12-HETE-forming activity (that would be typical of other epithelial lipoxygenases) led us to speculate that epidermal cells did not express significant lipoxygenase activity (in contrast to mucosal epithelial cells from lung and gut).[11,24] However, several features of epidermal 12-HETE-forming activity led us to question this conclusion: (a) the difference in 12*S*-/12*R*-HETE ratios in mitochondrial versus microsomal fractions from epidermal cells raised the possibility that the cells contained two distinct 12-HETE-forming enzymes with differing stereopreferences; (b) the absence of complete dependence of epidermal 12-HETE-forming activity on NADPH supplementation or complete inhibition by carbon monoxide raised the possibility that a non-monooxygenase activity was responsible at least in part for 12-HETE formation; and (c) the subsequent discovery of microsomal 12-lipoxygenase activity in other types of cultured epithelial cells[36,37] suggested that keratinocytes might also contain such an enzyme. Thus, studies were designed to reexamine the possibility that epidermal cells contain an arachidonate 12-lipoxygenase, to localize the enzyme to specific cell type(s) in the epidermis, and to determine whether the enzyme may have a distinct pattern of expression in inflamed versus normal skin.

As noted above, human epidermal cell preparations exhibited none of the cytosolic lipoxygenase activity that is prominent in mucosal epithelial cells, but instead contained a microsomal activity that converted arachidonic acid to 12-HETE at a high rate (13 nmol/mg protein/15 min in detergent extracts of microsomes). Identification of the extractable 12-HETE-forming activity as a 12-lipoxygenase (distinct from cytochrome P450 activity) included 12*S*-stereospecificity of product formation, trapping of 12-HPETE as an intermediate reaction product, and lack of NADPH dependence for activity.[38] Northern blot analysis of epidermal cell poly(A)$^+$RNA indicated high levels of a single 2.3-kb mRNA band that selectively hybridized with human platelet 12-lipoxygenase cDNA, and partial cDNA sequence of the band indicated identity to platelet 12-lipoxygenase. Immunoprecipitation and immunostaining experiments showed that epidermal 12-lipoxygenase was not recognized by anti-12-lipoxygenase antibodies that react with animal and human leukocyte-type 12- and 15-lipoxygenases (found in leukocytes, reticulocytes, and mucosal epithelial cells), but was recognized by an anti-12-lipoxygenase antibody raised against purified, recombinant platelet 12-lipoxygenase. The epidermal 12-lipoxygenase antigen was selectively expressed in keratinocytes within the germinal layers of the epidermis of healthy and psoriatic skin, and these cell layers exhibited hyperplasia and markedly increased immunostaining for antigen in inflamed skin from psoriatic subjects. Taken together with previous results, these observations indicate that: (a) human epidermis generates 12-HETE by either cytochrome P450- or lipoxygenase-based mechanisms depending on reaction conditions; and (b) 12-lipoxygenases (originally described in hematopoietic cell types) may be expressed in at least two distinct isoforms in epithelial barriers in humans; and in the case of the skin, a microsomal (platelet-type) 12-lipoxygenase is selectively overexpressed in germinal layer keratinocytes during psoriatic inflammation. The selective expression in germinal layer keratinocytes naturally raises questions of a role for the enzymatic activity in epithelial cell growth and differentiation.

Lipoxygenase Function in the Epithelial Barrier: Lessons from Cultured Epithelial Cells

Perhaps the most published and/or popular theories of 12- and 15-lipoxygenase function are focused on the generation of biologically active mediators. The enzymes may (with or without help from adjacent enzymes) lead to the generation of a diverse array of hydroxylated fatty acid products (as noted above), any one of which may be the "active ingredient" in a given biologic system. Because 12- and 15-HPETE and their respective reduction products 12- and 15-HETE are generally the most abundant products of these lipoxygenases, much of the work for determining functional significance has concentrated on the possibility that these compounds have biologically significant effects in epithelial (and other) tissues. Evidence for proinflammatory and antiinflammatory functions of 15-lipoxygenase products (especially 15-HETE) has been reviewed,[5] with the impression that the concentrations required often appear too high to be biologically significant. In addition to direct effects on the inflammatory response, others have proposed that products of the 15-lipoxygenase pathway (at lower concentrations) may: (i) promote cell growth;[39] (ii) participate in signal transduction events in neurons, endocrine cells and, perhaps mucus-secreting cells;[40–42] and (iii) alter integrin expression and cell migration.[43] Taken together, however, there is little consensus for a biologic role of 15- (or related 12-) lipoxygenase products as mediators of biologic function in humans (let alone for epithelial cell function), and the view that the products are often biologically inactive has also been proposed.[44,45]

The alternative role for the 12- and 15-lipoxygenases is suggested by their capacity to alter membrane lipid composition. This line of reasoning begins with the observations that 15-lipoxygenase products are rapidly incorporated into membrane phospholipids,[46–49] and (perhaps more importantly) the fact that at least some isoforms may actively utilize intact phospholipid as a substrate *in vitro* and *in vivo*.[50–54] This functional possibility has been further characterized in reticulocytes, a cell type that must degrade cellular membranes during maturation. Thus, reticulocyte mitochondrial membranes contain increased levels of the oxygenated derivatives of linoleic and arachidonic acids,[52] and the formation of these products may destabilize membrane structure and predispose it to proteolysis during erythrocyte maturation.

With this background, we had also postulated that lipoxygenase activities might be responsible for changes in membrane lipid composition in epithelial cells, but any linkage of this idea to physiologic regulatory mechanisms was not apparent. The absence of data is due in part to the difficulty in establishing cell culture systems that maintain the lipoxygenase phenotype of the original tissue. For example, cultured epithelial cells (in contrast to freshly isolated cells) often exhibit little or no evidence of 12- or 15-lipoxygenase activities under a variety of tissue culture conditions.[5,55–57] In the course of studying oxygenation pathways in cultured ovine tracheal epithelial cells (set up for our concurrent studies of PGH synthases), we noticed that the intact cells converted arachidonic acid only to prostaglandin E_2 (PGE_2), but microsome-containing subcellular fractions prepared from these cells under calcium-free conditions converted arachidonic acid to PGE_2 and to 12-HETE at a high rate (2-4 nmol/mg protein/15 min).[36] Identification of the membrane-bound 12-HETE-forming activity as a 12-lipoxygenase included 12*S*-stereospecificity of product formation and trapping of 12-HPETE as a reaction product. The 12-lipoxygenase activity was extracted from cell membranes only with detergent (1% Triton X-100), and the activity (membrane bound or detergent solubilized) was completely inactivated by mixing with the cytosol-containing subcellular fraction.

The inhibitory effect of the cytosolic fraction was reversed by treating the cytosol with GSH-depleting agents (2-cyclohexene-1-one or N-ethylmaleimide) or by mixing it with lipid hydroperoxide (13-hydroperoxyoctadecadienoic acid) at a concentration that had little direct effect on enzyme activity. Inhibition of 12-lipoxygenase activity could also be achieved by treatment of enzyme preparations with GSH at levels (0.1-10 mM) found in epithelial cell cytosol. In addition, treatment of cultured epithelial cells with a GSH-depleting agent (buthionine sulfoximine) and lipid hydroperoxide restored cellular 12-lipoxygenase activity (FIG. 1). In comparative experiments, the PGH synthase/PGE isomerase pathway of cultured ovine tracheal epithelial cells exhibited increased (rather than decreased) activity in response to GSH and decreased (not increased) activity after lipid hydroperoxide. Little or no 12-lipoxygenase activity was detected in freshly isolated ovine tracheal epithelial cells, but the cytosolic 12-lipoxygenase found in freshly isolated bovine tracheal epithelial cells was relatively insensitive to regulation by GSH or lipid hydroperoxide. These observations indicate that a 12-lipoxygenase is expressed in a cryptic, membrane-bound form in primary-culture epithelial cells and that this form of the enzyme may be selectively regulated by changes in cellular oxidation-reduction conditions dependent on the cytosolic levels of GSH versus lipid hydroperoxide.

These findings in cultured epithelial cells serve to establish a link between the microsomal-type 12-lipoxygenase and the chain of enzymes that regulate cellular oxidation-reduction conditions (FIG. 2). The capacity of endogenous GSH to inhibit

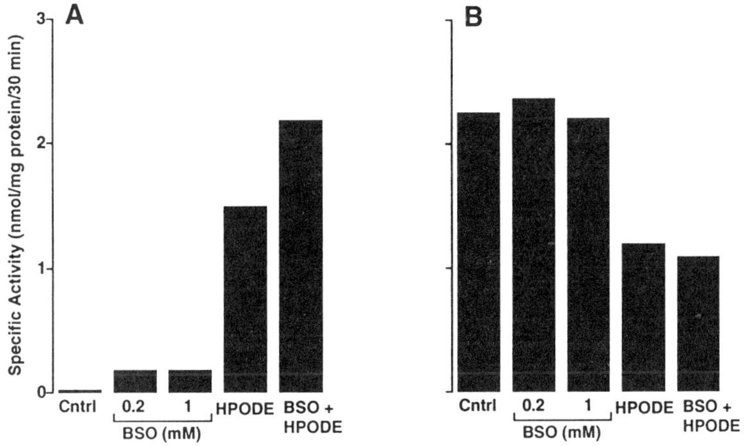

FIGURE 1. Decreased GSH and increased lipid hydroperoxide levels can restore epithelial cell 12-lipoxygenase (**A**) and inhibit PGH synthase/PGE isomerase (**B**) activities. Cultured ovine tracheal epithelial cells were treated with GSH-depleting agent buthionine sulfoximine (BSO) for 24 h, lipid hydroperoxide 13-HPODE for 5 min, or with both reagents and then assayed for enzymatic activity. For activity assays, 2×10^6 cells were incubated with 10 μM [^{3}H]arachidonic acid (2.2×10^5 dpm/ml) in 3 ml of Hepes-buffered HBSS for 15 min at 37°C. Enzymatic products were resolved and quantified using a reverse-phase HPLC/radioactivity flow detector system as described previously.[36] Values represent the mean for two experiments.

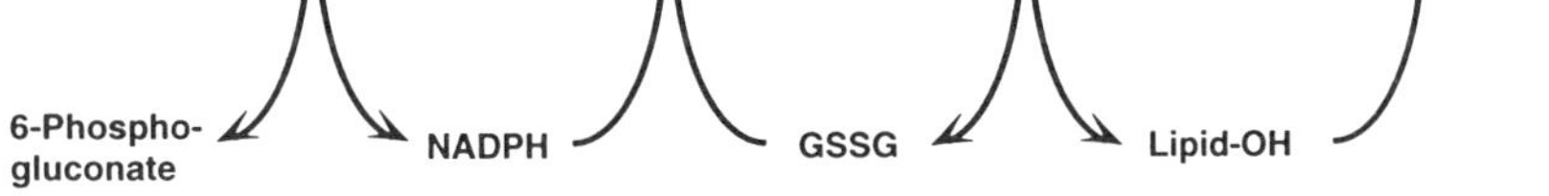

FIGURE 2. Pathways for reduced glutathione (GSH) generation and 12-lipoxygenase converge to regulate the levels of lipid hydroperoxide. In this scheme, GSH generated by the serial activities of glucose-6-phosphate dehydrogenase (G-6-P) and oxidized glutathione (GSSG) reductase may then regulate lipid hydroperoxide levels by acting both as a substrate for GSH peroxidase and an inhibitor of 12-lipoxygenase. For example, high GSH levels would favor decreased lipid hydroperoxide levels by promoting GSH peroxidase activity and inhibiting 12-lipoxygenase activity. By contrast, low GSH levels would favor increased lipid hydroperoxide levels by promoting GSH peroxidase activity and inhibiting 12-lipoxygenase activity.

microsomal-type 12-lipoxygenase may offer a distinct mechanism for the resistance of epithelial barrier cells to oxidative damage. Alternatively, the ability of lipid hydroperoxide to activate microsomal-type 12-lipoxygenase may offer a means to directly augment the peroxidative membrane damage that takes place in response to the release of hydroperoxides during inflammation. Activation of the airway epithelial 12-lipoxygenase by oxidant stress with peroxidation of membrane phospholipids (and consequent epithelial damage) may be analogous to the membrane-degrading role of the 15-lipoxygenase in erythrocyte maturation (and perhaps in the autolysis of eosinophils during inflammation). The marked increases in microsomal-type 12-lipoxygenase in cultured versus freshly isolated ovine tracheal epithelial cells may also suggest that the enzyme is inducible by cell culture conditions. Resolving these types of questions—that is, the regulation of cellular function by 12-lipoxygenase and the regulation of 12-lipoxygenase expression—will be facilitated by current studies aimed at determination of structure and development of more specific molecular probes. Such findings will also establish the relationship of the microsomal-type, redox-sensitive 12-lipoxygenase with the previously described leukocyte- and platelet-type enzymes and thereby help complete the genetic profile for the epithelial lipoxygenase family.

THE PGH SYNTHASE GENE FAMILY

PGH synthase catalyzes the conversion of arachidonic acid (and certain other polyunsaturated fatty acids) to prostaglandin endoperoxides. The cyclooxygenase activity of PGH synthase inserts two molecules of oxygen into arachidonic acid to yield prostaglandin G_2 (PGG_2), and a peroxidase activity of the enzyme reduces PGG_2 to PGH_2 (the 15-hydroxy analogue). Nonsteroidal antiinflammatory drugs inhibit the cyclooxygenase but not the hydroperoxidase activity of the enzyme. In particular, aspirin selectively acetylates and irreversibly inactivates the prostaglandin-forming activity of the enzyme, and this action is responsible for both therapeutic effects of the drug and idiosyncratic reactions to it. Identification of the aspirin-acetylation site and other structural features of PGH synthase were facilitated by purification of the enzyme (a homodimer of two 70-kDa polypeptides) from bovine and ovine seminal vesicle and by molecular cloning of the corresponding 2.7-3.0 kb cDNA encoding for the complete protein from ovine seminal vesicle, murine fibroblast, and human platelet cDNA libraries[58–62] and a human genomic library.[63] Thus, when we began our studies of epithelial prostaglandin-forming activity, the "dogma" for PGH synthase was that it consisted of a single gene-single polypeptide system with a ~2.8-kb mRNA encoding for a 70-kDa polypeptide and that its catalytic activity was completely inhibitable by aspirin acetylation.

Initial Evidence of PGH Synthase Heterogeneity

As noted above, we initiated experiments that took advantage of the newly developed reagents for ovine PGH synthase in order to study the regulation of the enzyme in cultured epithelial cells. In the first experiments, we found that ovine tracheal epithelial cells maintained in serum-free, growth-promoting conditions exhibited progressive increases in PGE_2-forming activity, and we anticipated that we would find corresponding increases in 2.8-kb PGH synthase mRNA and 70-kDa PGH synthase polypeptide. In fact, Northern analysis and immunoprecipitation experiments indicated that these two molecular species were coordinately expressed in cultured epithelial cells, but their levels were relatively stable in the face of increases in PGH synthase activity. These contradictory results were difficult to understand until it became apparent that a second, PGH synthase-related mRNA was also expressed in cultured ovine tracheal epithelial cells.[55] The existence of a PGH synthase-related mRNA (later designated PGH synthase-2) that might be responsible for increased PGH synthase activity (and increased PGE_2 formation) under these conditions was uncovered when rehybridization of Northern blots at lower stringency revealed the presence of a new 4.0-kb mRNA species exhibiting increased expression during cell culture. Hybridization of the 4.0-kb mRNA with two nonoverlapping PGH synthase-1 cDNA probes only at low stringency conditions suggested that it was derived from a distinct gene. Its relatedness to PGH synthase and its increase in parallel with enzymatic activity further suggested that the larger mRNA encoded for a PGH synthase that (in contrast to the 2.8-kb PGH synthase) might be more subject to regulation under growth factor stimulation.

Identification of a Pharmacologically Distinct PGH Synthase

Further evidence of PGH synthase heterogeneity came with experiments aimed at studying the regulation of enzymatic activity. This work was initiated to determine the recovery rate of PGH synthase activity in cultured ovine tracheal epithelial cells following aspirin treatment. In the course of the experiments, we noted that (as expected) aspirin treatment irreversibly blocked the enzymatic synthesis of PGE_2, but the generation of 15-HETE from arachidonic acid was invariably enhanced.[56] Previous work on ovine seminal vesicle and on other cell types demonstrated that 15-HETE (and 11-HETE) might be minor byproducts of arachidonate/PGH synthase interaction, but their formation (as well as PGH formation) was invariably inhibited by treatment with aspirin or indomethacin.[64,65]

Additional experiments on cultured tracheal epithelial cells were therefore performed to detect other enzymatic mechanisms of HETE formation; they led to the conclusion that the novel effect of aspirin in this cell culture system resulted from a PGH synthase-catalyzed reaction. The 15-HETE formed by aspirin-treated ovine tracheal epithelial cells was generated by a PGH synthase-dependent mechanism because: (i) the 15-HETE-forming activity was just as sensitive as PG-forming activity to selective inhibition by indomethacin; and (ii) both 15-HETE- and PG-forming activities were quantitatively immunoprecipitated (depleted from supernatants and recovered in immune complex pellets) by a specific anti-PGH synthase antiserum. Additional immunoprecipitation experiments indicated that anti-PGH synthase monoclonal antibodies (*cyo-1* and *cyo-5*) raised against the aspirin-inhibited form of the enzyme did not recognize the aspirin-stimulated 15-HETE-forming PGH synthase. Thus, sequential immunoprecipitation of cultured epithelial cell material first with excess *cyo-1* followed by anti-PGH synthase antiserum indicated that

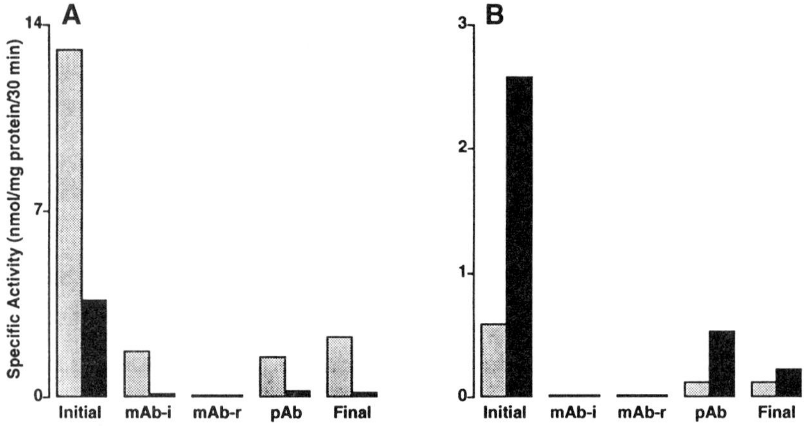

FIGURE 3. Two pharmacologically distinct PGH synthases can be demonstrated by sequential immunoprecipitation of PG-forming (**A**) and 15-HETE-forming (**B**) activities from cultured ovine tracheal epithelial cells. In each panel, detergent extracts from untreated (*gray bars*) or aspirin-treated (*black bars*) epithelial cell membranes were subjected to initial immunoprecipitation with anti-PGH synthase *cyo-1* monoclonal antibody (mAb-i), repeat immunoprecipitation with *cyo-1* mAb (mAb-r), then final immunoprecipitation with anti-PGH synthase polyclonal antiserum (pAb). The resulting supernatants and immune complex pellets were assayed for enzymatic activity using a reverse-phase HPLC/radioactivity flow detector system as described previously.[56] Values represent mean for two experiments. In concomitant experiments, SDS-PAGE analysis of immunoprecipitated material indicated that a 70-kDa PGH synthase that comigrates with ovine seminal vesicle PGH synthase was immunoprecipitated with mAb, and that a distinct 72-kDa PGH synthase was immunoprecipitated with pAb.

two isoforms of PGH synthase were expressed in these cells (FIG. 3). SDS-PAGE immunoprecipitated PGH synthase from cultured epithelial cells revealed distinct protein bands for each form of the enzyme (Mr = 70,000 and 72,000). The identification of a distinct PGH synthase that may be modified by aspirin so that selective oxygenation of fatty acid substrate is enhanced (while PG formation is inhibited) indicates that isozymes of PGH synthase exist that are pharmacologically distinct. By using two indices of PGH synthase-2 expression (aspirin-evoked catalytic activity and selective immunoreactivity), we demonstrated that a 72-kDa PGH synthase (nominally designated PGH synthase-2) was responsible for at least 50% of the prostaglandin-forming activity and at least 50% of the immunodetectable PGH synthase protein in cultured ovine tracheal epithelial cells. The findings implicate PGH synthase-2 as a major source of epithelial cell prostaglandins.

Our findings raised critical questions for possible differences in PGH synthase-1 and PGH synthase-2 catalytic properties. Previous studies utilizing 20-carbon substrates of PGH synthase (including those enriched in the 13S-^{3}H) indicated that the initial step in prostaglandin formation was abstraction of the pro-S hydrogen at carbon-13.[64,66,67] In the case of arachidonic acid as substrate, the initial enzyme-catalyzed abstraction of the 13S-hydrogen followed by typical insertion of molecular oxygen on the opposite face of the substrate molecule at carbon-11 or carbon-15

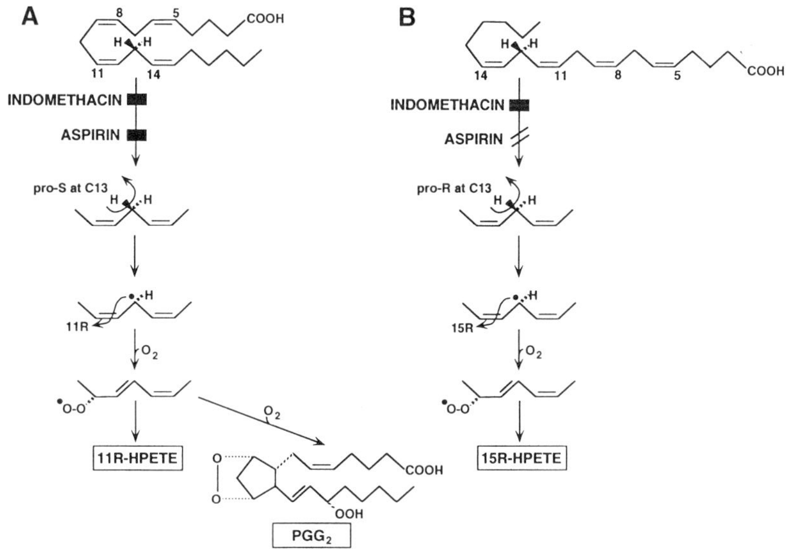

FIGURE 4. Scheme for arachidonic acid and inhibitor interaction with PGH synthases. In **A**, the initial step in prostaglandin formation is abstraction of the pro-*S* hydrogen at carbon-13 followed by formation of a peroxy radical in the pro-*R* position at carbon-11 and subsequent antarafacial addition of oxygen. This intermediate may be reduced to 11*R*-HPETE or may undergo further reactions (attack on carbon-9 with endoperoxide formation, cyclopentane ring generation with bond formation between carbons-8 and -12, and a second dioxygenation at carbon-15) to yield PGG_2. For both PGH synthase-1 and PGH synthase-2, the evidence indicates that the initial step in this scheme is inhibited by aspirin or indomethacin. In **B**, an alternative scheme is presented in which the substrate is in the reverse orientation with respect to the enzyme. In this case, the initial step involves abstraction of the pro*R*-hydrogen at carbon-13 followed by formation of a peroxy radical in the pro-*R* position at carbon-15. Subsequent steps would lead to the formation of 15*R*-HPETE that in turn is reduced to 15*R*-HETE. Indomethacin again blocks the initial step in this scheme, but aspirin does not block and may even augment this reaction. In addition, this reaction scheme may be more prominent in PGH synthase-2 catalyzed reactions especially when the enzyme is aspirin acetylated.

(antarafacial addition) would be expected to lead to the formation of 11*R*-HETE and 15*S*-HETE accompanying PGH_2 synthesis[66–69] (FIG. 4A). However, PGH synthase activity (and in particular, aspirin-acetylated PGH synthase-2) results predominantly in 15*R*-HETE formation.[56] This finding implies that an alternative catalytic scheme may exist that involves initial abstraction of the pro-*R* hydrogen at carbon-13 followed by antarafacial addition of oxygen in the pro-*R* position at carbon-15. If the same structure in the enzyme is involved in catalyzing this reaction, then the substrate needs to be in the reverse orientation with respect to the enzyme (FIG. 4B). The existence of this alternative scheme would imply that the catalytic site in the second PGH synthase-2 is more "promiscuous" in that it is capable of accepting substrate in reverse orientation even after acetylation, an event that would normally prevent substrate binding in the PGH synthase-1 catalytic site. This scheme may also (unfortunately) imply that a selective therapeutic advantage may be difficult to achieve if, for example, the end point is selective PGH synthase-2 inhibition in inflammatory cells. Further definition of PGH synthase catalytic mechanisms may be approached more exactly by using purified (recombinant) isoenzymes with and without inhibitor treatment.

Molecular Cloning and Expression of PGH Synthase-2 Using Epithelial Cells

Our findings indicated that isozymes of PGH synthase exist that are pharmacologically distinct, but these studies did not yet provide structural evidence linking the PGH synthase-related mRNA and the pharmacologically distinct PGH synthase polypeptide in cultured airway epithelial cells and did not allow direct comparison to results obtained in other cell types. Thus, while we were performing studies of PGH synthase catalytic activity, an additional form of PGH synthase mRNA and protein was also detected by molecular cloning from chicken and murine fibroblast and human endothelial cDNA libraries.[70–73] In each case, PGH synthase-2 mRNA is larger (4.0-4.7 kb) and contains an open-reading frame encoding for a polypeptide that is 59-61% homologous with PGH synthase-1. Each of the two types of recombinant polypeptides exhibit PGH synthase activity in assays of transfected cells,[59,72–74] but the possibility that the two PGH synthase isoforms might be catalytically or pharmacologically distinct was not deduced from these studies. Evidence of selective regulation of PGH synthase-2 mRNA levels by stimulatory factors and by glucocorticoids has reinforced the proposal for a regulated PGH synthase-2 gene (implicated in inflammation and/or cell growth) and a constitutive PGH synthase-1 gene (for physiologic activities).

To better define the distinct features of epithelial PGH synthase activity, we have cloned and sequenced a full-length 3.5-kb cDNA encoding for PGH synthase-2 (designated $PGHS_{ov}$-2) from cultured ovine tracheal epithelial cells (V. Zhang, H. Hussain, and M. J. Holtzman, unpublished observations). Analysis of PGH synthase-2 from ovine tracheal epithelial cells indicates that it contains unique sequence that is homologous with chicken, murine, and human PGH synthase-2. The ovine PGH synthase-2 is 78-82% homologous with PGH synthase-2 from mitogen-stimulated human vascular endothelial cells or transformed chicken and murine fibroblasts, and 60% homologous with ovine seminal vesicle PGH synthase-1 ($PGHS_{ov}$-1). The heme coordination site histidines, the peroxidase activity histidine, the active site tyrosine, the aspirin-sensitive serine, a putative *N*-glycosylation site, and the FSLKG sequence that appears critical for cyclooxygenase and peroxidase activities are each preserved in the ovine PGH synthase-2 gene.[75] Thus, these findings are quite similar to those reported for other PGH synthase-2 sequences.

Regulation of PGH Synthase Levels in Epithelial Cells

Prostaglandin-generating pathways may be regulated at several levels, but (as noted above) at least one critical control point appears to be the level of PGH synthase-2 activity. In studies designed to determine mechanisms governing the control of PGH synthase-2 levels in epithelial cells, two distinct mechanisms have been identified. The first relates directly to the most prominent structural difference between PGH synthase-2 and its homologue—that is, the large 3′-untranslated region (UTR). The large 3′-UTR in the ovine PGH synthase-2 (similar to the gene from other species) contains 11 AU-rich elements (AREs), and these elements have been shown to destabilize mRNAs derived from other genes.[76] To investigate whether such a mechanism might also regulate PGH synthase mRNA levels in an epithelial cell model system, we expressed ovine PGH synthase-2 constructs containing varying amounts of the AU-rich elements in a bronchial epithelial cell line that contained no endogenous PGH synthase mRNA or activity. In fact, increases in PGH synthase mRNA levels correlate with decreases in the number of AU-rich elements, suggesting that these elements serve to destabilize PGH synthase-2 mRNA under these experimental conditions (V. Zhang and M. J. Holtzman, unpublished observations). The results thereby offer a mechanism for posttranscriptional control of PGH synthase-2 levels in epithelial cells (and presumably other cell types). Whether this mechanism is influenced by PGH synthase regulatory factors (such as growth factors and cytokines) is under study.

A second interesting feature of PGH synthase-2 regulation is the apparent dependency of regulatory stimuli on cell type. For example, PGH synthase-2 mRNA levels are sensitive to interleukin-1β stimulation in several cell types (e.g., monocytes and endothelial cells) and dexamethasone inhibition in others (e.g., monocytes and fibroblasts).[77–79] However, we found little effect of these agents on PGH synthase levels in cultured ovine tracheal epithelial cells, even though these cells are highly capable of expressing PGH synthase-2 mRNA and protein. By contrast, our preliminary experiments indicate that interleukin-4 treatment of ovine tracheal epithelial cells caused fivefold increases in the rate of arachidonate conversion to PGE_2 and (in the presence of aspirin) to 15-HETE, and these increases in enzymatic activities correlate with increases in PGH synthase-2 mRNA levels (despite stable PGH synthase-1 mRNA levels). Cultured human tracheal epithelial cells also express a PGH synthase-2 mRNA species and exhibit the same aspirin evoked 15-HETE forming activity found in ovine cells as an index of PGH synthase-2 catalytic activity (V. Zhang, J. D. Wilson, and M. J. Holtzman, unpublished observations). Therefore, we have also begun to determine whether the same distinct profile for cytokine responsiveness of the PGH synthase-2 system is exhibited by mucosal epithelial cells from humans.

SUMMARY

Epithelial barrier cells (in skin, gut, and airway) are both active modulators and important targets of the inflammatory response, and some of these cellular events may be regulated at a molecular level by products of phospholipid-arachidonic acid metabolism. Accordingly, we have defined some of the characteristics of gene expression and enzyme regulation for distinct members of the PGH synthase and lipoxygenase gene families in normal and inflamed epithelial tissues and in epithelial cells isolated from mucosal and epidermal tissue (TABLE 1). A unifying scheme for our findings includes the following enzymatic systems: (i) a PGH synthase-1/PG iso-

TABLE 1. Lipoxygenase and PGH Synthase Gene Families in the Epithelial Cell Barrier in Humans

Enzyme Family	Epithelial Cell Type	Inducibility	GSH Sensitivity
Platelet-type			
12-Lipoxygenase	Keratinocytes	+[a]	–[b]
Leukocyte type			
15-Lipoxygenase	Airway mucosal cells	+[a]	±[c]
12-Lipoxygenase	Colonic mucosal cells	+	±[c]
PGH Synthase-1	Airway mucosal cells[d]	–	–[e]
PGH Synthase-2	Airway mucosal cells[d]	+[a]	–[e]

[a] This enzyme is expressed constitutively in normal subjects, but inducibility is suggested by overexpression during inflammation.

[b] Unpublished observations from J. D. Wilson and M. J. Holtzman.

[c] Based on work with the homologous 12-lipoxygenase from bovine tracheal mucosal epithelial cells.

[d] Expression may not be restricted to this type of epithelial cell.

[e] No significant effect on PGH synthase, but sensitivity of PGE isomerase may result in GSH-dependent decreases in PGE levels.

merase pathway responsible for constitutive generation of prostaglandins (e.g., PGE_2) and maintenance of physiologic epithelial function; (ii) a PGH synthase-2/PG isomerase and synthase pathway capable of producing additional prostaglandins (e.g., excess PGE_2 and/or $PGF_{2\alpha}$ and PGD_2) especially after stimulation by growth factors and cytokines; and (iii) a family of arachidonate 12- and 15-lipoxygenases that may serve to generate hydroxy acids (e.g., 12- and 15-HETE) as mediators of basal epithelial function and that (after overexpression and oxidant activation) may also catalyze membrane peroxidation that contributes to epithelial damage during inflammation. The regulatory mechanisms inherent in the control of this scheme provide a biochemical rationale for balancing constitutive and inducible oxygenation activities and maintaining epithelial barrier function.

REFERENCES

1. Hunter, J. A., W. E. Finkbeiner, J. A. Nadel, E. J. Goetzl & M. J. Holtzman. 1985. Predominant generation of 15-lipoxygenase metabolites of arachidonic acid by epithelial cells from human trachea. Proc. Natl. Acad. Sci. USA **82:** 4633.
2. Hansbrough, J. R., A. B. Atlas, J. Turk & M. J. Holtzman. 1989. Arachidonate 12-lipoxygenase and cyclooxygenase:PGE isomerase as predominant pathways for oxygenation in bovine tracheal epithelial cells. Am. J. Respir. Cell Mol. Biol. **1:** 237.
3. Alpert, S. E., C. M. Kramer, J. R. Brashler & M. K. Bach. 1990. Generation of lipoxygenase metabolites of arachidonic acid by monolayer cultures of tracheal epithelial cells and intact tracheal segments from rabbits. Exp. Lung Res. **16:** 207.
4. Holtzman, M. J., A. Pentland, N. L. Baenziger & J. R. Hansbrough. 1989. Heterogeneity of cellular expression of arachidonate 15-lipoxygenase: Implications for biological activity. Biochim. Biophys. Acta **1003:** 204.

5. HOLTZMAN, M. J. 1991. Epithelial cell regulation of arachidonic acid oxygenation. *In* The Airway Epithelium: Physiology, Pathophysiology, and Pharmacology, Lung Biology in Health and Disease. S. G. Farmer & D. W. P. Hay, Eds. Marcel Dekker. New York, NY.
6. HOLTZMAN, M. J., H. AIZAWA, J. A. NADEL & E. J. GOETZL. 1983. Selective generation of leukotriene B_4 by tracheal epithelial cells from dogs. Biochem. Biophys. Res. Commun. **114:** 1071.
7. HOLTZMAN, M. J., J. R. HANSBROUGH, G. D. ROSEN & J. TURK. 1988. Uptake, release, and novel species-dependent oxygenation of arachidonic acid in human and animal airway epithelial cells. Biochim. Biophys. Acta **963:** 401.
8. ELING, T. E., R. M. DANILOWICZ, D. C. HENKE, K. SIVARAJAH, J. R. YANKASKAS & R. C. BOUCHER. 1986. Arachidonic acid metabolism by canine tracheal epithelial cells. J. Biol. Chem. **261:** 12841.
9. GRAY, P. R., F. J. DERKSEN, N. E. ROBINSON, R. F. SLOCOMBE & M. L. PETERS-GOLDEN. 1992. Epithelial strips: An alternative technique for examining arachidonate metabolism in equine tracheal epithelium. Am. J. Respir. Cell Mol. Biol. **6:** 29.
10. HADA, T., N. UEDA, Y. TAKAHASHI & S. YAMAMOTO. 1991. Catalytic properties of human platelet 12-lipoxygenase as compared with the enzymes of other origins. Biochim. Biophys. Acta **1083:** 89.
11. HANSBROUGH, J. R., Y. TAKAHASHI, N. UEDA, S. YAMAMOTO & M. J. HOLTZMAN. 1990. Identification of a novel arachidonate 12-lipoxygenase in bovine tracheal epithelial cells distinct from leukocyte and platelet forms of the enzyme. J. Biol. Chem. **265:** 1771.
12. SHANNON, V. R., J. R. HANSBROUGH, Y. TAKAHASHI, N. UEDA, S. YAMAMOTO & M. J. HOLTZMAN. 1990. Biochemical and immunohistochemical evidence for selective expression of novel epithelial lipoxygenases. *In* Advances in Prostaglandin, Thromboxane, and Leukotriene Research. B. Samuelsson, R. Paoletti, P. W. Ramwell, G. Folco & E. Granstrom, Eds. 37. Raven Press. New York, NY.
13. SHANNON, V. R., E. C. CROUCH, N. UEDA, Y. TAKAHASHI, S. YAMAMOTO & M. J. HOLTZMAN. 1991. Related expression of arachidonate 12- and 15-lipoxygenases in animal and human lung tissue. Am. J. Physiol. (Lung Cell. Mol. Physiol.) **261:** L399.
14. HENKE, D., R. M. DANILOWICZ, J. F. CURTIS, R. C. BOUCHER & T. E. ELING. 1988. Metabolism of arachidonic acid by human nasal and bronchial epithelial cells. Arch. Biochem. Biophys. **267:** 426.
15. IZUMI, T., S. HOSHIKO, O. RADMARK & B. SAMUELSSON. 1990. Cloning of the cDNA for human 12-lipoxygenase. Proc. Natl. Acad. Sci. USA **87:** 7477.
16. FUNK, C. D., L. FURCI & G. A. FITZGERALD. 1990. Molecular cloning, primary structure, and expression of the human platelet/erythroleukemia cell 12-lipoxygenase. Proc. Natl. Acad. Sci. USA **87:** 5638.
17. YOSHIMOTO, T., H. SUZUKI, S. YAMAMOTO, T. TAKAI, C. YOKOYAMA & T. TANABE. 1990. Cloning and sequence analysis of the cDNA for arachidonate 12-lipoxygenase of porcine leukocytes. Proc. Natl. Acad. Sci. USA **87:** 2142.
18. DE MARZO, N., D. L. SLOANE, S. DICHARRY, E. HIGHLAND & E. SIGAL. 1992. Cloning and expression of an airway epithelial 12-lipoxygenase. Am. J. Physiol. (Lung Cell. Mol. Physiol.) **262:** L198.
19. DAHLEN, S., G. HANSSON, P. HEDQVIST, T. BJORCK, E. GRANSTROM & B. DAHLEN. 1983. Allergen challenge of lung tissue from asthmatics elicits bronchial contraction that correlates with the release of leukotrienes C_4, D_4, and E_4. Proc. Natl. Acad. Sci. USA **80:** 1712.
20. HAMBERG, M., P. HEDQVIST & K. RADEGRAN. 1980. Identification of 15-hydroxy-5,8,11,13-eicosatetraenoic acid (15-HETE) as a major metabolite of arachidonic acid in human lung. Acta Physiol. Scand. **110:** 219.

21. KUMLIN, M., M. HAMBERG, E. GRANSTROM, T. BJORCK, B. DAHLEN, H. MATSUDA, O. ZETTERSTROM & S.-E. DAHLEN. 1990. 15(S)-Hydroxyeicostetraenoic acid is the major arachidonic acid metabolite in human bronchi: Association with airway epithelium. Arch. Biochem. Biophys. **282:** 254.
22. MURRAY, J. J., A. B. TONNEL, A. R. BRASH, L. J. ROBERTS, P. GOSSET, R. WORKMAN, A. CAPRON & J. A. OATES. 1986. Release of prostaglandin D_2 into human airways during acute antigen challenge. N. Engl. J. Med. **315:** 800.
23. SHANNON, V. R., P. CHANEZ, J. BOUSQUET & M. J. HOLTZMAN. 1993. Histochemical evidence for induction of arachidonate 15-lipoxygenase in airway disease. Am. Rev. Respir. Dis. **147:** 1024.
24. SHANNON, V. R., W. F. STENSON & M. J. HOLTZMAN. 1993. Induction of epithelial arachidonate 12-lipoxygenase at active sites of inflammatory bowel disease. Am. J. Physiol. (Gastrointest. Liver Physiol.) **264:** G104.
25. HAMMARSTROM, S., M. HAMBERG, B. SAMUELSSON, E. A. DUELL, M. STAWISKI & J. J. VOORHEES. 1975. Increased concentrations of nonesterified arachidonic acid, 12L-hydroxy-5,8,10,14-eicosatetraenoic acid, prostaglandin E_2, and prostaglandin $F_{2\alpha}$ in epidermis of psoriasis. Proc. Natl. Acad. Sci. USA **72:** 5130.
26. BRAIN, S., R. CAMP, P. DOWD, A. KOBZA BLACK & M. GREAVES. 1984. The release of leukotriene B_4-like material in biologically active amounts from the lesional skin of patients with psoriasis. J. Invest. Dermatol. **83:** 70.
27. CAMP, R. D. R., A. I. MALLET, P. M. WOOLARD, S. D. BRAIN, A. KOBZA BLACK & M. W. GREAVES. 1983. The identification of hydroxy fatty acids in psoriatic skin. Prostaglandins **26:** 431.
28. DUELL, E. A., C. N. ELLIS & J. J. VOORHEES. 1988. Determination of 5,12, and 15-lipoxygenase products in keratomed biopsies of normal and psoriatic skin. J. Invest. Dermatol. **91:** 446.
29. HAMMARSTROM, S., J. A. LINDGREN, C. MARCELO, E. A. DUELL, T. F. ANDERSON & J. J. VOORHEES. 1979. Arachidonic acid transformations in normal and psoriatic skin. J. Invest. Dermatol. **73:** 180.
30. ZIBOH, V. A., T. L. CASEBOLT, C. L. MARCELO & J. J. VOORHEES. 1984. Biosynthesis of lipoxygenase products by enzyme preparations from normal and psoriatic skin. J. Invest. Dermatol. **83:** 426.
31. SOTER, N. A., R. A. LEWIS, E. J. COREY & K. F. AUSTEN. 1983. Local effects of synthetic leukotrienes (LTC_4, LTD_4, LTE_4, and LTB_4) in human skin. J. Invest. Dermatol. **80:** 115.
32. CHAN, C. C., L. DUHAMEL & A. FORD-HUTCHISON. 1985. Leukotriene B_4 and 12-hydroxyeicosatetraenoic acid stimulate epidermal proliferation in vivo in the guinea pig. J. Invest. Dermatol. **85:** 333.
33. WOOLLARD, P. M. 1986. Stereochemical difference between 12-hydroxy-5,8,10,14-eicosatetraenoic acid in platelets and psoriatic lesions. Biochem. Biophys. Res. Commun. **136:** 169.
34. CAPDEVILA, J., P. YADAGIRI, S. MANNA & J. R. FALCK. 1986. Absolute configuration of the hydroxyeicosatetraenoic acids (HETEs) formed during catalytic oxygenation of arachidonic acid by microsomal cytochrome P-450. Biochem. Biophys. Res. Commun. **141:** 1007.
35. HOLTZMAN, M. J., J. TURK & A. PENTLAND. 1989. A regiospecific monooxygenase with novel stereopreference is the major pathway for arachidonic acid oxygenation in isolated epidermal cells. J. Clin. Invest. **84:** 1446.
36. SHORNICK, L. P. & M. J. HOLTZMAN. 1993. A cryptic, microsomal-type arachidonate 12-lipoxygenase is tonically inactivated by oxidation-reduction conditions in cultured epithelial cells. J. Biol. Chem. **268:** 371.

37. CHANG, W.-C., C.-C. NING, M. T. LIN & J.-D. HUANG. 1992. Epidermal growth factor enhances a microsomal 12-lipoxygenase activity in A431 cells. J. Biol. Chem. **267:** 3657.
38. HUSSAIN, H., L. P. SHORNICK, V. R. SHANNON, J. D. WILSON, C. D. FUNK, A. P. PENTLAND & M. J. HOLTZMAN. 1994. Epidermis contains platelet-type 12-lipoxygenase that is overexpressed in germinal layer keratinocytes in psoriasis. Am. J. Physiol. (Cell Physiol.) **266:** C243.
39. SETTY, B. N. Y., J. E. GRAEBER & M. J. STUART. 1987. The mitogenic effect of 15- and 12-hydroxyeicosatetraenoic acid on endothelial cells may be mediated by diacylglycerol kinase inhibition. J. Biol. Chem. **262:** 17613.
40. SHANNON, V. R., S. RAMANADHAN, J. TURK & M. J. HOLTZMAN. 1992. Selective expression of an arachidonate 12-lipoxygenase by pancreatic islet β-cells. Am. J. Physiol. (Endocrinol. Metab.) **263:** E828.
41. PIOMELLI, D., A. VOLTERRA, N. DALE, S. A. SIEGELBAUM, E. R. KANDEL, J. H. SCHWARTZ & F. BALARDETTI. 1987. Lipoxygenase metabolites of arachidonic acid as second messengers for presynaptic inhibition of Aplysia sensory cells. Nature **328:** 38.
42. MAROM, Z., J. H. SHELHAMER, F. SUN & M. KALINER. 1983. Human airway monohydroxyeicosatetraenoic acid generation and mucus release. J. Clin. Invest. **72:** 122.
43. TANG, D. G., Y. Q. CHEN, C. A. DIGLIO & K. V. HONN. 1993. Protein kinase C-dependent effects of 12(S)-HETE on endothelial cell vitronectin receptor and fibronectin receptor. J. Cell Biol. **121:** 689.
44. HOLTZMAN, M. J. 1992. Arachidonic acid metabolism in airway epithelial cells. Annu. Rev. Physiol. **54:** 303.
45. BRASH, A. R. 1985. Review of the possible roles of the platelet 12-lipoxygenase. Circulation **72:** 702.
46. STENSON, W. F. & C. W. PARKER. 1979. Metabolism of arachidonic acid in ionophore-stimulated neutrophils. J. Clin. Invest. **64:** 1457.
47. FUNK, C. D. & W. S. POWELL. 1985. Release of prostaglandins and monohydroxy and trihydroxy metabolites of linoleic and arachidonic acids by adult and fetal aortae and ductus arteriosus. J. Biol. Chem. **260:** 7481.
48. BREZINSKI, M. E. & C. N. SERHAN. 1990. Selective incorporation of (15S)-hydroxyeicosatetraenoic acid in phosphatidylinositol of human neutrophils: Agonist-induced deacylation and transformation of stored hydroxyeicosanoids. Proc. Natl. Acad. Sci. USA **87:** 6248.
49. LEGRAND, A. B., J. A. LAWSON, B. O. MEYRICK, I. A. BLAIR & J. A. OATES. 1991. Substitution of 15-hydroxyeicostetraenoic acid in the phosphoinositide signaling pathway. J. Biol. Chem. **266:** 7570.
50. BRASH, A. R., C. D. INGRAM & T. M. HARRIS. 1987. Analysis of a specific oxygenation reaction of soybean lipoxygenase with fatty acids esterified in phospholipids. Biochemistry **26:** 5465.
51. JUNG, G., D. C. YANG & A. NAKAO. 1985. Oxygenation of phosphatidylcholine by human polymorphonuclear leukocyte 15-lipoxygenase. Biochem. Biophys. Res. Commun. **130:** 559.
52. KUHN, H. & A. R. BRASH. 1990. Occurrence of lipoxygenase products in membranes of rabbit reticulocytes. J. Biol. Chem. **265:** 1454.
53. KUHN, H., J. BELKNER, R. WIESNER & A. R. BRASH. 1990. Oxygenation of biological membranes by the pure reticulocyte lipoxygenase. J. Biol. Chem. **265:** 18351.
54. MURRAY, J. J. & A. R. BRASH. 1988. Rabbit reticulocyte lipoxygenase catalyzes specific 12(S) and 15(S) oxygenation of arachidonoyl-phosphatidylcholine. Arch. Biochem. Biophys. **265:** 514.
55. ROSEN, G. D., T. BIRKENMEIER, A. RAZ & M. J. HOLTZMAN. 1989. Identification of a cyclooxygenase-related gene and its potential role in prostaglandin formation. Biochem. Biophys. Res. Commun. **164:** 1358.

56. Holtzman, M. J., J. Turk & L. P. Shornick. 1992. Identification of a pharmacologically-distinct prostaglandin H synthase in cultured epithelial cells. J. Biol. Chem. **267:** 21438.
57. Pentland, A. P., M. Mahoney, S. C. Jacobs & M. J. Holtzman. 1990. Enhanced prostaglandin synthesis after UV injury is mediated by endogenous histamine stimulation: A mechanism for irradiation erythema. J. Clin. Invest. **86:** 566.
58. DeWitt, D. L. & W. L. Smith. 1988. Primary structure of prostaglandin G/H synthase from sheep vesicular gland determined from the complementary DNA sequence. Proc. Natl. Acad. Sci. USA **85:** 1412.
59. DeWitt, D. L., E. A. El-Harith, S. A. Kraemer, M. J. Andrews, E. F. Yao, R. L. Armstrong & W. L. Smith. 1990. The aspirin and heme-binding sites of ovine and murine prostaglandin endoperoxide synthases. J. Biol. Chem. **265:** 5192.
60. Funk, C. D., L. B. Funk, M. E. Kennedy, A. S. Pong & G. A. FitzGerald. 1991. Human platelet/erythroleukemia cell prostaglandin G/H synthase: cDNA cloning, expression, and gene chromosomal assignment. FASEB J. **5:** 2304.
61. Merlie, J. P., D. Fagan, J. Mudd & P. Needleman. 1988. Isolation and characterization of the complementary DNA for sheep seminal vesicle prostaglandin endoperoxide synthase (cyclooxygenase). J. Biol. Chem. **263:** 3550.
62. Yokoyama, C., T. Takai & T. Tanabe. 1988. Primary structure of sheep prostaglandin endoperoxide synthase deduced from cDNA sequence. FEBS Lett. **231:** 347.
63. Yokoyama, C. & T. Tanabe. 1989. Cloning of the human gene encoding prostaglandin endoperoxide synthase and primary structure of the enzyme. Biochem. Biophys. Res. Commun. **165:** 888.
64. Hemler, M. E., C. G. Crawford & W. E. M. Lands. 1978. Lipoxygenation activity of purified prostaglandin-forming cyclooxygenase. Biochemistry **17:** 1772.
65. Setty, B. N. Y., M. J. Stuart & R. W. Walenga. 1985. Formation of 11-hydroxyeicosatetraenoic acid and 15-hydroxyeicosatetraenoic acid in human umbilical arteries is catalyzed by cyclooxygenase. Biochim. Biophys. Acta **833:** 484.
66. Hamberg, M. & B. Samuelsson. 1967. On the mechanism of the biosynthesis of prostaglandins E_1 and $F_{1\alpha}$. J. Biol. Chem. **242:** 5336.
67. Hamberg, M. & B. Samuelsson. 1967. Oxygenation of unsaturated fatty acids by the vesicular gland of sheep. J. Biol. Chem. **242:** 5344.
68. Maas, R. L., C. D. Ingram, D. F. Taber, J. A. Oates & A. R. Brash. 1982. Stereospecific removal of the D_R hydrogen atom at the 10-carbon of arachidonic acid in the biosynthesis of leukotriene A_4 by human leukocytes. J. Biol. Chem. **257:** 13515.
69. Maas, R. L. & A. R. Brash. 1983. Evidence for a lipoxygenase mechanism in the biosynthesis of epoxide and dihydroxyl leukotrienes from 15(S)-hydroperoxyicosatetraenoic acid by human platelets and porcine leukocytes. Proc. Natl. Acad. Sci. USA **80:** 2884.
70. Xie, W., J. G. Chipman, D. L. Robertson, R. L. Erikson & D. L. Simmons. 1991. Expression of a mitogen-responsive gene encoding prostaglandin synthase is regulated by mRNA splicing. Proc. Natl. Acad. Sci. USA **88:** 2692.
71. Kujubu, D. A., B. S. Fletcher, B. C. Varnum, R. W. Lim & H. R. Herschman. 1991. TIS10, a phorbol ester tumor promoter-inducible mRNA from Swiss 3T3 cells, encodes a novel prostaglandin synthase/cyclooxygenase homologue. J. Biol. Chem. **266:** 12866.
72. O'Banion, M. K., H. B. Sadowski, V. Winn & D. A. Young. 1991. A serum- and glucocorticoid-regulated 4-kilobase mRBA encodes a cyclooxygenase-related protein. J. Biol. Chem. **266:** 23261.
73. Hla, T. & K. Neilson. 1992. Human cyclooxygenase-2 cDNA. Proc. Natl. Acad. Sci. USA **89:** 7384.
74. Fletcher, B. S., D. A. Kujubu, D. M. Perrin & H. R. Herschman. 1992. Structure of the mitogen-inducible TIS10 gene and demonstration that the TIS10-encoded protein is a functional prostaglandin G/H synthase. J. Biol. Chem. **267:** 4338.

75. SHIMOKAWA, T. & W. L. SMITH. 1993. Prostaglandin endoperoxide synthase. J. Biol. Chem. **267:** 12387.
76. JACKSON, R. J. 1993. Cytoplasmic regulation of mRNA function: The importance of the 3′ untranslated region. Cell **74:** 9.
77. O'BANION, M. K., V. D. WINN & D. A. YOUNG. 1992. cDNA cloning and functional activity of a glucocorticoid-regulated inflammatory cyclooxygenase. Proc. Natl. Acad. Sci. USA **89:** 4888.
78. KUJUBU, D. A. & H. R. HERSCHMAN. 1992. Dexamethasone inhibits mitogen induction of the TIS10 prostaglandin synthase/cyclooxygenase gene. J. Biol. Chem. **267:** 7991.
79. JONES, D. A., D. P. CARLTON, T. M. MCINTYRE, G. A. ZIMMERMAN & S. M. PRESCOTT. 1993. Molecular cloning of human prostaglandin endoperoxide synthase type II and demonstration of expression in response to cytokines. J. Biol. Chem. **268:** 9049.

Leukotriene C_4 Synthase and 5-Lipoxygenase Activating Protein

Regulators of the Biosynthesis of Sulfido-Leukotrienes

A. W. FORD-HUTCHINSON[a]

Merck Frosst Centre for Therapeutic Research
Kirkland, Quebec
Canada H9H 3L1

INTRODUCTION

Leukotrienes are products of arachidonic acid metabolism derived through the 5-lipoxygenase enzyme pathway. Leukotrienes have been implicated as mediators of inflammation and immediate hypersensitivity reactions—in particular, human bronchial asthma—and thus a considerable effort has been expended by the pharmaceutical industry to produce either inhibitors of the production or blockers of the action of these mediators. Various therapeutic approaches have been used including direct 5-lipoxygenase inhibitors, indirect leukotriene biosynthesis inhibitors that bind to 5-lipoxygenase activating protein and leukotriene D_4 receptor antagonists. In particular, there have been a number of clinical trials utilizing leukotriene D_4 receptor antagonists such as MK-571 and its individual enantiomer MK-0679, and ICI 204219 (Acculate).[1] These studies with receptor antagonists have indicated a role for leukotriene D_4 receptor activation in antigen-, exercise-, and aspirin-induced bronchoconstriction and antigen-induced hyperresponsiveness. In addition, leukotriene D_4 receptor antagonists have been shown to produce acute bronchodilatation in moderate to severe asthmatic subjects, suggesting that leukotrienes may contribute to baseline abnormal bronchomotor tone. In chronic studies, these agents have been shown to produce improvements in objective parameters, such as FEV_1 and peak expiratory flow rates as well as patient-assessed parameters such as reductions in β-agonist usage, improvements in symptom scores, and reductions in nocturnal awakenings due to asthma symptomology. In addition to leukotriene D_4 receptor antagonists other therapeutic approaches are possible—in particular, blockade of key steps in the biosynthesis of sulfido-leukotrienes. Two such targets will be described in the present article, 5-lipoxygenase activating protein (FLAP) and leukotriene C_4 synthase.

[a] Address for correspondence: A. W. Ford-Hutchinson, Merck Frosst Centre for Therapeutic Research, 16711 TransCanada Highway, Kirkland, Quebec, Canada H9H 3L1.

5-LIPOXYGENASE ACTIVATING PROTEIN

The enzyme 5-lipoxygenase catalyzes the first committed step in the biosynthesis of leukotrienes from arachidonic acid. The enzymic reaction is a two-step process involving oxygenation of arachidonic acid to produce the intermediate, 5-hydroperoxyeicosatetraenoic acid, followed by a dehyration step to produce the epoxide intermediate, leukotriene A_4. Subsequent metabolism of leukotriene A_4 can occur through either leukotriene A_4 hydrolase to produce the dihydroxy fatty acid leukotriene B_4 or through conjugation with glutathione by leukotriene C_4 synthase to produce the first of the sulfido-leukotrienes, leukotriene C_4. Leukotriene B_4 has been implicated as a mediator of inflammation and interacts with high-affinity, G protein-coupled receptors, which mediate a number of leukocyte and lymphocyte functions. Leukotriene C_4 can be further metabolized to leukotrienes D_4 and E_4, and the three compounds can interact with another high-affinity G protein-coupled receptor, the leukotriene D_4 receptor. Leukotriene D_4 receptor activation can result in bronchoconstriction, vascular permeability changes, induction of bronchial hyperreactivity in asthmatic subjects, modulation of mucus hypersecretion, and mediation of eosinophil migration into the lung. In addition, increased amounts of leukotrienes are excreted in asthmatic subjects, particularly following antigen and aspirin provocation. The above observations form the original rationale for testing leukotriene D_4 receptor antagonists in man.

During the search for inhibitors of leukotriene biosynthesis, compounds were discovered that were active in whole cells at preventing leukotriene biosynthesis but had no effect on the enzyme 5-lipoxygenase. One such compound was MK-886 (3-[(1-(4-chlorobenzyl-3-*t*-butyl-thio-5-isopropyl-2-yl)2,2-dimethyl-propanoicacid].[2-4] In addition to having no effect on 5-lipoxygenase activity on either the purified enzyme or in broken-cell preparations, MK-886 had no effect on either arachidonic acid release or other oxygenases such as cyclooxygenase and 12- and 15-lipoxygenases.[5] One suggestion was that compounds such as MK-886 blocked the activation of 5-lipoxygenase. Evidence for an activation step was obtained in studies that showed that activation of neutrophils to produce leukotrienes resulted in translocation of 5-lipoxygenase from the cytosol to a membrane fraction, and this translocation could be blocked by inhibitors such as MK-886.[2] The fact that MK-886 was active in whole-cell preparations at nanomolar concentrations suggested that it might be binding to a target site with high affinity. Such a target protein was identified through the use of photoaffinity labels and purified through the use of affinity columns based around the structure of MK-886.[3] Sequence was obtained on this novel 18-kDa protein, which was present only in the cells known to produce leukotrienes, and the enzyme was cloned and expressed in various cell lines.[4] This protein has been termed 5-lipoxygenase activating protein (FLAP), and its structure has been hypothesized to consist of three transmembrane spanning regions, with the C and N termini on opposite sides of the membrane.

Through the use of a series of transfection studies it became apparent that FLAP was essential for cellular leukotriene biosynthesis.[4] The current literature indicates that all cells known to synthesizes leukotriene contain both 5-lipoxygenase and FLAP. This has raised a question as to the biological function of FLAP. The original hypothesis was that FLAP might act as a "docking" protein for 5-lipoxygenase in the membrane. This hypothesis would require that a stable complex be formed at the membrane between active 5-lipoxygenase and FLAP, as well as other possible components of the leukotriene biosynthetic pathway, such as cytosolic arachidonic acid-specific phospholipase A_2. This complex would then regulate the action of the enzyme on its substrate, arachidonic acid, resulting in efficient leukotriene biosynthesis. This hypothesis suggests that drugs such as MK-886 bind to a site on FLAP that mediates the interaction of FLAP with 5-lipoxygenase, thus blocking a protein-protein reaction. This would require that there be a good correlation between inhibition of leukotriene biosynthesis by such compounds and inhibition of the translocation of 5-lipoxygenase from the cytosol to the membrane. A number of recent studies have suggested that this may not be the case. Thus, Coffey *et al.* have shown that in resting alveolar macrophages, all 5-lipoxygenase is found in the membrane fraction.[6] MK-886 was unable to reverse this membrane association, although it was a very efficient inhibitor of leukotriene biosynthesis in these cells. In addition, FLAP-independent translocation of 5-lipoxygenase has been shown in osteosarcoma cells transfected with the 5-lipoxygenase gene alone, this translocation being unaffected by MK-886.[7] Finally, a series of studies on analogs BAY X1005, a quinoline-based leukotriene biosynthesis inhibitor, have shown a lack of correlation between inhibition of leukotriene biosynthesis and inhibition of 5-lipoxygenase translocation, although a good correlation was observed between binding to FLAP and inhibition of leukotriene biosynthesis.[8,9]

These results have suggested that the MK-886 binding site on FLAP directly mediates an interaction with 5-lipoxygenase. The current hypothesis is that this site may be an arachidonic acid binding site that allows for the transfer of arachidonic acid to 5-lipoxygenase to permit the enzymic reaction to proceed in a more efficient fashion. Evidence for this has been obtained using a novel photoaffinity analog of arachidonic acid that binds to FLAP, this binding being inhibited by drugs such as MK-886.[10] Other evidence has been obtained in polymorphonuclear leukocytes, where Hill *et al.* have obtained active 5-lipoxygenase on the membranes by stimulating the cells with ionophore A23187 in the presence of a direct 5-lipoxygenase inhibitor, Zileuton.[11] When the experiments are carried out under these conditions, accumulation of active, membrane-associated 5-lipoxygenase was inhibited and reversed by MK-886. This membrane-associated 5-lipoxygenase was more efficient in producing leukotriene A_4 from arachidonic acid-derived 5-hydroperoxyeicosatetraenoic acid than the cytosolic enzyme and in addition showed a different substrate specificity profile, being able to metabolize additional arachidonic acid metabolites. Finally, a series of studies have been carried out in sf-9 insect cells infected with recombinant baculoviruses to express either the human 5-lipoxygenase gene or human 5-lipoxygenase and human FLAP together. In these experiments, FLAP was shown to stimulate utilization of arachidonic acid by 5-lipoxygenase and increase the efficiency with which 5-lipoxygenase converted 5-hydroperoxyeicosatetraenoic acid to leukotriene A_4.[12]

It was obviously of considerable interest to study the cellular localization of FLAP. Woods *et al.* have used immunoelectronmicroscopic labeling of ultrathin frozen sections to directly visualize the intracellular location of both 5-lipoxygenase and FLAP in resting and ionophore-activated human polymorphonuclear leukocytes and monocytes.[13] Interestingly, in activated cells both 5-lipoxygenase and FLAP were found to be localized to the cisternum of the nuclear envelope and the associated endoplasmic reticulum. In contrast, in resting cells the distribution of FLAP was identical to that observed in activated cells, but 5-lipoxygenase was not present in the nuclear envelope. These results are the first direct demonstration that FLAP is present in the membrane following stimulation of the cell and that 5-lipoxygenase moves to the same membrane site as FLAP. These results were confirmed by subcellular fractionation studies monitoring the presence of FLAP in various fractions by immunoblot and in a FLAP binding assay.[13] A similar study in ionophore-stimulated rat peritoneal macrophages has shown that both 5-lipoxygenase and cytosolic phospholiphase A_2 redistribute from the high-speed supernatant to a nuclear fraction. FLAP was localized predominantly to the nuclear fraction in both resting and activated cells.[14] This study points to the coordinated subcellular localization of the proteins involved in leukotriene synthesis from endogenous arachidonic acid.

In conclusion, it is now clear that FLAP is localized to the nuclear envelope and the adjacent endoplasmic reticulum. Following activation of the cell, 5-lipoxygenase and cytosolic phospholipase A_2 appear to translocate to the nuclear fraction, and FLAP may act as an arachidonic acid transfer protein allowing for the efficient synthesis of leukotrienes from the substrate, arachidonic acid. The fact that 5-lipoxygenase can function as an effective enzyme independently of FLAP both in broken-cell preparations and as a pure recombinant protein suggests that FLAP does not simply activate 5-lipoxygenase in the cell but may also counteract some inhibition or constraint for leukotriene biosynthesis.

LEUKOTRIENE C_4 SYNTHASE

Another possible site for therapeutic intervention in the synthesis of sulfido-leukotrienes is the enzyme leukotriene C_4 synthase. This enzyme catalyzes the conjugation of glutathione to leukotriene A_4 and can thus be considered as a glutathione S-transferase-like enzyme. The purification and characterization of this enzyme has been carried out in two human monocytic cell lines, THP-1 and differentiated U937 cells.[15–17] Leukotriene C_4 synthase enzymic activity could be solubilized from microsomal pellets by the anionic detergent taurocholate. This solubilization procedure leaves all the microsomal glutathione S-transferase activity in the membrane. The enzyme could be further distinguished from α, μ, π, and θ glutathione S-transferases through chromatographic separations, substrate specificity, and Western blot analysis. The enzyme has been purified to homogeneity from THP-1 cells (≥25,000-fold purification) through a multistep purification scheme and can be visualized as a single 18-kDa polypeptide using SDS polyacrylamide gel electrophoresis. Evidence that this enzyme was indeed leukotriene C_4 synthase was obtained through the use of a leukotriene D_4 photoaffinity probe. In a series of competition experiments inhibition of binding of the photoaffinity probe was compared to enzymic activity. These studies showed no correlation with inhibition of microsomal glutathione S-transferase activity but did show a correlation with inhibition of leukotriene C_4 synthase activation, which has led to the conclusion that the 18-kDa polypeptide is in fact leukotriene C_4 synthase itself.[18] The enzyme may function as a homodimer, as evidenced by data obtained on calibrated gel filtration columns where in the presence of three detergents the

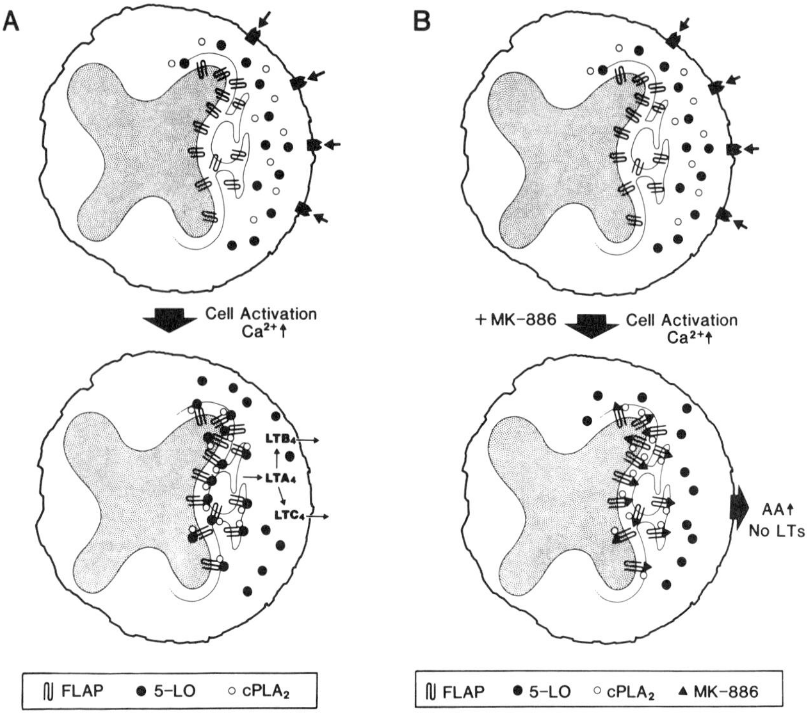

FIGURE 1. A. Proposed mechanism for leukotriene biosynthesis in intact cells showing translocation of 5-lipoxygenase (5-LO) and cytosolic phospholipase A_2 ($cPLA_2$) to FLAP. **B.** Inhibition of leukotriene biosynthesis in the presence of FLAP inhibitors such as MK-886. (Reproduced, with permission, from *The Annual Review of Biochemistry*, Volume 63, 1994, by Annual/Reviews Inc.)

enzyme eluted a volume corresponding to 39.2 ± 3.3 kDa.[17] Sequence data has been obtained on the N-terminal 35 amino acids, and this has indicated that leukotriene C_4 synthase is a unique protein, composed of predominantly hydrophobic amino acids and containing a consensus sequence for protein kinase C phosphorylation.[17] Phosphorylation may negatively down-regulate leukotriene C_4 synthase activity, as shown in DMSO-differentiated HL-60 cells, which resemble mature granulocytes. In these cells the synthesis of sulfido-leukotrienes was specifically inhibited following incubation with ionophore A23187 and the tumor promotor phorbol 12-myristate 12-acetate (Kargman *et al.*, submitted for publication). Under these conditions the syntheses of leukotrienes B_4 and A_4 were unaffected. This inhibition of leukotriene C_4 production could be reversed in the presence of protein kinase inhibitors such as staurosporine and bisindolymaleimide, whereas inhibitors of protein kinase A, tyrosine kinase, and the respiratory burst oxidase were ineffective.

These results indicate that human leukotriene C_4 synthase is a unique glutathione S-transferase, unrelated to known existing families of glutathione S-transferases, and is composed of an 18-kDa polypeptide that is enzymatically active as the homodimer and may be negatively phosphoregulated in cell systems. Because of the unique nature of this enzyme, it represents a potential therapeutic target for regulation of sulfido-leukotriene synthesis.

REFERENCES

1. Ford-Hutchinson, A. W. 1993. Springer Semin. Immunopathol. **15:** 37-50.
2. Rouzer, C. A., A. W. Ford-Hutchinson, H. E. Morton & J. W. Gillard. 1990. J. Biol. Chem. **265:** 1436-1442.
3. Miller, D. K., J. W. Gillard, P. J. Vickers, S. Sadowski, C. Léveillé *et al.* 1990. Nature **343:** 278-281.
4. Dixon, R. A. F., R. E. Diehl, E. Opas, E. Rands, P. J. Vickers *et al.* 1990. Nature **343:** 282-284.
5. Gillard, J. W., A. W. Ford-Hutchinson, C. Chan, S. Charleson, D. Denis *et al.* 1989. Can. J. Physiol. Pharmacol. **67:** 456-464.
6. Coffey, M., M. Peters-Golden, J. C. Fantone III & P. H. S. Sporn. 1992. J. Biol. Chem. **267:** 570-576.
7. Kargman, S., P. J. Vickers & J. F. Evans. 1992. J. Cell. Biol. **119:** 1701-1709.
8. Hatzelmann, A., R. Fruchtmann, K. H. Mohrs, S. Raddatz & R. Müller-Peddinghaus. 1993. Biochem. Pharmacol. **45:** 101-111.
9. Hatzelmann, A., R. Fruchtmann, K. H. Mohrs, M. Matzke, U. Pleiss *et al. In* Advances in Prostaglandin, Thromboxane and Leukotriene Research. B. Samuelsson, S. E. Dahlén, J. Fritsh & P. Hedqvist, Eds. Raven Press. New York, NY. In press.
10. Mancini, J. A., M. Abramovitz, M. E. Cox, E. Wong, S. Charleson *et al.* 1993. FEBS Lett. **318:** 277-281.
11. Hill, E., J. Maclouf, R. C. Murphy & P. M. Henson. 1992. J. Biol. Chem. **267:** 22048-22053.
12. Kennedy, B. P., R. E. Diehl, Y. Boie, M. Adam & R. A. F. Dixon. 1991. J. Biol. Chem. **266:** 8511-8516.
13. Woods, J. W., J. F. Evans, D. Ethier, S. Scott, P. J. Vickers *et al.* 1993. J. Exp. Med. **178:** 1935-1946.
14. Peters-Golden, M. & R. W. McNish. 1993. Biochem. Biophys. Res. Common. **196:** 147-153.
15. Nicholson, D. W., A. Ali, N. W. Klemba, N. A. Munday, R. J. Zamboni & A. W. Ford-Hutchinson. 1992. J. Biol. Chem. **267:** 17849-17857.
16. Nicholson, D. W., M. W. Klemba, D. M. Rasper, R. J. Zamboni & A. W. Ford-Hutchinson. 1992. Eur. J. Biochem. **209:** 725-734.
17. Nicholson, D. W., A. Ali, J. P. Vaillancourt, J. R. Calaycay, R. A. Mumford, R. J. Zamboni & A. W. Ford-Hutchinson. 1993. Proc. Natl. Acad. Sci. U.S.A. **90:** 2015-2019.
18. Ali, A., R. J. Zamboni, A. W. Ford-Hutchinson & D. W. Nicholson. 1993. FEBS Lett. **317:** 195-201.

Cytokine Regulation of Mast Cell Protease Phenotype and Arachidonic Acid Metabolism

MAKOTO MURAKAMI, JONATHAN P. ARM, AND K. FRANK AUSTEN[a]

Department of Medicine
Harvard Medical School
and
Department of Rheumatology and Immunology
Brigham and Women's Hospital
Boston, Massachusetts 02115

THE FUNCTION OF SECRETORY AND CYTOSOLIC PHOSPHOLIPASE A_2 DURING MAST CELL ACTIVATION

Heterogeneity of Phospholipase A_2 Enzymes

Phospholipase A_2 (PLA_2) is an enzyme that catalyzes liberation of free fatty acids from the *sn*-2 position of glycerophospholipids. This reaction is of particular significance when the fatty acid is arachidonic acid, since it serves as the rate-limiting step for synthesis of prostaglandins and leukotrienes. The other product of PLA_2 cleavage is lysophospholipid, the intermediate precursor of platelet-activating factor. Thus an understanding of the nature and regulation of PLA_2 is important to understanding the role of lipid mediators in inflammation and the potential mechanisms by which these products may be regulated.

Three classes of PLA_2 enzymes have been characterized in mammalian cells. There are two secretory forms ($sPLA_2$), each with a molecular mass of 14 kDa, termed group I (pancreatic) and group II (nonpancreatic) PLA_2.[1–4] They hydrolyze the *sn*-2 position of phospholipids without any selectivity for arachidonic acid in the presence of millimolar concentrations of Ca^{2+}. While this may appear inconsistent with selective arachidonate release upon cell activation, several reports suggest the participation of group II PLA_2 in eicosanoid generation under certain conditions.[5,6] The finding of a third family of PLA_2, namely 85-kDa cytosolic PLA_2 ($cPLA_2$), which is selective for arachidonate-containing phospholipids, has provided new insight into the regulation of arachidonate release in response to receptor-mediated signaling.[7–9]

Secretory PLA_2

Pancreatic PLA_2 is classified as group I PLA_2 based on its sequence homology with PLA_2 in snake venoms.[1] The main role of pancreatic group I PLA_2 is digestion

[a] To whom correspondence should be addressed.

of phospholipids in food. Recently it has become clear that group I PLA_2 is distributed in other tissues, such as spleen and lung,[2] although its function in these tissues is unknown. Hanasaki and Arita demonstrated that group I PLA_2 can regulate the proliferation of vascular smooth muscle cells and fibroblasts *in vitro* via a specific binding site on the cell surface.[10] This is the first demonstration that $sPLA_2$ has a function unrelated to its phospholipid-hydrolyzing activity.

The primary structure of an $sPLA_2$ obtained from human and rat reveals that it shares some typical characteristics with snake venom group II PLA_2.[3,4] Although the homology between mammalian group I and group II PLA_2 is no more than 30%, the residues essential for the hydrolysis of a phospholipid substrate, such as the catalytic site, Ca^{2+}-binding site, and hydrophobic surface-recognition site, are well conserved. One of the most significant differences between group I and group II PLA_2 is that group I PLA_2 is synthesized as the proenzyme (zymogen) manifesting only limited activity and is cleaved proteolytically to express its full activity. Group II PLA_2 does not exist as a pro-form, indicating that group II PLA_2 can exhibit full enzymatic activity once the signal sequence is properly processed. Regulatory mechanisms must therefore operate to avoid nonspecific cell membrane damage.

Inflammatory exudate contains group II PLA_2,[4] and platelets and mast cells store the enzyme in secretory granules for release by exocytosis upon stimulation.[3,11] Hence, the involvement of the PLA_2 in the process of inflammation has been postulated since its discovery. Injection of group II PLA_2 into the inflamed hindfoot of rats with adjuvant-induced arthritis exacerbates local tissue edema.[12] A series of reports demonstrated that inflammatory cytokines, such as interleukin (IL)-1, IL-6, and tumor necrosis factor (TNF)-α, up-regulated gene expression and secretion of group II PLA_2 in a wide variety of cells accompanied by the increased production of prostaglandins.[4,13] Dexamethasone, an antiinflammatory glucocorticoid, suppressed the induction of group II PLA_2 as well as prostaglandin synthesis stimulated by the cytokines.[14] PGI_2 synthesis in TNF-α-treated endothelial cells was enhanced significantly by exogenous group II PLA_2.[6] Moreover, an antibody neutralizing group II PLA_2 partly inhibited TNF-α-stimulated PGI_2 synthesis in human umbilical vein endothelial cells,[6] and antisense oligonucleotides specific for group II PLA_2 reduced the release of arachidonic acid from a macrophage-like cell line, $P388D_1$.[15]

Cytosolic PLA_2

The rapid and fairly selective release of arachidonic acid from membrane phospholipids following cellular activation with various kinds of agonists suggests the presence of a cellular PLA_2 that is activated during the process of stimulus-coupled signal transduction, so as to preferentially hydrolyze arachidonoyl residues. This novel type of PLA_2, now termed cytosolic PLA_2 ($cPLA_2$), was purified from the cytosolic fractions of various cells, including monocytes and platelets.[7,16] Purified $cPLA_2$ hydrolyzes phospholipids bearing arachidonic acid much more efficiently than those bearing other fatty acids. The catalytic activity of $cPLA_2$ increases sharply with an increase in Ca^{2+} concentration from 10^{-7} to 10^{-6} M, which corresponds to the range of change in cytosolic Ca^{2+} concentration when cells are activated. Thus, the activity of $cPLA_2$ may be regulated by the concentration of Ca^{2+} in the cytosol during stimulus-coupled signal transduction.

In 1991, the cDNA for human $cPLA_2$ was cloned from U937 cells.[7] The cDNA predicts a protein of 749 amino acids, with an estimated molecular mass of 85 kDa, showing no homology with 14-kDa group I and group II PLA_2 enzymes. Sequence analysis revealed the presence of a region showing high homology with a motif identified as the Ca^{2+}-dependent membrane-binding domain found in several proteins, such as protein kinase C, phospholipase C (PLC), GTPase-activating protein, and p53. The existence of such a motif in $cPLA_2$ is consistent with the observation that an increase in free Ca^{2+} from 10^{-7} to 10^{-6} M induced translocation of $cPLA_2$ from the cytosol to the cell membrane.[17]

The treatment of Chinese hamster ovary (CHO) cells overexpressing $cPLA_2$ with various agonists resulted in an increase of arachidonic acid release, demonstrating stimulus/receptor coupling of $cPLA_2$.[8] Interestingly, the activation of $cPLA_2$ with either G protein-coupled (e.g., ATP, thrombin) or tyrosine-kinase-coupled (e.g., epidermal growth factor, platelet-derived growth factor [PDGF]) ligands stimulated the phosphorylation of $cPLA_2$ on serine (not tyrosine) residues. The *in vitro* $cPLA_2$ activity was increased twofold by phosphorylation, and pretreatment of the cells with protein kinase inhibitors blocked agonist-induced phosphorylation as well as activation of $cPLA_2$. $cPLA_2$ contains a consensus phosphorylation site (around Ser 505) for the mitogen-activated protein (MAP) kinase, which is believed to serve as a convergence point that integrates diverse signal transduction pathways. Indeed, $cPLA_2$ is both phosphorylated and activated by MAP kinase.[9] It is reported that the pretreatment of homogenates of HL-60 cells with anti-$cPLA_2$ antibody attenuates the basal as well as GTP-γS- and Ca^{2+}-stimulated arachidonic acid release.[18] The activity of $cPLA_2$ might thus be regulated by G proteins either directly or indirectly.

Treatment of various cell lines for several hours with a range of cytokines including IL-1β, TNF-α, macrophage colony-stimulating factor (M-CSF), and transforming growth factor (TGF)-β stimulates eicosanoid synthesis, in association with increased $cPLA_2$ expression.[19,20] The cytokine-mediated increase in $cPLA_2$ activity and eicosanoid generation was suppressed by glucocorticoids. As discussed above, group II PLA_2 is also up-regulated by proinflammatory cytokines and inhibited by glucocorticoids. Thus, increased eicosanoid production in cytokine-stimulated cells may be mediated by both group II PLA_2 and $cPLA_2$.

There are several lines of evidence suggesting the presence of unidentified PLA_2 enzymes distinct from the well-characterized $sPLA_2$ and $cPLA_2$. These include a Ca^{2+}-independent PLA_2 widely distributed in mammalian cells, sheep erythrocyte PLA_2 tightly associated with the cell membrane, and both serum- and cell-associated acetylhydrolase, which inactivates PAF and oxidizes phospholipids. Recently a 30-kDa protein was cloned and considered to encode a novel PLA_2 that catalyzes the cleavage of the *sn*-2 fatty acid of phospholipids through the formation of a stable acylenzyme intermediate. Sequence homology suggests that the cloned enzyme mediating this transesterification is a member of a family of proteins collectively designated 14-3-3 proteins.[21] It is therefore evident that there are multiple distinct species of mammalian PLA_2, which likely play different roles under differing physiological and pathological conditions.

PLA_2 and Mast Cell Activation

Direct evidence for the participation of group II PLA_2 in mast cell activation has been obtained from the work of Inoue and his colleagues. They demonstrated that exogenously added group II PLA_2 alone showed modest direct effects on rat serosal mast cell prostaglandin synthesis.[5] However, exposure of rat serosal mast cells to group II PLA_2 augmented PGD_2 synthesis in response to suboptimal stimulation with IgE and antigen.[5] The mechanism by which $sPLA_2$ enhances eicosanoid generation by these mast cells has yet to be elucidated.

Several investigators have postulated the possible involvement of PLA_2 in degranulation in various secretory cells based on the observations that classical $sPLA_2$ inhibitors prevented various types of degranulation such as histamine release from mast cells, catecholamine release from chromaffin cells, and hormone release from pituitary cells. Release of histamine from rat serosal mast cells in response to IgE and antigen was suppressed by group II PLA_2-specific inhibitors.[11] Group II PLA_2 was expressed transiently on the cell surface and was secreted into the medium upon mast cell activation. Moreover, histamine release was induced by exposure of mast cells to exogenous group II PLA_2.[22] Binding of group II PLA_2 to the mast cell surface was inhibited by heparin, accompanied by suppression of the mast cell degranulating activity of group II PLA_2. These findings indicate that the ''ecto-form'' of group II PLA_2, released during exocytosis, might play a crucial role in the degranulation process. Localization of group II PLA_2 at the fusion sites of secretory granules and the plasma membrane is demonstrated by an immunocytochemical study of TNF-α-stimulated vascular smooth muscle cells, in which the apical membrane, where fusion with secretory granules occurred, is intensely stained with anti-group II PLA_2 antibody.[23] Since the catalytic activity of the enzyme appeared essential for the degranulation process, some ''fusogenic'' compounds derived from phospholipid hydrolysis, such as lysophospholipids, fatty acids, or their derivatives, might augment or mediate the fusion between the granule membrane and plasma membrane. The ''fusogenic'' compounds are not products of cyclooxygenase and 5-lipoxygenase pathways because inhibitors of these enzymes have no effect on degranulation of mast cells.

Arachidonic acid release from mast cells in response to cross-linking of FcϵRI is largely regulated by $cPLA_2$. The presence of $cPLA_2$ in mast cells is confirmed in mouse bone marrow-derived mast cells (mBMMC) and the rat mucosal mast cell line RBL-2H3.[24] FcϵRI cross-linking causes activation of several nonreceptor tyrosine kinases, resulting in activation of PLC-γ1 to induce phosphoinositide breakdown, leading to an increase in cytosolic Ca^{2+}, protein kinase C activation, and MAP kinase activation.[25] It is reasonable, therefore, to consider that the rapid activation of $cPLA_2$ observed in antigen-stimulated mast cells may result from its phosphorylation by MAP kinase together with its translocation to the membrane in response to the increase in cytosolic Ca^{2+}.

CYTOKINE REGULATION OF MAST CELL HETEROGENEITY AND FUNCTION

Mast Cell Heterogeneity

There are at least two distinct populations of mast cells in rats and mice. These are termed mucosal mast cells (MMC) and connective tissue mast cells (CTMC), based upon their anatomic distribution (reviewed in Ref. 26). The serosal mast cell (SMC) is generally studied as a CTMC surrogate but can be distinguished in the mouse by differences in the protease profile of the secretory granule.[27] Rat SMC stain with safranin after exposure to alcian blue and retain their staining characteristics after fixation in formalin; their secretory granules contain predominantly heparin proteoglycan, a chymotryptic serine protease termed RMCP-I, and a high content of histamine; they preferentially generate PGD_2 upon IgE-dependent stimulation.[28] In contrast, rat MMC stain with alcian blue, but not after fixation in formalin, and do not counterstain with safranin; they contain predominantly chondroitin sulfate-diB/E proteoglycan, a chymotryptic serine protease termed RMCP-II, and 12-fold less histamine than SMC; they preferentially generate LTC_4 upon IgE-dependent stimulation.[29] Mouse SMC resemble rat SMC in their staining characteristics, granule phenotype, and preferential generation of PGD_2 in response to immunologic stimulation.[30] There are limited direct data on mouse MMC due to their availability in such small numbers; their staining characteristics resemble those of rat MMC. However, phenotyping of neutral proteases by RNA blot analysis of *Trichinella spiralis*-infected mice has distinguished mouse MMC from SMC and CTMC. BALB/c mouse MMC preferentially express mouse mast cell protease (mMCP)-1 and mMCP-2.[31,32] SMC preferentially express mMCP-4, mMCP-5, mMCP-6, and mouse mast cell carboxypeptidase A (mMC-CPA).[33–36]

BALB/c mBMMC derived with IL-3 represent a relatively immature population of mast cells that are alcian blue$^+$/safranin$^-$; synthesize chondroitin sulfate E proteoglycan; are poor in histamine and mMC-CPA content; preferentially express mRNA for mMCP-5, mMCP-6, and mMC-CPA; and transiently express the mMCP-7 transcript.[26,34–37] Reconstitution of WBB6F_1-W/W^v mast-cell-deficient mice with mBMMC derived from congenic WBB6F_1-+/+ mice led to the appearance of cells in the intestine and in the peritoneal cavity with the staining characteristics of normal MMC and SMC, respectively.[38,39] Thus, mBMMC are pluripotent progenitor mast cells capable of differentiation into mature mast cells of either phenotype depending upon the influence of the local microenvironment.

In Vitro *Differentiation of Mouse Mast Cells*

c-*Kit* is a plasma membrane protein with tyrosine kinase activity.[40] WBB6F_1-W/W^v mice, which have a mutation of c-*kit*,[41] and WCB6F_1-Sl/Sl^d mice, which are deficient in functional c-*kit* ligand (KL),[42] share a similar phenotype with macrocytic anemia, altered coat color, sterility, and deficiency of mast cells. KL is a pluripotent fibroblast-derived cytokine[43,44] that acts as a growth factor for murine mast cells *in vitro* and *in vivo*.[45] It was shown to induce clonal proliferation of mouse CTMC *in vitro*; these cells retained the staining characteristics of CTMC. When IL-3 was added to the medium, colony size increased, but mast cells became uniformly berberine and safranin negative.[46] When BALB/c mBMMC were developed with recombinant KL or by sequential culture in recombinant IL-3 followed by KL, they counterstained with safranin, synthesized both heparin and chondroitin sulfate proteoglycan, and expressed high levels of transcripts for mMCP-4 (normally expressed in BALB/c SMC) as well as mMCP-5, mMCP-6, and mMC-CPA.[47] The continued presence of IL-3 suppressed the KL-induced accumulation of mMCP-4, but not the other protease transcripts. These cells also failed to counterstain with safranin and produced negligible amounts of ^{35}S-labeled heparin proteoglycan.

IL-9 is a 30-40-kDa glycoprotein produced by activated lymphocytes whose synthesis is IL-2 dependent. Receptors for IL-9 have been identified on mast cells.[48] IL-10 is a 35-40-kDa homodimer derived from T_H2 cells and B cells, and is a strong inhibitor of cytokine production by T_H1 cells in the presence of antigen-presenting cells.[49] IL-9 and IL-10 act as accessory growth factors for murine mast cells.[50–53] When BALB/c mBMMC were developed in the presence of IL-3 and then cultured for a further two weeks in IL-10, mRNA transcripts for mMCP-1 and mMCP-2 were induced (proteases normally found in MMC of *T. spiralis*-infected mice).[32,51] IL-9, in the presence of suitable growth factors, also induced transcription of mMCP-1 and mMCP-2. IL-3 suppressed the induction of transcripts for mMCP-1 and mMCP-2 by IL-9 and IL-10.[51] Further, IL-4 markedly inhibited the induction of transcripts for mMCP-1 and mMCP-2 by IL-9 and IL-10, and significantly inhibited the induction of mMCP-4 by KL in the presence of IL-9.[53]

When spleen cells of neonatal rats were primed with 2.5S nerve growth factor (NGF) and injected into the brain of developing rats, they differentiated into mast cells and were associated with increased local microvessel formation.[54] Spleen cells are known to contain progenitor populations of IL-3-dependent immature mast cells. Culture of mBMMC, derived from $WBB6F_1$-+/+ mice, in methyl cellulose with 2.5S NGF and suboptimal doses of IL-3 led to increased clonal proliferation of the mast cells. These mast cells stained with berberine sulfate, contained heparin proteoglycan, and had an increased histamine content.[55]

ARACHIDONIC ACID METABOLISM BY MAST CELLS

The Enzymes of Arachidonic Acid Metabolism

Arachidonic acid is a 20-carbon polyunsaturated fatty acid released from membrane phospholipids by the action of PLA_2—predominantly by the action of 85-kDa cytosolic, rather than 14-kDa secretory, PLA_2.[56] Metabolism by the 5-lipoxygenase (5-LO) and cyclooxygenase pathways leads to the generation of leukotrienes (LT) and prostanoids, respectively (reviewed in Refs. 57 and 58).

5-Lipoxygenase Pathway

5-LO generates the unstable intermediate 5*S*-hydroperoxy-eicosatetraenoic acid (5-HPETE), which is reduced to 5*S*-hydroxy-eicosatetraenoic acid (5-HETE) or is converted by the sequential action of 5-LO[59] to an epoxide, LTA_4. 5-LO is a cytosolic metalloenzyme that translocates to the nuclear membrane upon activation[60,61] and requires both ATP and Ca^{2+} for activity.[62] The capacity of 5-LO to subsequently dissociate from the cell membrane following cell activation depends upon the nature of the activating stimulus. Following stimulation of mouse mast cells with ionophore A23187 there is irreversible association of 5-LO with the cell membranes and permanent inactivation of the enzyme, whereas after IgE-mediated activation there is return of active 5-LO to the cytosol.[60]

The activity of 5-LO is dependent upon a second factor termed 5-LO-activating protein (FLAP). Thus, when osteosarcoma cells were transfected with 5-LO cDNA, the cell lysates had 5-LO activity, but the intact cells failed to synthesize leukotrienes in response to an activating stimulus.[62] A novel class of drug has been described that inhibits leukotriene generation by activated cells without affecting subcellular 5-LO activity. This class of drugs was shown to recognize an 18-kDa protein that was subsequently purified, cloned, and shown to be necessary for the 5-LO activity of intact cells.[63,64] The mechanism of action of FLAP, once believed to be involved in the translocation of 5-LO from the cytoplasm to a cell membrane compartment, is more likely central to the presentation of released arachidonic acid.[63,64]

LTA_4 is processed by a cytosolic epoxide hydrolase to LTB_4 or by a novel microsomal glutathione (GSH)-S-transferase, termed LTC_4 synthase, to LTC_4; each product is exported from the cell by a carrier system.[65–67] Extracellularly, the glutamic acid residue is released from the GSH of LTC_4 by γ-glutamyl-transpeptidase to generate LTD_4, from which the glycyl residue is cleaved by a dipeptidase to form LTE_4.

LTC_4 Synthase

The purification of LTC_4 synthase has been hindered by its presence in small quantities and its instability, especially in purified form. This laboratory has worked with several sources of this enzyme, including guinea pig lung,[68] and various mammalian hematopoietic cell lines, including THP-1, differentiated U-937, HL-60, RBL-1,[69] and KG-1.[70] LTC_4 synthase was characterized and partially purified from RBL-1 cells (10-fold)[69] and guinea pig lung (91-fold) by conventional techniques.[68] The cell source chosen for the further isolation of this enzyme was switched to the human leukemic cell line, KG-1, which could be grown to high density in large volumes without a requirement for defined growth factors. LTC_4 synthase was purified from detergent-solubilized microsomes by affinity chromatography over two S-hexyl-GSH agarose columns, eluted with probenecid, and further purified by electrophoresis in a nonreducing acrylamide gel. An 18-kDa protein was purified greater than 10,000-fold to homogeneity, and was associated with LTC_4 synthase enzymatic activity.[70] A novel and sensitive fluorescence-linked immunoassay for LTC_4 was used to isolate a cDNA encoding LTC_4 from a KG-1 cDNA expression library.[71] The 694-bp cDNA hybridized to a 0.7-kb mRNA transcript in RNA from human eosinophils and KG-1 cells. The nucleotide and deduced amino acid sequence of the LTC_4 synthase cDNA shows no significant homology to GSH-S-transferases but shows a 31% overall amino acid identity with FLAP. Moreover, LTC_4 synthase is inhibitable by a FLAP inhibitor, MK-886; and therefore LTC_4 synthase is distinct from known GSH-S-transferases and belongs to a novel gene family.

Other investigators have purified an 18-kDa protein, believed to be LTC_4 synthase, from solubilized microsomes of differentiated U-937 cells by two sequential MONO-Q anion exchange chromatographic steps followed by LTC_2 affinity chromatography.[72] A subsequent publication from the same group reported 35 amino acids from the NH_2-terminus of an 18-kDa homodimeric protein purified from THP-1 cells.[73]

Cyclooxygenase Pathway

Cyclooxygenase catalyzes the sequential bis-oxygenation of arachidonic acid to PGG_2, which is reduced to PGH_2 by the hydroperoxidase activity of the same enzyme. PGH_2 is metabolized by specific synthases, each of which has a restricted distribution, to thromboxane (TX) A_2, PGI_2 (prostacyclin), PGD_2, PGE_2, and $PGF_{2\alpha}$. The capacity of a cell to synthesize a particular prostanoid is determined by the specific synthase(s) it contains. There are two forms of cyclooxygenase. Cyclooxygenase 1 is constitutively expressed with minimal (threefold) induction of steady state mRNA levels in human HEL cells and human fibroblasts by phorbol myristate acetate (PMA) and by cytokines such as PDGF, TGF-β1, IL-1β, and TNF-α.[74,75] Cyclooxygenase 2 is inducible with ~30-fold increase in steady state mRNA levels following activation of human umbilical vein endothelial cells, monocytes, and fibroblasts with PMA, PDGF, TNF-α, and IL-1β,[76,77] and substantial inhibition of this induction by corticosteroids.[77]

PGD_2 Synthase

Two PGD_2 synthase enzymes have been described. One is brain specific, does not require GSH, and is a member of the lipocalin family of proteins, consisting of hydrophobic small-molecule transporters.[78] The second PGD_2 synthase was initially characterized and isolated from rat spleen. Like the brain enzyme, it is a 28 kDa protein. However, spleen- and brain-derived enzymes have been distinguished from one another both biochemically and immunologically.[79] The spleen enzyme is dependent upon glutathione for its activity, has a different pH optimum and kinetics from the brain enzyme, and possesses GSH-S-transferase activity.[79] Polyclonal antibodies raised against each enzyme showed no cross-reactivity, each antibody specifically neutralizing only the activity of the enzyme against which it was raised.[79] Splenic PGD_2 synthase was distinguished from other GSH-S-transferases in that it is anionic and monomeric, and by its substrate specificity.[79] The polyclonal antibody raised against spleen PGD_2 synthase recognized the enzyme on SDS-PAGE immunoblotting and has been used in immunohistochemical analysis to localize the enzyme to Langerhans cells, macrophages, and mast cells of rat skin.[80]

The activity of PGD_2 synthase, present in the cytosolic fraction of rat SMC, was dependent upon the presence of GSH and was insensitive to 1mM 1-chloro-2,4-dinitrobenzene, thus distinguishing the mast cell enzyme from the brain isoform of PGD_2 synthase and from other GSH-S-transferases. The activity was neutralized by specific antibody directed to the spleen form of the enzyme. The mast cell enzyme and the spleen enzyme showed identical patterns on SDS-PAGE immunoblotting after two-dimensional PAGE, and identical peptide maps following partial digestion with *S. aureus* V8 protease. Thus the PGD_2 synthase activity in rat mast cells is attributable to the splenic form of PGD_2 synthase.[81] The anti-spleen PGD_2 synthase antibody recognized rat CTMC in the septum of the thymus, the submucosa of the intestine, and the skin. No significant staining of MMC was demonstrated.[81]

Metabolism of Arachidonic Acid in Activated Mast Cells

As described above, SMC preferentially generate PGD_2,[28] whereas MMC preferentially generate LTC_4 upon IgE-dependent stimulation.[29] The immature progenitor BMMC generates LTC_4 in preference to PGD_2. Coculture of BALB/c mBMMC with 3T3 fibroblasts in the presence of IL-3 results in the appearance of cells that counterstain with safranin, synthesize both heparin and chondroitin sulfate E proteoglycan, and contain increased quantities of histamine and mMC-CPA.[82,83] BMMC cocultured with 3T3 fibroblasts released seven fold more PGD_2 than mBMMC cultured in WEHI-3-conditioned medium alone (as a source of IL-3), in response to cross-linking of FcϵRI. However, LTC_4 generation was still predominant.[83]

While mBMMC possess abundant transcripts for FcγRIIa, FcγRIIb, and FcγRIII, only the type II FcγR are expressed on the cell surface,[84] due to rapid degradation of FcγRIII in the endoplasmic reticulum.[85] mBMMC cannot be activated for mediator release through IgG-dependent stimuli.[30] When mBMMC are cocultured in the presence of 3T3 fibroblasts, they express FcγRIII on the cell surface in association with the γ chain of FcϵRI,[85] and may be activated via IgG-dependent mechanisms for release of granule-associated mediators and eicosanoids with the same profile as occurs with FcϵRI-mediated stimulation.[30] Thus the product profile is determined by the phenotype of the mast cell rather than by the specificity of the Fc receptor mediating activation.

Heterogeneity of human mast cells is not as clear-cut as that described in rodents, although human mast cells may be distinguished based upon their protease phenotype. With regard to arachidonic acid metabolism, all relatively mature classes of human mast cell examined to date synthesize PGD_2 in preference to LTC_4, with the possible exception of formalin-sensitive lung mast cells; these were reported to generate LTC_4 in the absence of PGD_2,[86] although this was not confirmed in a separate study.[87] Human skin mast cells synthesize PGD_2 in an approximate 10- to 20-fold excess over LTC_4 in response to both immunologic and nonimmunologic stimulation.[88,89] The data for human pulmonary mast cells is conflicting. In all reports substantial quantities of PGD_2 (~80 ng per 10^6 cells) are synthesized upon cross-linking of FcϵRI,[28,90] but the relative generation of LTC_4 varied from barely detectable[90] to around 40 ng/10^6 cells–one half of that seen for PGD_2.[91] Human uterine mast cells were reported to synthesize 89 ng of PGD_2 and 45 ng of LTC_4 per 10;6 cells.[92] In the report by Levi-Schaffer and associates[90] there was no detectable LTC_4 synthesis by freshly isolated and purified human pulmonary mast cells, although chopped human lung fragments had previously yielded significant quantities of LTC_4 in response to IgE-mediated events. Following coculture with fibroblasts LTC_4 generation recovered to one-half of the PGD_2 synthesis, which was augmented threefold compared to freshly purified mast cells. These data indicate that 5-lipoxygenase metabolism by

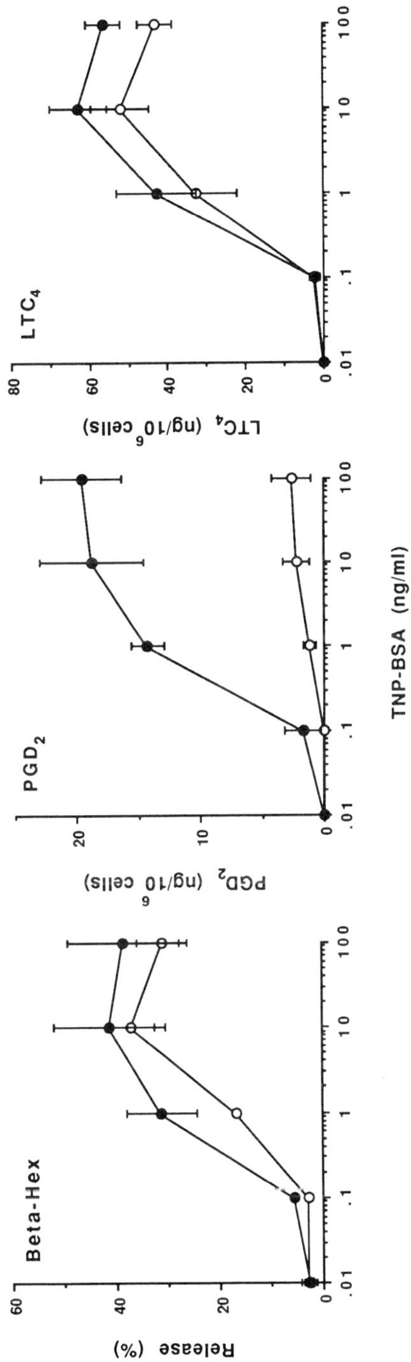

FIGURE 1. Dose-dependent release of β hexosaminidase (beta-HEX), prostaglandin D_2 (PGD_2), and leukotriene C_4 (LTC_4) from BALB/c mouse bone marrow culture-derived mast cells (BMMC). Three-week-old BMMC, derived in the presence of WEHI-3 conditioned medium as a source of interleukin-3, were cultured for a further week in the presence of 100 U/ml interleukin-3 (*open circles*), or 100 U/ml interleukin-3 plus 100 ng/ml c-*kit* ligand (*closed circles*), sensitized with IgE anti-trinitrophenyl (TNP), and activated with increasing doses of TNP conjugated to bovine serum albumin (BSA).

human pulmonary mast cells is particularly sensitive to isolation techniques, and perhaps to the microenvironment in which the cell resides.

THE EFFECT OF CYTOKINES ON IGE-MEDIATED DEGRANULATION AND RELEASE OF ARACHIDONATE METABOLITES BY MOUSE MAST CELLS

BALB/c mBMMC, developed in the presence of WEHI-3 cell-conditioned medium were cultured for a further week in the presence of 100 ng/ml KL in the presence or absence of 100 U/ml IL-3, in the presence of IL-3 alone, or in the continued presence of 50% WEHI-3 cell-conditioned medium. Each cell population was sensitized with 10 μg/ml monoclonal IgE anti-trinitrophenol (TNP) and activated with 1 to 100 ng/ml of TNP linked to bovine serum albumin (BSA). After 10 minutes, the reaction was stopped by centrifugation and cooling on ice, and generation of LTC_4, PGD_2, and the granule-associated mediator β-hexosaminidase was assessed.

Treatment with KL alone led to a marked diminution of granule mediator release without any significant reduction of eicosanoid synthesis and release. Indeed, there was a 2.5-fold increase in PGD_2 generation without an appreciable change in LTC_4 generation. In contrast, when mBMMC were cultured in the presence of both KL and IL-3, there was no inhibition of granule-associated mediator release, PGD_2 generation was markedly augmented, and the ratio of PGD_2:LTC_4 increased fourfold as compared to cells maintained in IL-3 alone (FIG. 1). These changes in KL-dependent PGD_2 generation were time dependent, reaching an optimum at 2 to 4 days. BALB/c mBMMC did not maintain viability when cultured for one week in the presence of KL alone at concentrations below 25 ng/ml. A dose- and time-dependent, selective enhancement of PGD_2 generation was demonstrated with respect to KL in the presence of IL-3. KL had minimal effects on LTC_4 generation. A nonselective dose-dependent effect of IL-3 upon augmented PGD_2, LTC_4, and β-hexosaminidase release was seen in cells cultured in the presence of 100 ng/ml KL, suggesting that IL-3 was required for optimal activation of mBMMC in terms of both granule exocytosis and eicosanoid synthesis and release.

These results demonstrate that KL, with the support of a growth factor such as IL-3, selectively augments IgE-dependent PGD_2 synthesis by BALB/c mBMMC in a dose-dependent fashion without appreciably altering exocytosis. This priming by KL for increased IgE-dependent PGD_2 synthesis is dependent on constitutive cyclooxygenase 1. In contrast, culture of mBMMC with KL in combination with IL-1 and IL-10 leads to substantial generation of PGD_2 over the initial 10 hours of culture, an event that is dependent on inducible cyclooxygenase 2.[93]

REFERENCES

1. HEINRIKSON, R. L., E. T. KRUEGER & P. S. KEIM. 1977. J. Biol. Chem. **252:** 4913-4921.
2. SAKATA, T., E. NAKAMURA, Y. TSURUTA, M. TAMAKI, H. TERAOKA, H. TOJO, T. ONO & M. OKAMOTO. 1989. Biochim. Biophys. Acta **1007:** 124-126.
3. KOMADA, M., I. KUDO, H. MIZUSHIMA, N. KITAMURA & N. INOUE. 1989. J. Biochem. **106:** 545-557.

4. KRAMER, R. M., C. HESSION, B. JOHANSEN, G. HAYES, P. MCGRAY, E. P. CHOW, R. TIZARD & R. B. PEPINSKY. 1989. J. Biol. Chem. **264:** 5768-5775.
5. MURAKAMI, M., I. KUDO & K. INOUE. 1991. FEBS Lett. **294:** 247-251.
6. MURAKAMI, M., I. KUDO & K. INOUE. 1993. J. Biol. Chem. **268:** 839-844.
7. CLARK, J. D., L. L. LIN, R. W. KRIZ, C. S. RAMESHA, L. A. SULTZMAN, A. Y. LIN, N. MILONA & J. L. KNOPF. 1991. Cell **65:** 1043-1051.
8. LIN, L. L., A. Y. LIN & J. L. KNOPF. 1992. Proc. Natl. Acad. Sci. USA **89:** 6147-6151.
9. LIN, L. L., M. WARTMANN, A. Y. LIN, J. L. KNOPF, A. SETH & R. J. DAVIS. 1993. Cell **72:** 269-278.
10. HANASAKI, K. & H. ARITA. 1992. J. Biol. Chem. **267:** 6414-6420.
11. MURAKAMI, M., I. KUDO, Y. SUWA & K. INOUE. 1992. Eur. J. Biochem. **209:** 257-265.
12. MURAKAMI, M., I. KUDO, H. NAKAMURA, Y. YOKOYAMA, H. MORI & K. INOUE. 1990. FEBS Lett. **268:** 113-116.
13. SCHALKWIJK, C., J. PFEILSCHIFTER, F. MARKI & H. VAN DEN BOSCH. 1992. J. Biol. Chem. **267:** 8846-8851.
14. NAKANO, T., O. OHARA, H. TERAOKA & H. ARITA. 1990. J. Biol. Chem. **265:** 12745-12748.
15. BARBOUR, S. E. & E. A. DENNIS. 1993. J. Biol. Chem. **268:** 21875-21882.
16. KIM, D. K., I. KUDO & K. INOUE. 1991. Biochim. Biophys. Acta **1083:** 80-88.
17. CHANNON, J. Y. & C. C. LESLIE. 1990. J. Biol. Chem. **265:** 5409-5413.
18. XING, M. & R. MATTERA. 1992. J. Biol. Chem. **267:** 25966-25975.
19. LIN, L. L., A. Y. LIN & D. L. DEWITT. 1992. J. Biol. Chem. **267:** 23451-23454.
20. HOECK, W. G., C. S. RAMESHA, D. J. CHANG, N. FAN & R. A. HELLER. 1993. Proc. Natl. Acad. Sci. USA **90:** 4475-4479.
21. ZUPAN, L. A., D. L. STEFFENS, C. A. BERRY, M. LANDT & R. W. GROSS. 1992. J. Biol. Chem. **267:** 8707-8710.
22. MURAKAMI, M., N. HARA, I. KUDO & K. INOUE. 1993. J. Immunol. **151:** 5675-5684.
23. KURIHARA, H., T. NAKANO, N. TAKASU & H. ARITA. 1991. Biochim. Biophys. Acta **1082:** 285-292.
24. MURAKAMI, M., I. KUDO, M. UMEDA, A. MATSUZAWA, M. TAKEDA, M. KOMADA, Y. FUJIMORI, K. TAKAHASHI & K. INOUE. 1992. J. Biochem. (Tokyo) **111:** 175-181.
25. EISEMAN, E. & J. B. BOLEN. 1992. Nature **355:** 78-80.
26. KATZ, H. R., R. E. KAYE & K. F. AUSTEN. 1991. Transplantation Proc. **23:** 2900-2904.
27. STEVENS, R. L., D. S. FRIEND, H. P. MCNEIL, V. SCHILLER, N. GHILDYAL & K. F. AUSTEN. 1994. Proc. Natl. Acad. Sci. USA **91:** 128-132.
28. LEWIS, R. A., N. A. SOTER, P. T. DIAMOND, K. F. AUSTEN, J. A. OATES & L. J. ROBERTS. 1982. J. Immunol. **129:** 1627-1631.
29. HEAVEY, D. J., P. B. ERNST, R. L. STEVENS, A. D. BEFUS, J. BIENENSTOCK & K. F. AUSTEN. 1988. J. Immunol. **140:** 1953-1957.
30. KATZ, H. R., M. B. RAIZMAN, C. S. GARTNER, H. C. SCOTT, A. C. BENSON & K. F. AUSTEN. 1992. J. Biol. Chem. **148:** 868-871.
31. SERAFIN, W. E., D. S. REYNOLDS, S. ROGELJ, W. S. LANE, G. A. CONDER, S. S. JOHNSON, K. F. AUSTEN & R. L. STEVENS. 1990. J. Biol. Chem. **265:** 423-429.
32. GHILDYAL, N., H. P. MCNEIL, S. STECHSCHULTE, K. F. AUSTEN, D. SILBERSTEIN, M. F. GURISH, L. L. SOMERVILLE & R. L. STEVENS. 1992. J. Immunol. **149:** 2123-2129.
33. SERAFIN, W. T., T. P. SULLIVAN, G. A. CONDER, A. EBRAHIMI, P. MARCHAM, S. S. JOHNSON, K. F. AUSTEN & D. S. REYNOLDS. 1991. J. Biol. Chem. **266:** 1934-1941.
34. MCNEIL, H. P., K. F. AUSTEN, L. L. SOMERVILLE, M. F. GURISH & R. L. STEVENS. 1991. J. Biol. Chem. **266:** 20316-20322.
35. REYNOLDS, D. S., D. S. GURLEY, K. F. AUSTEN & W. E. SERAFIN. 1991. J. Biol. Chem. **266:** 3847-3853.

36. Reynolds, D. S., R. L. Stevens, D. S. Gurley, W. S. Lane & K. F. Austen. 1989. J. Biol. Chem. **264:** 20094-20099.
37. McNeil, H. P., D. S. Reynolds, V. Schiller, N. Ghildyal, D. S. Gurley, K. F. Austen & R. L. Stevens. 1992. Proc. Natl. Acad. Sci. USA **89:** 11174-11178.
38. Otsu, K., T. Nakano, Y. Kanakura, H. Asai, H. R. Katz, K. F. Austen, R. L. Stevens, S. J. Galli & Y. Kitamura. 1987. J. Exp. Med. **165:** 615-627.
39. Nakano, T., T. Sonoda, C. Hayashi, A. Yamatodani, Y. Kanayama, T. Yamamura, H. Asai, T. Yonezawa, Y. Kitamura & S. J. Galli. 1985. J. Exp. Med. **162:** 1025-1043.
40. Yarden, Y., W. J. Kuang, T. Yang-Feng, L. Coussens, S. Munemitsu, T. J. Dull, E. Chen, J. Schlessinger, U. Francke & A. Ullrich. 1987. EMBO J. **6:** 3341-3351.
41. Nocka, K., J. C. Tan, E. Chiu, T. Y. Chu, P. Ray, P. Traktman & P. Besmer. 1990. EMBO J. **9:** 1805-1813.
42. Flanagan, J. G., D. C. Chan & P. Leder. 1991. Cell **64:** 1025-1035.
43. Zsebo, K. M., J. Wypych, I. K. McNiece, H. S. Lu, K. A. Smith, S. B. Karkare, R. K. Sachdev, V. N. Yuschenkoff, N. C. Birkett, L. R. Williams *et al.* 1990. Cell **63:** 195.
44. Martin, F. H., S. V. Suggs, K. E. Langley, H. S. Lu, J. Ting, K. H. Okino, C. F. Morris, I. K. McNiece, F. W. Jacobsen, E. A. Mediaz *et al.* 1990. Cell **63:** 203.
45. Anderson, D. M., S. D. Lyman, A. Baird, J. M. Wignall, J. Eisenman, C. Rauch, C. J. March, H. S. Boswell, S. D. Gimpel & D. Cosman. 1990. Cell **63:** 235-243.
46. Tsuji, K., K. M. Zsebo & M. Ogawa. 1991. J. Clin. Pathol. **148:** 362-369.
47. Gurish, M. F., N. Ghildyal, H. P. McNeil, K. F. Austen, S. Gillis & R. L. Stevens. 1992. J. Exp. Med. **175:** 1003-1012.
48. Druez, C., P. Coulie, C. Uyttenhove & J. Van Snick. 1990. J. Immunol. **145:** 2494-2499.
49. Moore, K. W., P. Vieira, D. F. Fiorentino, M. L. Trounstine, T. A. Khan & T. R. Mosmann. 1990. Science **248:** 1230-1234.
50. Thomson-Snipes, L., V. Dhar, M. W. Bond, T. R. Mosmann & K. W. Moore. 1991. J. Exp. Med. **1735:** 507-510.
51. Ghildyal, N., H. P. McNeil, M. F. Gurish, K. F. Austen & R. L. Stevens. 1992. J. Biol. Chem. **267:** 8473-8477.
52. Hultner, L., C. Druez, J. Moeller, C. Uyttenhove, E. Schmitt, E. Rüde, P. Dormer & J. Van Snick. 1990. Eur. J. Immunol. **20:** 1413-1416.
53. Eklund, K. K., N. Ghildyal, K. F. Austen & R. L. Stevens. 1993. J. Immunol. **151:** 4266-4273.
54. Aloe, L. & R. Desimone. 1989. Int. J. Dev. Neurosci. **7:** 565-573.
55. Matsuda, H., Y. Kannan, H. Ushio, Y. Kiso, T. Kanemoto, H. Suzuki & Y. Kitamura. 1991. J. Exp. Med. **174:** 7-14.
56. Murakami, M., I. Kudo, Y. Fujimori, H. Suga & K. Inoue. 1991. Biochem. Biophys. Res. Commun. **181:** 714-721.
57. Samuelsson, B., M. Goldyne, E. Granstrom, M. Hamberg, S. Hammarstrom & C. Malmsten. 1978. Annu. Rev. Biochem. **47:** 997-1029.
58. Samuelsson, B. 1983. Science **220:** 568-575.
59. Rouzer, C. A., T. Matsumoto & B. Samuelsson. 1986. Proc. Natl. Acad. Sci. USA **83:** 857-861.
60. Malaviya, R. & B. A. Jakschik. 1993. J. Biol. Chem. **268:** 4939-4944.
61. Rouzer, C. A. & S. Kargman. 1988. J. Biol. Chem. **263:** 10980-10988.
62. Rouzer, C. A., E. Rands, S. Kargman, R. E. Jones, R. B. Register & R. A. Dixon. 1988. J. Biol. Chem. **263:** 10135-10140.

63. Dixon, R. A. F., R. E. Diehl, E. Opas, E. Rands, P. J. Vickers, J. F. Evans, J. W. Gillard & D. K. Miller. 1990. Nature **343:** 282-284.
64. Miller, D. K., J. W. Gillard, P. J. Vickers, S. Sadowski, C. Léveillé, J. A. Mancini, P. Charleson, R. A. F. Dixon, A. W. Ford-Hutchinson, R. Fortin, J. Y. Gauthier, J. Rodkey, R. Rosen, C. Rouzer, I. S. Sigal, C. D. Strader & J. F. Evans. 1990. Nature **343:** 278-281.
65. Lam, B. K., W. F. Owen, Jr., K. F. Austen & R. J. Soberman. 1989. J. Biol. Chem. **264:** 12885-12889.
66. Lam, B. K., L. Gagnon, K. F. Austen & R. J. Soberman. 1990. J. Biol. Chem. **265:** 13438-13441.
67. Lam, B. K., K. Xu, M. B. Atkins & K. F. Austen. 1992. Proc. Natl. Acad. Sci. USA **89:** 11598-11602.
68. Yoshimoto, T., R. J. Soberman, B. Spur & K. F. Austen. 1988. J. Clin. Invest. **81:** 866-971.
69. Yoshimoto, T., R. J. Soberman, R. A. Lewis & K. F. Austen. 1985. Proc. Natl. Acad. Sci. USA **82:** 8399-8403.
70. Penrose, J. F., L. Gagnon, M. Goppelt-Struebe, P. Myers, B. K. Lam, R. M. Jack, K. F. Austen & R. J. Soberman. 1992. Proc. Natl. Acad. Sci. USA **89:** 11603-11606.
71. Lam, B. K., J. F. Penrose, G. J. Freeman & K. F. Austen. 1994. Proc. Natl. Acad. Sci. USA. In press.
72. Nicholson, D. W., A. Ali, M. W. Klemba, N. A. Munday, R. J. Zamboni & A. W. Ford-Hutchinson. 1992. J. Biol. Chem. **267:** 17849-17857.
73. Nicholson, D. W., A. Ali, J. P. Vaillancourt, J. R. Calaycay, R. A. Mumford, R. J. Zamboni & A. W. Ford-Hutchinson. 1993. Proc. Natl. Acad. Sci. USA **90:** 2015-2019.
74. Funk, C. D., L. B. Funk, M. E. Kennedy, A. S. Pong & G. A. Fitzgerald. 1991. FASEB J. **5:** 2304-2312.
75. Diaz, A., A. M. Reginato & S. A. Jimenez. 1992. J. Biol. Chem. **267:** 10816-10822.
76. Hla, T. & K. Neilson. 1992. Proc. Natl. Acad. Sci. USA **89:** 7384-7388.
77. O'Banion, M. K., V. D. Winn & D. A. Young. 1992. Proc. Natl. Acad. Sci. USA **89:** 4888-4892.
78. Nagata, A., Y. Suzuki, M. Igarashi, N. Eguchi, H. Toh, Y. Urade & O. Hayaishi. 1991. Proc. Natl. Acad. Sci. USA **88:** 4020-4024.
79. Urade, Y., N. Fujimoto, M. Ujihara & O. Hayaishi. 1987. J. Biol. Chem. **262:** 3820-3825.
80. Ujihara, M., Y. Horiguchi, K. Ikai & Y. Urade. 1988. J. Invest. Dermatol. **90:** 448-451.
81. Urade, Y., M. Ujihara, Y. Horiguchi, M. Igarashi, A. Nagata, K. Ikai & O. Hayaishi. 1990. J. Biol. Chem. **265:** 371-375.
82. Levi-Schaffer, F., K. F. Austen, P. M. Gravallese & R. L. Stevens. 1986. Proc. Natl. Acad. Sci. USA **83:** 6485-6488.
83. Levi-Schaffer, F., E. T. Dayton, K. F. Austen, A. Hein, J. P. Caulfield, P. M. Gravallese, F-T. Liu & R. L. Stevens. 1987. J. Immunol. **139:** 3431-3441.
84. Katz, H. R., J. P. Arm, A. C. Benson & K. F. Austen. 1990. J. Immunol. **145:** 3412-3417.
85. Lobell, R. B., J. P. Arm, M. B. Raizman, K. F. Austen & H. R. Katz. 1993. J. Biol. Chem. **268:** 1207-1212.
86. Van Overveld, F. J., L. A. Houben, F. E. Schmitz Du Moulin, P. L. Bruijnzeel, J. A. Raaijmakers & G. K. Terpstra. 1989. Clin. Sci. **77:** 297-304.
87. Schulman, E. S., R. B. Pollack, T. J. Post & S. P. Peters. 1990. J. Immunol. **144:** 4195-4201.

88. LAWRENCE, I. D., J. A. WARNER, V. L. COHAN, W. C. HUBBARD, A. KAGEY-SOBOTKA & L. M. LICHTENSTEIN. 1987. J. Immunol. **139:** 3062-3069.
89. ROBINSON, C., C. BENYON, S. T. HOLGATE & M. K. CHURCH. 1989. J. Invest. Dermatol. **93:** 397-404.
90. LEVI-SCHAFFER, F., K. F. AUSTEN, J. P. CAULFIELD, A. HEIN, P. M. GRAVALLESE & R. L. STEVENS. 1987. J. Immunol. **139:** 494-500.
91. MACGLASHAN, D. W., R. P. SCHLEIMER, S. P. PETERS, E. S. SCHULMAN, K. ADAMS III, H. H. NEWBALL & L. M. LICHTENSTEIN. 1982. J. Clin. Invest. **70:** 747-751.
92. MASSEY, W. A., C. B. GUO, A. M. DVORAK, W. C. HUBBARD, B. S. BHAGAVAN, V. L. COHAN, J. A. WARNER, A. KAGEY-SOBOTKA & L. M. LICHTENSTEIN. 1991. J. Immunol. **147:** 1621-1627.
93. MURAKAMI, M., R. MATSUMOTO, K. F. AUSTEN & J. P. ARM. 1994. J. Biol. Chem. In press.

On the Positional Specificity of 15-Lipoxygenase[a]

DAVID L. SLOANE AND ELLIOTT SIGAL

Syntex Discovery Research
3401 Hillview Avenue
Palo Alto, California 94303

The lipoxygenases constitute a family of lipid peroxidating enzymes that catalyze the formation of inflammatory mediators such as leukotrienes from arachidonic acid.[1] Lipoxygenases have been implicated in the pathogenesis of a variety of inflammatory conditions such as arthritis, psoriasis, and bronchial asthma. 15-Lipoxygenase is unique among the human lipoxygenases in that it is capable of oxidizing complex membrane lipids.[2,3] Furthermore, it has been demonstrated that the 15-lipoxygenase is induced in the inflammatory reaction associated with human atherosclerotic lesions.[4] This may be relevant to the pathogenesis of this disease because there is some evidence that cellular lipoxygenases are capable of oxidizing low-density lipoprotein to its atherogenic form.[5,6]

Despite the increasing interest in the biological actions of lipoxygenase, little is known about the structural basis of the enzymatic reaction. All lipoxygenases catalyze the stereospecific dioxygenation of fatty acids that contain a 1,4-*cis,cis*-pentadiene. The lipoxygenases differ in the position on the fatty acid substrate at which they catalyze this reaction. The enzymes are named for their positional (or regio-) specificity on arachidonic acid. Insight into the structural basis for this specificity will provide information about the substrate binding pocket of the enzyme. Such information could be useful in the design of inhibitors that are specific for certain isoforms of lipoxygenase. Our current knowledge has relied on the activity of the enzymes on various fatty acids, as well as differences and conservations in the primary structures of the different lipoxygenases. Recently, the three-dimensional structure of the soybean lipoxygenase isoform-1 has been reported.[7] Although a substrate binding cavity was inferred from this structure, in the absence of substrate or substrate analogues bound to the enzyme, the structural features of the enzyme-substrate interaction cannot be deduced with certainty.

This article reviews our current knowledge of the structural basis for the regiospecificity of 15-lipoxygenase. Earlier studies on the substrate specificity are reviewed, as well as our own studies on mutant forms of 15-lipoxygenase. Finally, we discuss what insights the results of mutagenesis and the structure of the soybean enzyme can give us about the regiospecificities of 15- and 12-lipoxygenases.

The paradigm for our current models of the lipoxygenase reaction mechanism comes from experiments done by Hamberg and Samuelsson.[8,9] By analyzing the activity of the soybean lipoxygenase on a total of 14 unsaturated fatty acids, which

[a]This work was supported by Grant RO1 HL48591 from the National Institutes of Health.

differed in their lengths, positions, and degrees of unsaturation, these researchers were able to demonstrate that the soybean enzyme requires a *bis*-allylic methylene at the ω-8 position (i.e., the eighth carbon from the methyl terminus of the substrate), with oxygen introduced at the ω-6 position (FIG. 1A). The hydroperoxy acid product formed from 8,11,14-eicosatrienoic acid was demonstrated to be 15-hydroperoxy-8(*cis*), 11(*cis*), 13(*trans*)-eicosatrienoic acid by gas-liquid chromatography and mass spectrometry. The product was shown to be in the L configuration by its optical rotatory dispersion spectrum. The stereospecificity of the oxygen insertion led these researchers to examine the stereospecificity of hydrogen abstraction from the *bis*-allylic methylene. By using stereospecific tritium-labeled 8,11,14-eicosatrienoic acid, it was demonstrated that hydrogen is abstracted from the ω-8 methylene in a stereospecific manner. Furthermore, a kinetic isotope effect was observed with the tritium-labeled fatty acid, implying that the hydrogen abstraction is rate limiting. Hamberg and Samuelsson concluded that the first, rate-limiting, step of the reaction is the abstraction of hydrogen from the ω-8 carbon, followed by redistribution of the π electrons of the ω-6 and ω-9 double bonds to form a conjugated pentadienyl free radical. Oxygen is then introduced stereospecifically at the ω-6 carbon. One key feature of this mechanism is that the enzyme counts from the methyl terminus of the substrate. It can be imagined that the substrate binds to the enzyme with its methyl end inside the binding cavity of the enzyme. This binding cavity would contain structural features specific for an aliphatic hydrocarbon. The substrate would be bound in such a way as to align the *bis*-allylic ω-8 methylene with the reaction center of the enzyme, enabling the enzyme to abstract hydrogen from this carbon.

The activity of the 15-lipoxygenase from rabbit reticulocytes has been concluded to obey a similar mechanism, although this enzyme is not as strict in its positional specificity, forming both 15(S)- and 12(S)-hydroperoxyeicosatetraenoic acid from arachidonic acid, in a ratio of approximately 9 : 1[10,11] (FIG. 1A). This ratio of product formation is also seen with the human 15-lipoxygenase. This implies that the mammalian 15-lipoxygenases abstract hydrogen predominantly from the ω-8 position, and to a lesser extent from the ω-11 position (FIG. 1A). Results from a direct comparison of the soybean enzyme with the rabbit enzyme on various fatty acids demonstrated that the rabbit enzyme reaction is consistent with the mechanism proposed by Hamberg and Samuelsson. In this case, however, the reaction center of the rabbit enzyme is slightly shifted toward the carboxy terminus of the substrate, relative to the soybean enzyme. Thus, the optimum position of a *bis*-allylic methylene for hydrogen abstraction on an ideal substrate for the soybean enzyme is at ω-8, whereas for the rabbit enzyme it is at ω-9.[11]

A particularly interesting substrate is 6Z,9Z,12Z,15Z-eicosatetraenoic acid (6,9,12,15-20 : 4ω5, FIG. 1B). Like arachidonic acid (5,8,11,14-20 : 4ω6), this is a 20-carbon-long fatty acid with 4 *cis*-unsaturated positions; however, relative to arachidonic acid, these four double bonds are shifted by one carbon toward the methyl terminus of the fatty acid. The rabbit enzyme performs roughly equal (54 : 46) oxygenation at the C-13 and C-16 positions on this substrate, whereas the soybean enzyme preferentially (35 : 65) oxygenates the C-16 position. When the substrate binds in the active site of the enzyme, the site of hydrogen abstraction is near C-13; in the case of the rabbit enzyme the hydrogen abstraction site is shifted slightly toward the carboxyl terminus of the substrate; in the soybean enzyme it is directly

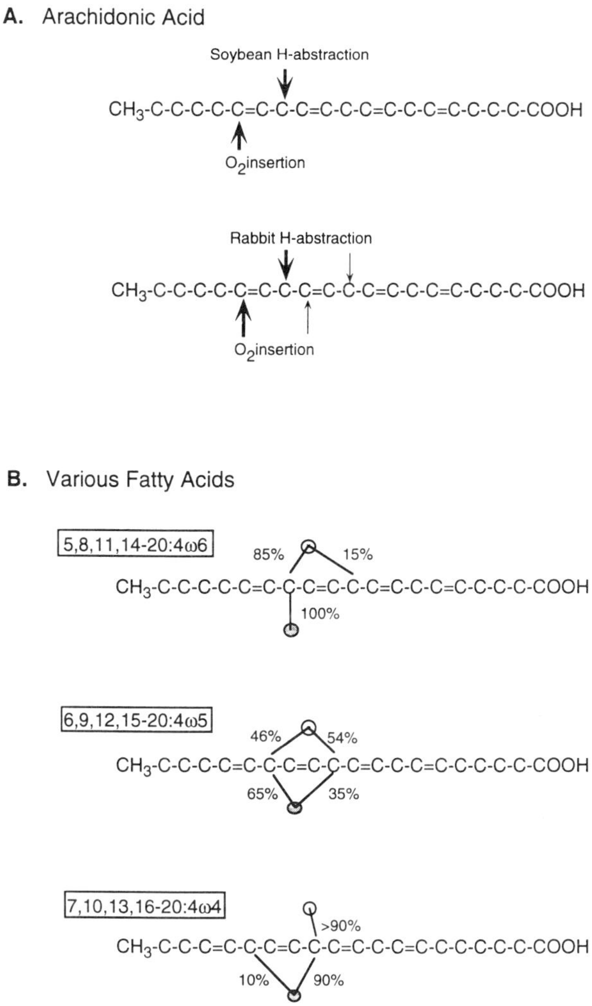

FIGURE 1. A. The sites of hydrogen abstraction and oxygen insertion catalyzed by the soybean 15-lipoxygenase isoform-1 and the rabbit reticulocyte 15-lipoxygenase on arachidonic acid. For simplification, the fatty acid is depicted linearly. The soybean enzyme catalyzes hydrogen abstraction at the ω-8 carbon and oxygen insertion at the ω-6 position, exclusively. The rabbit enzyme shows dual positional specificity, abstracting hydrogen predominantly from the ω-8 carbon, or (10% of the time) from the ω-11 carbon, leading to oxygen insertion primarily at the ω-6 position, and to a lesser extent at the ω-9 position. **B.** A comparison of the activities of the soybean 15-lipoxygenase isoform-1 and the rabbit reticulocyte 15-lipoxygenase on three different fatty acids. The site of hydrogen abstraction for the rabbit enzyme is depicted as an *open circle* and for the soybean enzyme as a *shaded circle.* The three different fatty acids differ in the position of the four double bonds. (After References 9 and 11.)

adjacent to C-13. This carbon of arachidonic acid is a *bis*-allylic methylene, and these enzymes abstract hydrogen from it as the first step in the reaction. In the case of the ω5 substrate, C-13 is an alkene, and the nearest sites for hydrogen abstraction are C-11 and C-14. These two *bis*-allylic methylenes straddle the reaction center in the rabbit enzyme, leading to equal product formation. In the case of the soybean enzyme, C-14 is closer to the site of hydrogen abstraction than C-11, explaining why the soybean enzyme preferentially oxygenates C-16 of this substrate. Using another 20-carbon-long fatty acid substrate that has its double bonds shifted by one more carbon toward the methyl end (an ω4 fatty acid), greater than 90% of the products formed by the rabbit enzyme are at the ω-7 position, demonstrating preferential hydrogen abstraction at the ω-9 position (FIG. 1B).

Further strong evidence in support of the rabbit enzyme binding the methyl terminus of the substrate comes from the observation that this enzyme can oxygenate arachidonyl-phosphatidylcholine[2] and biological membranes.[12] In other words, structural information on the substrate at the carboxyl terminus does not influence the enzyme-substrate interaction, consistent with the model in which the enzyme binds the methyl terminus of the substrate in such a way that the reaction center is aligned adjacent to the appropriate part of the substrate. The ability of the enzyme to oxygenate biological membranes may be relevant to its biological role in the maturation of the reticulocyte by participating in the degradation of the mitochondrial membranes. It may also be relevant to intracellular processing of low-density lipoprotein in atherogenesis.

To understand the structural features of the protein that define the methyl terminus binding region, we searched for conservations and differences in the sequences of the 12- and 15-lipoxygenases. The 12- and 15-lipoxygenases are very similar in amino acid sequence. Furthermore, the mammalian 15-lipoxygenases perform 12-lipoxygenation, and some 12-lipoxygenases perform 15-lipoxygenation. The three 12-lipoxygenase sequences that were known at the time of the experiment share 414 identical amino acids out of 661.[13–15] Of these, 47 differ from 15-lipoxygenase sequences. In this set of 47 differences between 12- and 15-lipoxygenases, only four amino acids are conserved between the two known mammalian 15-lipoxygenase sequences.[16,17] These four conserved differences, or a subset of them, were candidates for positioning the substrate relative to the active site and hence may determine either 12- or 15-lipoxygenation.

By site-directed mutagenesis, these four amino acids in human 15-lipoxygenase were changed to the four amino acids found in the 12-lipoxygenase sequences. The resulting quadruple mutant performs equal amounts of 12- and 15-lipoxygenation.[18] Next, all the permutations of single, double, and triple mutations at these four positions were generated, and we demonstrated that Met 418 is a primary determinant of positional specificity for 15-lipoxygenase. When Met 418 is replaced with a Val (15LOX M418V) either alone or in combination with any of the other changes, the resulting enzyme carries out equal 12- and 15-lipoxygenation. This result is reminiscent of the activity of the rabbit enzyme on the 6Z,9Z,12Z,15Z-eicosatetraenoic acid (20 : 4ω5) substrate, implying that 15LOX M418V binds arachidonic acid in such a way that the *bis*-allylic methylenes at positions 10 and 13 straddle the reaction center of the enzyme. By mutating the methionine at position 418 to a valine, presumably the substrate shifts deeper into the enzyme by a distance roughly equal to one C-C

hum15LO	:	414	FDQI**M**STGGGGHV	426
rab15LO	:	415	-------------	427
bov12LO	:	415	---V**V**--------	427
porc12LO	:	415	---V**V**--------	427
hum12LO	:	414	--KA**V**--------	426
rat12LO	:	414	--KA**M**--------	426
hum5LO	:	422	--KA**N**A-------	434
rat5LO	:	421	--KA**N**A-------	433
soy15LO	:	552	IETT**F**LPSKYSVE	564

FIGURE 2. Sequence comparisons and conserved differences in the region containing methionine 418 of human 15-lipoxygenase. The sequences are the human reticulocyte 15-lipoxygenase (hum15LO),[16] rabbit reticulocyte 15-lipoxygenase (rab15LO),[17] bovine tracheal epithelial 12-lipoxygenase (bov12LO),[15] porcine leukocyte 12-lipoxygenase (porc12LO),[14] human platelet 12-lipoxygenase (hum12LO),[13] rat brain 12-lipoxygenase (rat12LO),[19] human leukocyte 5-lipoxygenase (hum5LO),[22] rat leukocyte 5-lipoxygenase (rat5LO),[23] and soybean 15-lipoxygenase isoform-1 (soy15LO).[24] The *numbers* flanking each sequence correspond to the number of the initial and final amino acid in that region, where the initial methionine of the protein is number 1. A *dash* indicates amino acid identity with human reticulocyte 15-lipoxygenase. The single-letter amino acid code is used. The methionine at position 418 in the human reticulocyte 15-lipoxygenase sequence is shown in *bold-face*, as are the corresponding amino acids in the sequence alignment of the other lipoxygenases.

bond. The prediction from this is that 15LOX M418V would have greater preference for oxygenation at carbon 13 of the ω5 substrate. When 15LOX M418V was incubated with the ω5 substrate, it performed predominantly (75 : 25) 13-oxygenation,[18] supporting the conclusion that the M418V mutation allows the substrate to bind deeper into the enzyme.

The activity of 15LOX M418V implicates a particular region of the enzyme in the active site. In the primary structure, the region immediately surrounding amino acid 418 is highly conserved in all mammalian lipoxygenases (FIG. 2). The human platelet 12-lipoxygenase is known to perform exclusively 12-lipoxygenation. Therefore, we further mutated 15LOX M418V to contain the platelet 12-lipoxygenase amino acids. This variant, 15LOX Q416K, I417A, M418V generated 12-lipoxygenase and 15-lipoxygenase products in the proportions of 15:1.[18] When the double variant, 15LOX I417V, M418V was generated, it also had activity similar to the bovine 12-lipoxygenase (D. Sloane and R. Leung, unpublished observations).

These results demonstrate the importance of amino acids 417 and 418 in determining the regiospecificity of 15-lipoxygenase. One interpretation that is consistent with the data is that amino acids 417 and 418 define the binding region for the methyl terminus of the substrate. When these two bulky hydrophobic amino acids are replaced by smaller hydrophobic amino acids (e.g., alanine, valine), the binding region is deeper, allowing the substrate to shift relative to the reaction center of the enzyme so that the site of hydrogen abstraction shifts from carbon 13 to carbon 10 of arachidonic acid.

The binding determinants of 12-lipoxygenase have also been examined by mutagenesis. When the valine at position 419 of the bovine 12-lipoxygenase (the analogous position to 418 of human 15-lipoxygenase) was replaced with a methionine, no change in the enzyme's regiospecificity was measured. Further mutagenesis to replace both valines (positions 418 and 419 of the bovine 12-lipoxygenase sequence) with an isoleucine and a methionine results in a variant that performs 12-lipoxygenation and 15-lipoxygenation in a ratio of 3 : 1, indicating that this region is involved in determining the regiospecificity of this 12-lipoxygenase (D. Sloane and R. Leung, unpublished observations). Recently, a 12-lipoxygenase that contains a methionine at position 418 has been cloned from the rat brain (FIG. 2).[19] This suggests that, contrary to 15-lipoxygenase, position 418 in 12-lipoxygenase is not crucial for regiospecificity. When this methionine was mutated to a valine, the enzyme remained predominantly a 12-lipoxygenase.[20] Further mutations introduced a glutamine at position 416 and an isoleucine at position 417, making this region the same sequence as wild-type 15-lipoxygenase. This double mutant enzyme remained a 12-lipoxygenase.[20] These results may indicate that there are other structures in 12-lipoxygenases that define its regiospecificity. However, one must be cautious when interpreting the mutagenesis results without a three-dimensional structure of the mutants. In converting the 15-lipoxygenase to a 12-lipoxygenase, one is removing amino acid side chain bulk, whereas larger bulk must be introduced to convert a 12-lipoxygenase to a 15-lipoxygenase. Structurally, a protein will be more amenable to side chain removal than to the introduction of larger bulk.

The only known three-dimensional structure of any lipoxygenase is that of the soybean isoform-1.[7] The structure is composed of two domains. Domain I, comprising the 146 NH_2-terminal residues, is an eight-stranded antiparallel β-barrel "jelly roll," which has no counterpart in the mammalian structures, as predicted by sequence homology. Domain II is the major domain of the protein and contains the regions of highest homology to the mammalian sequences (overall, 22% identity to the human 15-lipoxygenase sequence[16]). Domain II is a bundle of 23 helices and two antiparallel β sheets. A particularly interesting aspect of the structure is the coordination of the catalytically essential iron in domain II. The coordination is a highly distorted octahedron with two unoccupied positions. Three histidine imidazoles and one carboxyl group (the carboxyl terminus of the polypeptide chain) serve as the iron ligands. Another recent crystallographic study of the iron coordination site of the soybean enzyme is slightly different, with the addition of an asparagine ligand and, thus, only one unoccupied position.[21] The other ligands are the same in this study as in the previous one. The reasons for this difference are not clear.

The structure has two cavities in it, which were interpreted to be an oxygen channel and a substrate binding cavity, based on the dimensions of each cavity. Cavity I, the presumed oxygen tunnel, is a conical tunnel that connects to one of the unoccupied positions of the iron. Cavity II, the presumed substrate binding cavity, is a 40-Å-long narrow cavity with a bend adjacent to the other unoccupied position of the iron. The amino acid in the primary structure of the soybean enzyme that corresponds to the human methionine 418, phenylalanine 557, lies in this cavity. Therefore, the structure supports the mutagenesis data that illustrated a role for this amino acid; however, its precise function remains to be determined. It is interesting to note that there is very little sequence identity between the mammalian lipoxygenases and the soybean lipoxygenase in this region (FIG. 2), even though the mammalian lipoxygenases all share a high degree of sequence identity in this region.

Since the structure was solved in the absence of substrate or substrate analogues, it is difficult to make any conclusions about the binding of the substrate in the enzyme. Boyington *et al.* state that arachidonic acid can fit snugly into the end of the bent region of the cavity;[7] however, the cavity is so long that one could propose many different binding modes of the substrate. Clearly, more structural determinations of both substrate-bound and of mammalian lipoxygenases are needed in order to fully understand the structural basis of substrate recognition.

In conclusion, several approaches have shown that 15-lipoxygenase recognizes its substrate by binding the methyl terminus of the fatty acid. By sequence alignments and mutagenesis, structural features of human 15-lipoxygenase that can affect its regiospecificity have been identified. Sequence information and mutagenesis on 12-lipoxygenases indicate that this enzyme uses other structural features to determine its regiospecificity. The region of the soybean enzyme that is analogous to amino acid 418 of the human enzyme has been found to be in a potential substrate binding cleft; however, in the absence of bound substrate it is difficult to determine how the substrate binds. Further mutagenesis and x-ray crystallography will more fully elucidate the structural basis of substrate recognition.

SUMMARY

The lipoxygenases comprise a family of fatty acid dioxygenases that are involved in a variety of inflammatory conditions. Various approaches have been taken in order to understand the different regiospecificities of the different lipoxygenases. Here we have reviewed the current knowledge of the structural features of the substrate and of the enzyme that form the basis of the regiospecificity of 15-lipoxygenase. Earlier experiments on the structural features of the substrate were reviewed, as well as more recent results of site-directed mutagenesis studies. The structure of the soybean lipoxygenase isoform-1 was also briefly reviewed.

ACKNOWLEDGMENTS

The authors would like to thank Dr. Charles S. Craik of the University of California at San Francisco and Regina Leung, who participated in the mutagenesis work described here. We thank Drs. Chakk Ramesha and John Wu for review of the manuscript.

REFERENCES

1. SAMUELSSON, B., S. E. DAHLEN, J. A. LINDGREN, C. A. ROUZER & C. N. SERHAN. 1987. Science **237:** 1171-1176.
2. MURRAY, J. J. & A. R. BRASH. 1988. Arch. Biochem. Biophys. **263:** 514-523.
3. KUHN, H. & A. R. BRASH. 1990. J. Biol. Chem. **265:** 1454-1458.
4. YLA-HERTTUALA, S., M. E. ROSENFELD, S. PARTHASARATHY, C. K. GLASS, E. SIGAL, J. L. WITZTUM & D. STEINBERG. 1990. Proc. Natl. Acad. Sci. U.S.A. **87:** 6959-6963.

5. STEINBERG, D., S. PARTHASARATHY, T. E. CAREW, J. C. KHOO & J. L. WITZTUM. 1989. N. Engl. J. Med. **320:** 915-924.
6. PARTHASARATHY, S., E. WIELAND & D. STEINBERG. 1989. Proc. Natl. Acad. Sci. U.S.A. **86:** 1046-1050.
7. BOYINGTON, J. C., B. J. GAFFNEY & L. M. AMZEL. 1993. Science **260:** 1482-1486.
8. HAMBERG, M. & B. SAMUELSSON. 1965. Biochem. Biophys. Res. Commun. **21:** 531.
9. HAMBERG, M. & B. SAMUELSSON. 1967. J. Biol. Chem. **242:** 5329-5335.
10. BRYANT, R. W., J. M. BAILEY, T. SCHEWE & S. M. RAPOPORT. 1982. J. Biol. Chem. **257:** 6050-6055.
11. KUHN, H., H. SPRECHER & A. R. BRASH. 1990. J. Biol. Chem. **265:** 16300-16305.
12. KUHN, H., J. BELKNER, R. WIESNER & A. R. BRASH. 1990. J. Biol. Chem. **265:** 18351-18361.
13. FUNK, C. D., L. FURCI & G. A. FITZGERALD. 1990. Proc. Natl. Acad. Sci. U.S.A. **87:** 5638-5642.
14. YOSHIMOTO, T., H. SUZUKI, S. YAMAMOTO, T. TAKAI, C. YOKOYAMA & T. TANABE. 1990. Proc. Natl. Acad. Sci. U.S.A. **87:** 2142-2146.
15. DE MARZO, N., D. L. SLOANE, S. DICHARRY, E. HIGHLAND & E. SIGAL. 1992. Am. J. Physiol. **6:** L198-L207.
16. SIGAL, E., C. S. CRAIK, E. HIGHLAND, D. GRUNBERGER, L. L. COSTELLO, R. A. F. DIXON & J. A. NADEL. 1988. Biochem. Biophys. Res. Commun. **157:** 457-464.
17. FLEMING, J., B. J. THIELE, J. CHESTER, J. O'PREY, S. JANETZKI, A. AITKEN, I. A. ANTON, S. M. RAPOPORT & P. R. HARRISON. 1989. Gene **79:** 181-188.
18. SLOANE, D. L., R. LEUNG, C. S. CRAIK & E. SIGAL. 1991. Nature **354:** 149-152.
19. WATANABE, T., J. F. MEDINA, J. Z. HAEGGSTRÖM, O. RÅDMARK & B. SAMUELSSON. 1993. Eur. J. Biochem. **212:** 605-612.
20. WATANABE, T. & J. Z. HAEGGSTRÖM. 1993. Biochem. Biophys. Res. Commun. **192:** 1023-1029.
21. MINOR, W., J. STECZKO, J. T. BOLIN, Z. OTWINOWSKI & B. AXELROD. 1993. Biochemistry **32:** 6320-6323.
22. DIXON, R. A. F., R. E. JONES, R. E. DIEHL, C. D. BENNETT, S. KARGMAN & C. A. ROUZER. 1988. Proc. Natl. Acad. Sci. U.S.A. **85:** 416-420.
23. BALCAREK, J. M., T. W. THEISEN, M. N. COOK, A. VARRICHIO, S.-M. HWANG, M. W. STROHSACKER & S. T. CROOKE. 1988. J. Biol. Chem. **263:** 13937-13941.
24. SHIBATA, D., J. STECZKO, J. E. DIXON, M. HERMODSON, R. YAZDANPARAST & B. AXELROD. 1987. J. Biol. Chem. **262:** 10080-10085.

Activation of Mitogen-Activated Protein Kinase and Arachidonate Release via Two G Protein–Coupled Receptors Expressed in the Rat Hippocampus[a]

MIKIRO MORI, HARUHIKO BITO, CHIE SAKANAKA, ZEN-ICHIRO HONDA, KAZUHIKO KUME, TAKASHI IZUMI, AND TAKAO SHIMIZU[b]

Department of Biochemistry
Faculty of Medicine
The University of Tokyo
7-3-1 Hongo, Bunkyo-ku
Tokyo 113, Japan

INTRODUCTION

Arachidonate and its related compounds have been considered to play various physiological and pathological roles in the central nervous system (CNS) including modulation of synaptic transmission possibly as retrograde messengers[1–3] and ischemic brain damage.[4] Several studies reported that long-term potentiation in the hippocampus is induced by arachidonate itself or its metabolites[5–7] as well as its related compound, platelet-activating factor (PAF : 1-*O*-alkyl-2-acetyl-*sn*-glycero-3-phosphocholine).[8] PAF is a derivative of alkylether phospholipids that has diverse biological effects[9–12] including expression of growth-related oncogenes[13,14] and neurite extension in PC-12 pheochromocytoma cells.[15]

Most of the effects of arachidonate-related compounds are considered to be mediated by their specific cell surface receptors.[9,10,16] All of the receptors identified so far have seven putative membrane-spanning domains,[17–21] a characteristic of receptors coupled to heterotrimeric guanine nucleotide-binding proteins (G proteins). On the other hand, several lines of evidence indicate that some of these compounds also act as intracellular second messengers.[22] Arachidonate lipoxygenase products were shown to directly regulate opening of S channel (serotonin-sensitive K^+ channel) in *Aplysia* sensory neurons,[23] G protein-sensitive K^+ channel in the atrial cells,[24,25] and somato-

[a]This work was supported in part by grants-in-aid from the Ministry of Education, Sciences, and Culture and the Ministry of Health and Welfare of Japan and by grants from the Uehara Memorial Foundation, the Toray Science Foundation, the Japan Cardiovascular Research Foundation, and the Yamada Science Foundation.

[b]To whom correspondence should be addressed.

statin (SST)-stimulated K^+ current (M current) in the rat hippocampus.[26,27] Intracellular receptors have been proposed for PAF.[28] Here we describe the signaling pathways downstream of G protein-coupled receptors by comparing the pathways mediated by the PAF receptor and those by a somatostatin receptor, another member of this superfamily. We chose the SST receptor (SSTR) because the SSTR-mediated signaling pathways involve no mobilization of intracellular Ca^{2+} (see below), which is advantageous as a model for analyzing the pathways regardless of the Ca^{2+} effects. SST is a 14-amino acid peptide distributed widely in the CNS and various other organs including pancreas and gastrointestinal tract.[29,30] Although this peptide was first isolated from the hypothalamus as a growth hormone-inhibitory substrate,[31] it has been considered to have diverse actions[29,30,32] including regulation of membrane conductance with direct relevance to arachidonate-related substances as described above. We have paid special attention to the signals to activation of mitogen-activated protein (MAP) kinase, which is considered to play a substantial role in cell differentiation/proliferation[33–39] and activation of arachidonate-specific phospholipase A_2 ($cPLA_2$).[40] The results led us to propose that these two receptors mediate different signals that converge at MAP kinase cascade leading to arachidonate release and to postulate these pathways in the rat hippocampus, the region closely related to neuronal plasticity and delayed cell death.

CLONING, STRUCTURE, AND DISTRIBUTION OF RECEPTORS FOR PAF AND SOMATOSTATIN

PAF Receptor

The PAFR cDNA was originally isolated by expression cloning using *Xenopus* oocytes from a cDNA library constructed from size-fractionated mRNA of the guinea pig lung.[18] The oocytes were microinjected with cRNA synthesized *in vitro* with T7 RNA polymerase, and the functional expression of the receptor was determined by measuring the inward current through the Ca^{2+}-dependent Cl^- channel opened by ligand-induced PI hydrolysis and intracellular Ca^{2+} mobilization. Three × 10^6 independent clones were screened, and the pool of the positive fractions was subdivided until a single clone with a 3-kb insert was obtained. The cDNA for the human[41] and rat[110] were obtained by homology screening from the leukocyte and spleen libraries, respectively. The hydropathy analyses indicated that PAFR proteins have seven putative transmembrane domains (FIG. 1), a finding compatible with previous reports that PAFR couples with G proteins[42,43] and activates phospholipase C.[44] Northern blot analysis revealed that this receptor mRNA is expressed in various tissues including the leukocytes, spleen, small intestine, kidney, lung, liver, and almost all regions of the brain, although the pattern of expression in various tissues differs among the species.[18,45] We also found that PAF elevates intracellular Ca^{2+} concentration ($[Ca^{2+}]_i$) in rat hippocampal culture, evidence that PAFR is functionally expressed in the rat hippocampal neurons.[45] *In situ* hybridization study revealed expression of PAFR mRNA in almost all regions of the adult rat brain with relatively higher levels in the hippocampus and cerebellar cortex (Mori *et al.*, unpublished data).

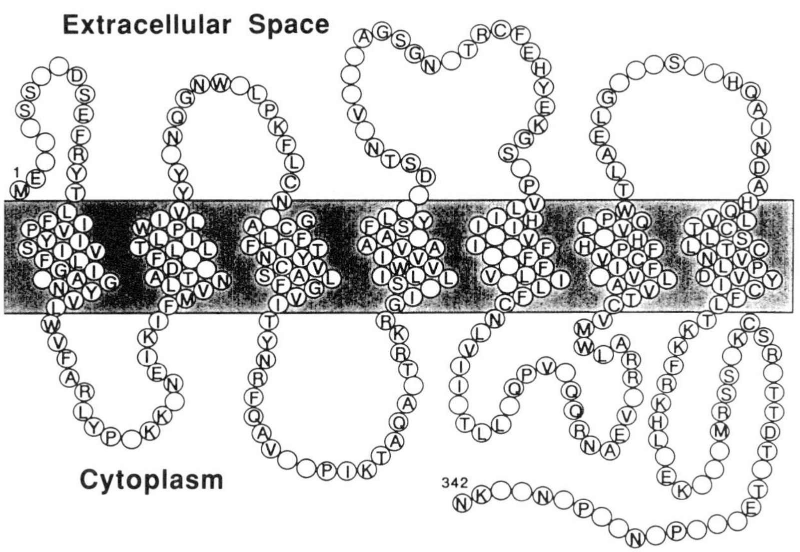

FIGURE 1. A putative structure of PAF receptors. The conserved amino acids between the guinea pig lung[18] and human leukocyte[41] receptors are indicated by *single letters.*

Somatostatin Receptor

The recent cloning of the multiple somatostatin receptor genes has revealed the cDNAs of at least five receptor subtypes (SSTR1-5),[46–53] which all indicate the presence of seven putative transmembrane domains. SSTR1-5 have been shown to inhibit adenylate cyclase,[51,53,54] although other studies report the absence of inhibitory actions of SSTR1 and 2.[55,56] We recently isolated a cDNA of the major SSTR subtype ($SSTR_h$, found to be identical to SSTR4[47]) expressed in the rat hippocampus.[57] Two distinct products were obtained from polymerase chain reaction (PCR) using a double-stranded rat hippocampal cDNA as a template and oligonucleotide primers based on the published sequences for mouse and human SSTR1 and SSTR2.[52] A 1.9-kb-long cDNA encoding $SSTR_h$ was isolated from a λZAPII rat hippocampus cDNA library by plaque hybridization using the $SSTR_h$ PCR fragments as a probe and was fully sequenced. The deduced primary structure of $SSTR_h$ consisted of 384 amino acid residues and seven hydrophobic, putative transmembrane domains (FIG. 2). The amino acid sequence homology with the rat homologues of SSTR1, SSTR2, SSTR3, and a pituitary-specific SSTR (SSTR5) was 66.1%, 56%, 46%, and 52.8%, respectively. Northern blot analysis revealed that $SSTR_h$ mRNA is prominently expressed in the hippocampus and moderately in the cerebral cortex, striatum, hypothalamus, and thalamus.[57] *In situ* hybridization study showed remarkable expression of this receptor in the CA1-3 regions of the hippocampus and in the lateral habenula (FIGS. 3,4). Moderate expression was found in the cerebral cortex, piriform cortex, olfactory nucleus, and amygdala; and lower levels of expression were observed in the striatum, medial thalamic nuclei, and hypothalamic nuclei.

```
rSSTRh       1                   MNTPATLPLGGEDTTWTPGINASWAPDEEEDAVRSDGTGTAGM  43
rSSTR1       1  MFPNGTAPSPTSSPSSSPGGCGEGVCSRGPGSGAADGMEEPGRNSSQNGTLSEGQGSAIL  60
rSSTR2       1                   MELTSEQFNGSQVWIPSPFDLNGSLGPSNGSNQTEPYYDMTSN  43
rSSTR3       1            MAAVTYPSSVPTTLDPGNASSAW-PLDT--SLGNASAGTSLAGLAVSGIL  46
rSSTRpit     1                      MEPLSLAS-TPSWNAS--AASSGNHNWSLVGSA-SPMGAR  36

                           TM1                                      TM2
rSSTRh      44  VTI-QCIYALVCLVGLVGNALVIFVILRYAKMKTATNIYLLNLAVADELFMLSVPFVASA  102
rSSTR1      61  ISF---IYSVVCLVGLCGNSMVIYVILRYAKMKTATNIYILNLAIADELLMLSVPFLVTS  117
rSSTR2      44  AVL-TFIYFVVCVVGLCGNTLVIYVILRYAKMKTITNIYILNLAIADELFMLGLPFLAMQ  102
rSSTR3      47  ISL---VYLVVCVVGLLGNSLVIYVVLRHTSSPSVTSVYILNLALADELFMLGLPFLAAQ  104
rSSTRpit    38  AVLVPVLYLLVCTVGLSGNTLVIYVVLRHAKMKTVTNVYILNLAVADVLFMLGLPFLATQ  96
                       *  ** *** **  ** * **       *  * **** ** * **  **

                                          TM3
rSSTRh     103  AA-LRHWPFGAVLCRAVLSVDGLNMFTSVFCLTVLSVDRYVAVVHPLRAATYRRPSVAKL  161
rSSTR1     118  TL-LRHWPFGALLCRLVLSVDAVNMFTSIYCLTVLSVDRYVAVVHPIKAARYRRPTVAKV  176
rSSTR2     103  VA-LVHWPFGKAICRVVMTVDGINQFTSIFCLTVMSIDRYLAVVHPIKSAKWRRPRTAKM  161
rSSTR3     105  NA-LSYWPFGSLMCRLVMAVDGINQFTSIFCLTVMSVDRYLAVVHPTRSARWRTAPVARM  163
rSSTRpit    98  NAVVSYWPFGSFLCRLVMTLDGINQFTSIFCLMVMSVDRYLAVVHPLRSARWRRPRVAKM  156
                      ****   ** *   *  * ***  ** * * *** *****   *  *    *

                         TM4                                               TM5
rSSTRh     162  INLGVWLASLLVTLPIAVFADTRPARGGEAVACNLHWPHPA--WSAVFVIYTFLLGFLLP  219
rSSTR1     177  VNLGVWVLSLLVILPIVVFSRT-AANSDGTVACNMLMPEPAQRWLVGFVLYTFLMGFLLP  235
rSSTR2     162  INVAVWGVSLLVILPIMIYAGLRSNQWGR-SSCTINWPGESGAWYTGFIIYAFILGFLVP  220
rSSTR3     164  VSAAVWVASAVVVLPVVVFSGV----PRGMSTCHMQWPEPAAAWRTAFIIYTAALGFFGP  219
rSSTRpit   158  ASAAVWVFSLLMSLPLLVFADVQEGW----GTCNLSWPEPVGLWGAAFITYTSVLGFFGP  212
                    **  *    **                 *    *     *   *  *    **  *

                                                                         TM6
rSSTRh     220  VLAIGLCYLLIVGKMRAVA--LRAGW---------QQRRRSEKKITRLVLMVVTVFVLCW  268
rSSTR1     236  VGAICLCYVLIIAKMRMVA--LKAGW---------QQRKRSERKITLMVMMVVMVFVICW  284
rSSTR2     221  LTIICLCYLFIIIKVKSSG--IRVGS---------SKRKKSEKKVTRMVSIVVAVFIFCW  269
rSSTR3     220  LLVICLCYLLIVVKVRSTTRRVRAPSCQWVQAPACQRRRRSERRVTRMVVAVVALFVLCW  279
rSSTRpit   214  LLVICLCYLLIVVKVKAAG--MRVGS---------SRRRRSEPKVTRMVVVVVLVFVGCW  261
                   * ***  *  *                       *  **   *  *  **  *  **

                                                  TM7
rSSTRh     269  MPFYVVQLLNLFVTS----LDATVNHVSLILSYANSCANPILYGFLSDNFRRSFQRVLCL  324
rSSTR1     285  MPFYVVQLVNVFAEQ----DDATVSQLSVILGYANSCANPILYGFLSDNFKRSFQRILCL  340
rSSTR2     270  LPFYIFNVSSVSVAISPTPALKGMFDFVVILTYANSCANPILYAFLSDNFKKSFQNVLCL  329
rSSTR3     280  MPFYLLNIVNVVCPLPEEPAFFGLYFLVVALPYANSCANPILYGFLSYRFKQGFRRIL-L  338
rSSTRpit   263  LPFFIVNIVNLAFTLPEEPTSAGLYFFVVVLSYANSCANPLLYGFLSDNFRQSFRKVLCL  321
                 **                           * ******** ** ***  *   *   * *

rSSTRh     325  RCCLLETTGGAEEEPLDYYATALKSRGGPGCICPPLPCQQEPMQAEPACKRVPFTKTTTF  384
rSSTR1     341  -----SWMDNAAEEPVDYYATALKSRAYSVEDFQPENLESGGVFRNGTCASRISTL      391
rSSTR2     330  VKVSGAEDGERSDSKQDKSRLNETTETQRTLLNGDLQTSI                      369
rSSTR3     339  RPSRRVRSQEPGSGPPEKTEEEEDEEEEERREEEERRMQRGQEMNGRLSQIAQPGPSGQQ  398
rSSTRpit   323  RRGYGMEDADAIEPRPDKSGRPQAHCPHAAARPMGSCRPAGFECPCNTLGVLQASTVLSS  381

rSSTR3     399  QRPCTGTAKEQQLLPQEATAGDKASTLSHL                                428
rSSTRpit   382  GI                                                            383
```

FIGURE 2. Alignment of deduced amino acid sequences of rat SSTRs. The cDNA for the major SSTR subtype in the hippocampus ($SSTR_h$ or SSTR4) was isolated as follows. A pair of specific primers was designed on the basis of published sequences of SSTR1 and SSTR2,[52] and reverse transcriptase (RT) polymerase chain reaction (PCR) was carried out with rat hippocampal poly(A)$^+$ RNA. A fragment corresponding to a novel receptor subtype was obtained from the PCR products with two distinct sequences. A λ ZapII (Stratagene) cDNA library was constructed from size-fractionated rat hippocampal poly(A)$^+$ RNA and screened by hybridization using the PCR fragment. The sequences of rat SSTR1-3 and $SSTR_{pit}$ (or SSTR5) are cited from the published literature.[48-51] *Asterisks* indicate invariant residues. TM1-7 represent seven putative transmembrane domains.

SIGNAL TRANSDUCTION PATHWAYS FOR PAFR AND SSTR4[57,58]

Chinese hamster ovary (CHO) cells that stably express the transfected gene for PAFR or SSTR4 were established for analyzing various G protein-mediated signals and activation of MAP kinase and arachidonate release. MAP kinase is activated

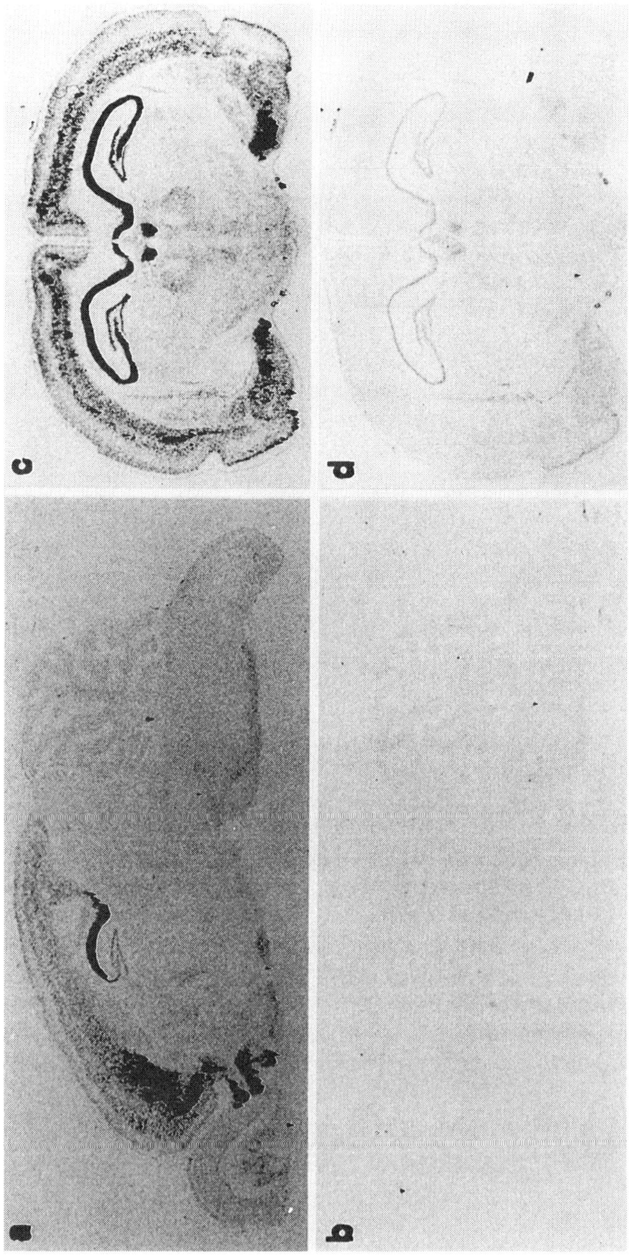

FIGURE 3. Positive film images of *in situ* hybridization of (**a, b**) parasagittal and (**c, d**) coronal sections showing distribution of SSTR4 mRNA in the adult rat brain. Sections were hybridized with (**a,c**) an antisense probe corresponding to the ~1.9-kb BamHI-HindIII fragment of pSS3,[57] (**b**) a sense probe, or (**d**) the antisense probe containing 100-fold excess of cold transcript.

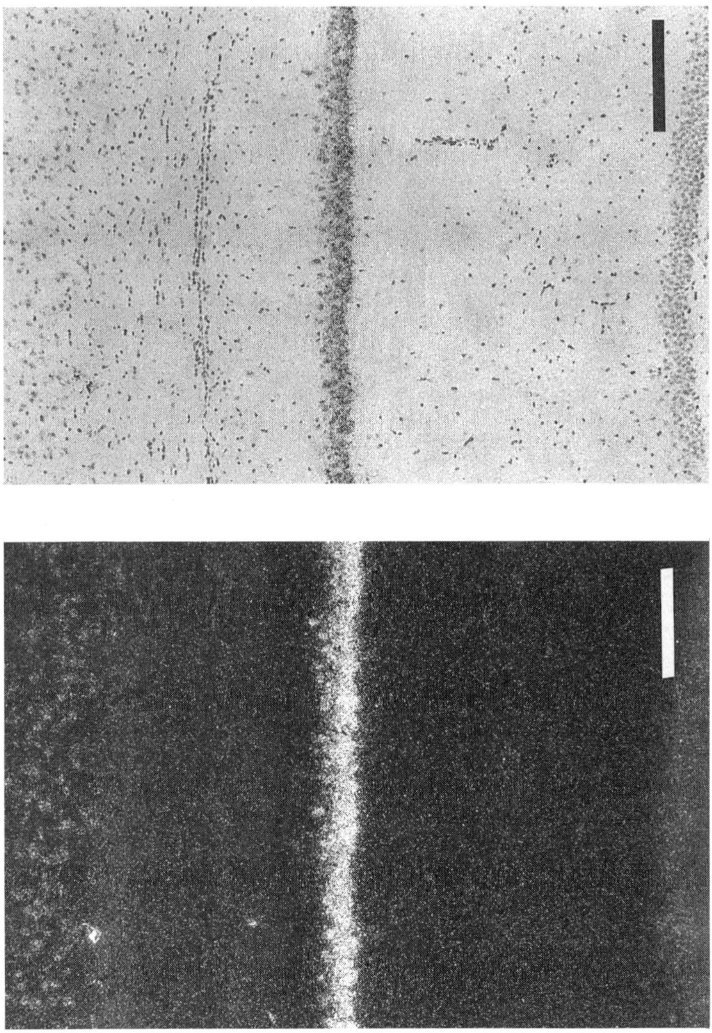

FIGURE 4. Dark (**left**)-and bright (**right**)-field photomicrographs showing the localization of SSTR4 mRNA in the CA1 region of the hippocampus. Pyramidal cell layer is intensely labeled. *Bars* = 200 μm.

during cell differentiation and cell cycle transition[33–39] and considered to play a key role in the kinase cascade.[59–61] MAP kinase is also known to activate $cPLA_2$,[40] which is a key enzyme involved in arachidonate release, one of the important consequences of stimulation by PAF[9,11,12] and SST.[26,27,62] Recent studies showed that MAP kinase activation is mediated by heterotrimeric G proteins[63–67] as well as receptor tyrosine kinases.[68] In fact, PAF triggers activation of MAP kinase in human B cell lines[69] and in sheep platelets.[70] MAP kinase is reported to be functionally expressed in the CNS: its mRNA is expressed at high levels in the hippocampus, neocortex, cerebellum;[71] and its immunostaining is prominent in the neocortex, hippocampus, dentate gyrus, and cerebellar Purkinje cells.[72] Further, *N*-methyl-D-aspartate (NMDA) receptor stimulation of rat hippocampal culture induces the tyrosyl phosphorylation of MAP kinase,[73] and similar phosphorylation occurs in the hippocampi of electroconvulsion-treated rats,[74] transient ischemia-treated gerbils,[75] and in hippocampal slices stimulated with muscarinic receptor agonists or phorbol esters.[76,77]

Binding of PAF and SST in the CHO Cells

The binding parameters for CHO cells stably expressing PAFR (CHOPR1) and SSTR4 (CHOS3-7) were determined by Scatchard analyses using [^{3}H]PAF and [^{125}I-Tyr^{11}]SST, respectively. The values of the dissociation constant (K_d) and maximal binding (B_{max}) for PAF in the CHOPR1 were 400 pM and 6.2 pmol/mg protein, respectively. The K_d and B_{max} values for SST in the CHOS3-7 cells were 2.6 nM and 1.3-3.5 pmol/mg protein, respectively. In the CHOPR1 cells, GTPγS (100 μM) altered the K_d and B_{max} values to 800 pM and 3.9 pmol/mg protein, respectively, supporting the notion that this receptor couples with G protein. In contrast, the same concentration of GTPγS caused much less reduction in SST binding to the CHOS3-7 cells (Sakanaka *et al.*, unpublished data). Although GTPγS may well be less effective in reducing ligand binding to cells overexpressing the receptor, no definite explanation is presently available.

Inhibition of cAMP Accumulation

PAF and SST dose dependently inhibited the forskolin-induced accumulation of cAMP in the CHO cells (FIG. 5a). These results are compatible with the inhibition by PAF[9,11,12] and SST[32] of adenylate cyclase system in native cells (e.g., platelets[78] and brain homogenates[79]).

Production of Inositol Phosphates and Calcium Mobilization

PAF dose dependently induced an increase in production of the inositol phosphates (IPs: IP_1+IP_2+IP_3) with the EC_{50} value of ~1 nM. PAF caused a transient rise in $[Ca^{2+}]_i$ up to ~1 μM in the first few minutes. In contrast, 1 μM SST elevated neither IP_3 (FIG. 5b) nor $[Ca^{2+}]_i$ in the CHOS3-7 cells.

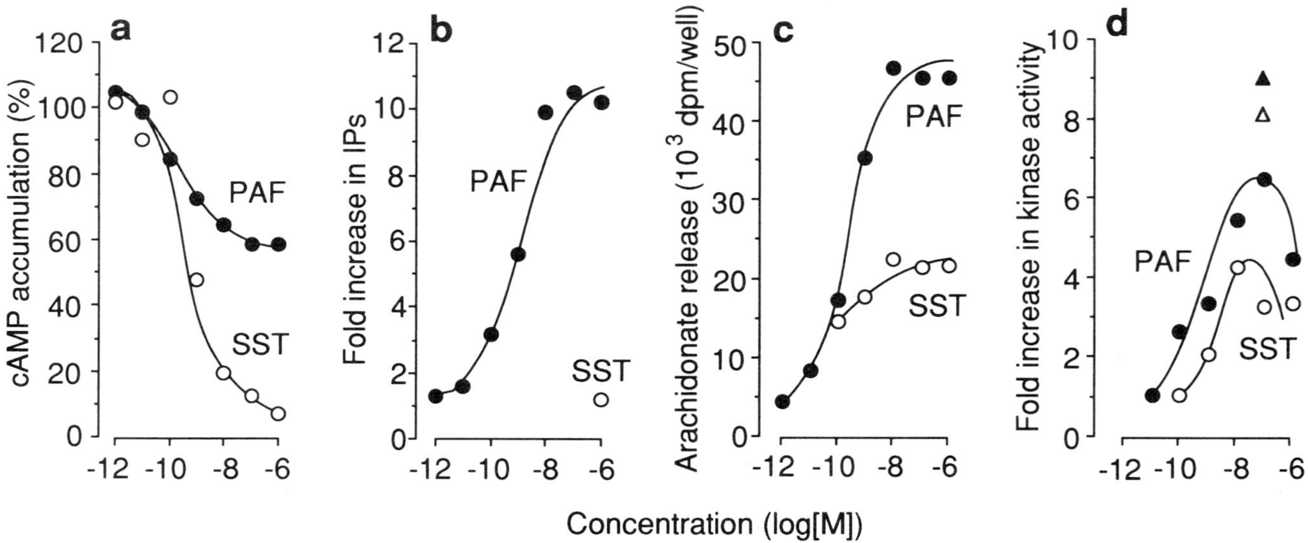

FIGURE 5. PAF- or SST-induced inhibition of forskolin-stimulated cAMP accumulation (**a**), inositol phosphates production (**b**), arachidonate release, (**c**), and MAP kinase/MAP kinase kinase activation (**d**). **a**, The CHO cells stably expressing PAFR (CHOPR1)[58] and SSTR4 (CHOS3-7)[57] were stimulated with 10 μM forskolin only or with forskolin plus various concentrations of PAF and SST, respectively. cAMP accumulation in the cells was determined by radioimmunoassay (Yamasa, Chiba). **b**, CHOPR1 cells prelabeled with *myo*-[^{3}H] inositol were stimulated with various concentrations of PAF, and the production of $IP_1 + IP_2 + IP_3$ was determined by anion exchange column chromatography. CHOS3-7 cells were stimulated with 1 μM SST, and IP_3 production was determined by a D-*myo* [^{3}H] IP_3 assay kit (Amersham). **c**, CHOPR1 and CHO SS3-7 cells were prelabeled with [^{3}H] arachidonate and stimulated with various concentrations of PAF and SST, respectively. Arachidonate release was determined from the radioactivity in the medium. **d**, MAP kinase activity was determined as follows: CHOPR1 and CHOS3-7 cells were stimulated with various concentrations of PAF and SST, respectively, and the MAP kinase in the cell lysate was immunoprecipitated with αC92,[81] anti-MAP kinase antibody. The sample was incubated with 1 mg/ml myelin basic protein (MBP) containing [γ-^{32}P] ATP and spotted on a p81 phosphocellulose paper. After extensive washing, ^{32}P incorporation into MBP was measured by Cerenkov counting. MAP kinase kinase activity was determined as follows: CHOPR1 and CHOS3-7 cells were stimulated with 100 nM PAF and SST, respectively, and MAP kinase kinase was partially purified with batch treatment of Q-Sepharose (Pharmacia). The sample was incubated with a purified kinase-negative mutant of a recombinant *Xenopus* MAP kinase[82,83] (50 μg/ml) containing [γ-^{32}P] ATP and subjected to SDS polyacrylamide gel electrophoresis. ^{32}P incorporation into the recombinant MAP kinase was measured with a Fuji image analyzer (Fujix BAS 2000). The *closed and open triangles* indicate increases in MAP kinase kinase activity induced by PAF and SST, respectively.

Release of Arachidonate

PAF and SST dose dependently induced arachidonate release from the transfected CHO cells (FIG. 5c). Less remarkable increase by SST may have resulted from the absence of SST-induced increase in $[Ca^{2+}]_i$, because translocation of $cPLA_2$ to the membranes is Ca^{2+} dependent.[80] In fact, the addition of calcium ionophore (A23187) caused a severalfold increase in arachidonate release from SST-stimulated CHOS3-7 cells (Sakanaka *et al.*, unpublished data).

Activation of MAP Kinases

PAF and SST dose dependently increased MAP kinase activity in the CHO cells as determined with myelin basic protein as a substrate (FIG. 5d). The maximum activation was elicited by 100 nM PAF or 10 nM SST. The activated MAP kinases included both 42- and 44-kDa isoforms that were identified by anti-MAP kinase antibody (αY91[81]).

Activation of MAP Kinase Kinase

PAF and SST activated MAP kinase kinase at 100 nM by about 9-fold and 8-fold, respectively (FIG. 5d) when kinase-negative mutant of *Xenopus* recombinant MAPK[82,83] was used as a substrate.

Inhibition of Various Signals by Pertussis Toxin

To analyze the nature of G proteins coupled to each receptor, effects of pertussis toxin (PTX) treatment were examined for various signals (FIG. 6). All the SST-induced signals (except IPs production that was *not* induced by SST) were almost completely abolished by PTX. In contrast, the extents of suppression by PTX of PAF-induced signals were ~100% for adenylate cyclase inhibition, less than 10% for IPs production, and ~50% for MAP kinase and MAP kinase kinase activation and arachidonate release. These results suggest that: (1) SSTR4 couples only with PTX-sensitive Gi, whereas PAFR couples with both PTX-sensitive Gi and PTX-insensitive G protein(s) (e.g., Gq); and (2) both Gi and Gq mediate the signal from PAFR to the MAP kinase kinase-MAP kinase cascade, whereas only Gi mediates that from SSTR4 to this cascade.

Absence of Ras Activation

It is well established that Ras mediates signals from receptor tyrosine kinases through Raf kinase activation to MAP kinase.[68] However, G protein-mediated pathways are less clearly understood: they are reported to be Ras dependent[84–86] or partially Ras independent.[67] In our system using the CHO-K1 cells, no significant accumulation of Ras-GTP accompanied PAF- or SST-induced activation of MAP kinases, sug-

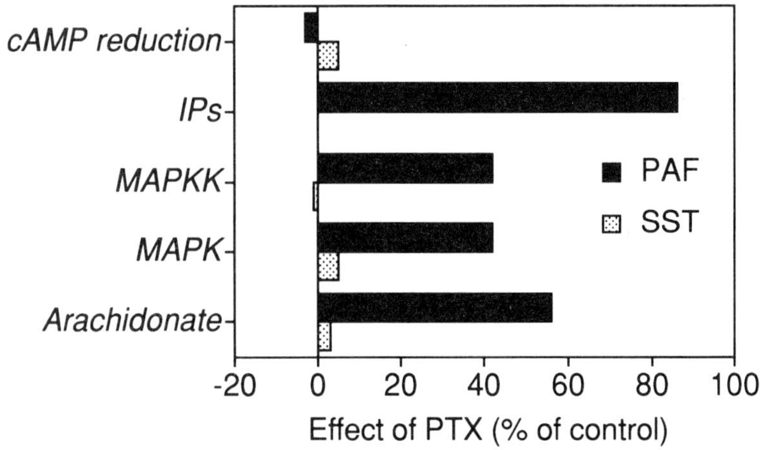

FIGURE 6. Effects of pertussis toxin (PTX) on PAF- or SST-induced signals in CHO cells. PAF (10 nM)- or SST (1 μM)-induced inhibition of forskolin-stimulated increase in cAMP (*cAMP reduction*), arachidonate release (*Arachidonate*), MAP kinase activation (*MAPK*), MAP kinase kinase activation (*MAPKK*), and PAF-induced IP formation (*IPs*) were determined in CHOPR1 and CHOS3-7 cells as described in the legend for FIGURE 5 with or without pretreatment with 100 ng/ml PTX for 24 h.

gesting that the Ras-mediated cascade is not important in PAF- or SST-induced activation of MAP kinases, at least in these cells. The participation of Ras and/or Raf in MAP kinase activation in various systems is now being scrutinized by expression of dominant inhibitory mutants of Ras and Raf.

PROPOSED MODEL FOR SIGNAL TRANSDUCTION PATHWAYS FROM G PROTEIN-COUPLED RECEPTORS TO MAP KINASE ACTIVATION

Among the divergent pathways upstream of MAP kinase (MAPK), the receptor tyrosine kinase (RTK)-mediated pathways are best understood. Interaction of Src homology (SH)2-SH3 proteins (GRB2/Ash)[87–89] and a guanine nucleotide-releasing factor of Ras[90,91] transduces RTK-originated signals to activation of Ras by GTP/GDP exchange.[92–96] The Ras-GTP then activates Raf kinase,[97–99] which phosphorylates MAP kinase kinase (MAPKK).[100,101] On the other hand, the G protein-mediated pathways are still in the course of elucidation. As described above, it is controversial whether Ras activation is always required in these pathways or not.[67,84–86] In a recent study, an additional activator of MAPKK (MEK kinase, MEKK) was identified as a mammalian homologue of *S. cerevisiae* STE11 and proposed to mediate signals from G protein-coupled receptors to MAPKK.[102] Although direct evidence of MEKK activation was not provided, our results suggest that Gi-MEKK cascade mediates the SSTR4-originated signal to MAP kinase kinase and that both the Gi-MEKK and Gq-

MAPKK cascades mediate PAFR-originated signals to MAPKK (FIG. 7). The Gi-MAPKK pathway involves activation of Ras and Raf in some systems[84] but not in ours, suggesting that this pathway is rather complex. It is also suggested that Gi activates MEKK through as-yet-unknown pathways(s), or that it activates MAPKK through other Ras-independent pathway(s). The Gq-MAPKK cascade appears to depend on protein-kinase C (PKC), because down-regulation of PKC inhibits activation of MAPK by endothelin[103] and by PAF (Nozawa *et al.*, unpublished data). The PKC-MAPKK pathway may be mediated by Raf, which was recently shown to be phosphorylated directly by PKC_{α}.[104] However, sequential activation of Gq-PLC_{β}-PKC-Raf-MAPKK is unlikely unless Ras is activated through as-yet-unidentified pathways. Direct activation of MEKK by PKC is a possible pathway independent of Ras activation. The activated MAP kinase induces expression of early oncogenes—for example, *c-jun,*[105] *c-myc,*[106] and $p62^{TCF}$[107] and phosphorylates various types of protein kinases[60] and $cPLA_2$.[40] Phosphorylated $cPLA_2$ is translocated to membrane by the action of Ca^{2+} through the Ca^{2+}-dependent lipid-binding domain and catalyzes synthesis of PAF and release of arachidonate,[40,80,108] which is further metabolized to prostaglandins (PGs), leukotrienes (LTs), and other derivatives[109] that exert various biological actions. The signaling pathways observed in the CHO cells overexpressing the receptor are compatible with the findings in the human B cell lines[69] but may not necessarily be identical to those in native cells. Although there has been no evidence that the physiological and pathological events induced by PAF or SST are the consequences of MAP kinase-mediated gene expression and/or activation of various enzymes and proteins, it seems likely that MAP kinase-mediated pathways are involved in at least some of the mechanisms underlying the long-term effects of stimuli to PAFR, SSTR4, and other G protein-coupled receptors. The molecules involved in these pathways are functionally present in the rat hippocampus, suggesting that these pathways found in the CHO cell systems might be responsible for the effects of PAF and SST in the CNS.

SUMMARY

Platelet-activating factor and somatostatin receptors, two G protein-coupled receptors expressed in the rat hippocampus, were analyzed for the downstream signaling pathways in Chinese hamster ovary cells stably expressing each receptor. Ligand stimulation to each CHO cell line induced (1) inhibition of forskolin-induced accumulation of cAMP, (2) arachidonate release, and (3) activation of mitogen-activated protein kinase and MAP kinase kinase. In contrast, inositol phosphate breakdown was seen only in the PAF-stimulated CHO cells. The induction of these signals accompanied no detectable Ras activation. Suppression of the signals by pertussis toxin was almost complete for the somatostatin receptor but partial for the PAF receptor, suggesting that the somatostatin receptor couples only with PTX-sensitive G protein, while the PAF receptor couples with both PTX-sensitive and -insensitive G proteins. A model of G protein-mediated signaling pathways was proposed in which the signals from Gi and those from Gq converge at MAP kinase kinase and lead to arachidonate release. The present system using CHO cells is useful for analyzing signaling pathways from G proteins to MAP kinase kinase and will thereby

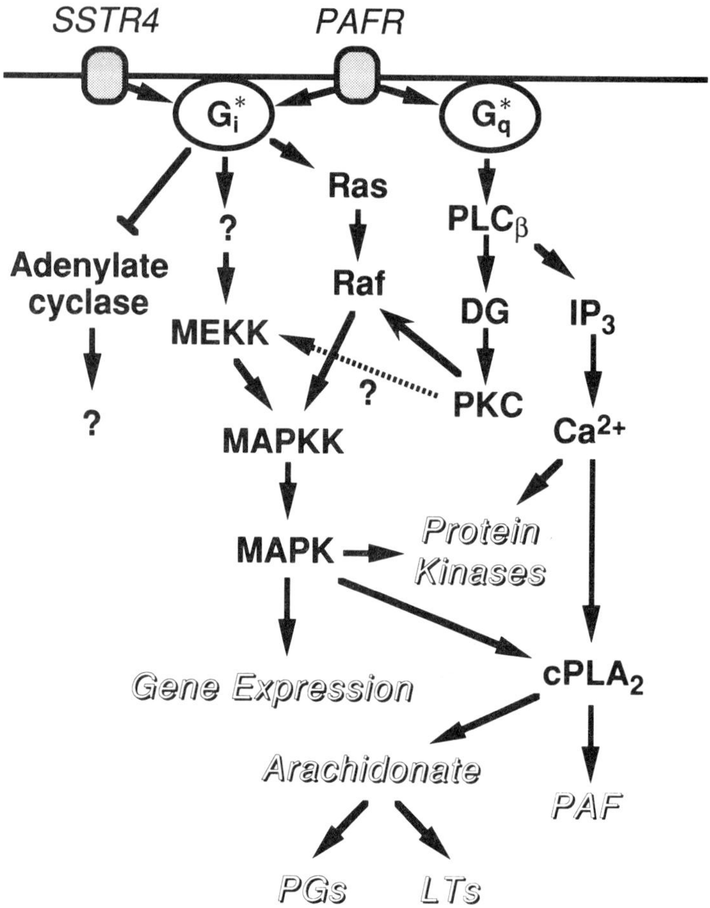

FIGURE 7. A proposed model for signal transduction pathways mediated by heterotrimeric G proteins. Gi mediates signals from both PAFR and SSTR4 to MAP kinase kinase (MAPKK) through Ras-Raf and/or MEK kinase (MEKK). Gq mediates signals from PAFR to phospholipase Cβ (PLCβ), which produces inositol triphosphate (IP_3) and diacylglycerol (DG). IP_3 induces release of Ca^{2+} from intracellular Ca^{2+} stores, and DG activates protein kinase C (PKC). PKC then activates MAPKK through Raf and possibly through MEKK. Signals through Gi and Gq thus converge at MAPKK, and MAPKK then activates MAP kinase (MAPK). MAPK activates various protein kinases and cytosolic phospholipase A_2 ($cPLA_2$) and induces growth-related gene expression. $cPLA_2$ is translocated to membrane by Ca^{2+} (mediated by IP_3 in some cases) and catalyzes synthesis of platelet-activating factor (PAF) and release of arachidonate, which exerts biological actions by itself and/or is metabolized to prostaglandins (PGs), leukotrienes (LTs), and other derivatives.[22,109] *Asterisks*: unidentified PTX-sensitive and -insensitive G proteins are putatively shown as Gi and Gq, respectively.

provide clues for understanding the mechanisms underlying the physiological and pathological events mediated by PAF, somatostatin, and other G protein-coupled receptors in the central nervous system and other tissues.

REFERENCES

1. BASHIR, Z. I. & G. L. COLLINGRIDGE. 1992. Synaptic plasticity: Long-term potentiation in the hippocampus. Curr. Opin. Neurobiol. **2:** 328-335.
2. FAZELI, M. S. 1992. Synaptic plasticity: On the trail of the retrograde messenger. Trends Neurosci. **15:** 115-117.
3. WILLIAMS, J. H., M. L. ERRINGTON, Y. G. LI, M. A. LYNCH & T. V. P. BLISS. 1993. The search for retrograde messengers in long-term potentiation. Semin. Neurosci. **5:** 149-158.
4. SIESJÖ, B. K. & K. KATSURA. 1992. Ischemic brain damage: Focus on lipids and lipid mediators. Adv. Exp. Med. Biol. **318:** 41-56.
5. WILLIAMS, J. H. & T. V. P. BLISS. 1988. Induction but not maintenance of calcium-induced long-term potentiation in dentate gyrus and area CA1 of the hippocampal slice is blocked by nordihydroguaiaretic acid. Neurosci. Lett. **88:** 81-85.
6. WILLIAMS, J. H. & T. V. P. BLISS. 1989. An in intro study of the effect of lipoxygenase and cyclooxygenase inhibitors of arachidonic acid on the induction and maintenance of long-term potentiation in the hippocampus. Neurosci. Lett. **107:** 301-306.
7. OKADA, D., S. YAMAGISHI & H. SUGIYAMA. 1989. Differential effects of phospholipase inhibitors in long-term potentiation in the rat hippocampal mossy fiber synapses and Schaffer/commissural synapses. Neurosci. Lett. **100:** 141-146.
8. WIERASZKO, A., G. LI, E. KORNECKI, M. V. HOGAN & Y. H. EHRLICH. 1993. Long-term potentiation in the hippocampus induced by platelet-activating factor. Neuron **10:** 553-557.
9. CHAO, W. & M. S. OLSON. 1993. Platelet-activating factor: Receptors and signal transduction. Biochem. J. **292:** 617-629.
10. SHIMIZU, T., Z. HONDA, M. NAKAMURA, H. BITO & T. IZUMI. 1992. Platelet-activating factor receptor and signal transduction. Biochem. Pharmacol. **44:** 1001-1008.
11. SHUKLA, S. D. 1992. Platelet-activating factor receptor and signal transduction mechanisms. FASEB J. **6:** 2296-2301.
12. VENABLE, M. E., G. A. ZIMMERMAN, T. M. MCINTYRE & S. M. PRESCOTT. 1993. Platelet-activating factor: A phospholipid autocoid with diverse actions. J. Lipid Res. **34:** 691-702.
13. TRIPATHI, Y. B., J. C. KANDALA, R. V. GUNTAKA, R. W. LIM & S. D. SHUKLA. 1991. Platelet activating factor induces expression of early response genes c-fos and TIS-1 in human epidermoid carcinoma A-431 cells. Life Sci. **49:** 1761-1767.
14. SQUINTO, S. P., A. L. BLOCK, P. BRAQUET & N. G. BAZAN. 1989. Platelet-activating factor stimulates a fos/jun/AP-1 transcriptional signaling system in human neuroblastoma cells. J. Neurosci. Res. **24:** 558-566.
15. KORNECKI, E. & Y. H. EHRLICH. 1988. Neuroregulatory and neuropathological actions of the ether-phospholipid platelet-activating factor. Science **240:** 1792-1794.
16. HARUSHKA, P. V., D. E. MAIS, P. R. MAYEUX & T. A. MORINELLI. 1989. Thromboxane, prostaglandin and leukotriene receptors. Annu. Rev. Pharmacol. Toxicol. **29:** 213-239.
17. HIRATA, M., Y. HAYASHI, Y. USHIKUBI, Y. YOKOTA, R. KAGEYAMA, S. NAKANISHI & S. NARUMIYA. 1991. Cloning and expression of cDNA for a human thromboxane A2 receptor. Nature **349:** 617-620.

18. HONDA, Z., M. NAKAMURA, I. MIKI, M. MINAMI, T. WATANABE, Y. SEYAMA, H. OKADO, H. TOH, K. ITO, T. MIYAMOTO & T. SHIMIZU. 1991. Cloning by functional expression of platelet-activating factor receptor from guinea-pig lung. Nature **349:** 342-346.

19. HONDA, A., Y. SUGIMOTO, T. NAMBA, T. WATABE, A. IRIE, M. NEGISHI, S. NARUMIYA & M. ICHIKAWA. 1993. Cloning and expression of a cDNA for mouse prostaglandin E receptor EP2 subtype. J. Biol. Chem. **268:** 7759-7762.

20. SUGIMOTO, Y., T. NAMBA, A. HONDA, Y. HAYASHI, M. NEGISHI, A. ICHIKAWA & S. NARUMIYA. 1992. Cloning and expression of a cDNA for mouse prostaglandin E receptor EP3 subtype. J. Biol. Chem. **267:** 6463-6466.

21. WATABE, A., Y. SUGIMOTO, A. HONDA, A. IRIE, T. NAMBA, M. NEGISHI, S. ITO, S. NARUMIYA & A. ICHIKAWA. 1993. Cloning and expression of cDNA for a mouse EP(1) subtype of prostaglandin-E receptor. J. Biol. Chem. **268:** 20175-20178.

22. SHIMIZU, T. & L. S. WOLFE. 1990. Arachidonic acid cascade and signal transduction. J. Neurochem. **55:** 1-15.

23. PIOMELLI, D., A. VOLTERRA, N. DALE, S. A. SIEGELBAUM, E. R. KANDEL, J. H. SCHWARTZ & F. BELADETTI. 1987. Lipoxygenase metabolites of arachidonic acid are second messengers for presynaptic inhibition of Aplysia sensory cells. Nature **328:** 38-43.

24. KURACHI, Y., H. ITO, T. SUGIMOTO, T. SHIMIZU, T. MIKI & M. UI. 1989. Arachidonic acid metabolites as intracellular modulators of the G protein-gated cardiac K^+ channel. Nature **337:** 555-557.

25. KIM, D., D. L. LEWIS, L. GRAZIADEI, E. J. NEER, D. BAR-SAGI & D. E. CLAPHAM. 1989. G-protein beta gamma-subunits activate the cardiac muscarinic K^+-channel via phospholipase A2. Nature **337:** 557-560.

26. SCHWEITZER, P., S. MADAMBA & G. R. SIGGINS. 1990. Arachidonic acid metabolites as mediators of somatostatin-induced increase of neuronal M-current. Nature **346:** 464-467.

27. SCHWEITZER, P., S. MADAMBA, J. CHAMPAGNAT & G. R. SIGGINS. 1993. Somatostatin inhibition of hippocampal CA1 pyramidal neurons: Mediation by arachidonic acid and its metabolites. J. Neurosci. **13:** 2033-2049.

28. HWANG, S. B. 1991. Function and regulation of extracellular and intracellular receptors of platelet activating factor. Ann. N.Y. Acad. Sci. **629:** 217-226.

29. PATEL, Y. C. 1992. Somatostatin. *In* Basic and Clinical Aspects of Neuroscience. E. E. Muller, M. O. Thorner & C. Weil., Eds.: 1-16. Springer-Verlag. Berlin.

30. REICHLIN, S. 1983. Somatostatin. N. Engl. J. Med. **309:** 1495-1501, 1556-1563.

31. BRAZEAU, P., W. VALE, R. BURGUS, N. LING, M. BUTCHER, J. RIVER & R. GUILLEMIN. 1973. Hypothalamic polypeptide that inhibits secretion on immunoreactive pituitary growth hormone. Science **179:** 77-79.

32. RENS-DOMIANO, S. & T. REISINE. 1992. Biochemical and functional properties of somatostatin receptors. J. Neurochem. **58:** 1987-1996.

33. PELECH, S. L., R. M. TOMBERS, L. MEIJER & E. G. KREBS. 1988. Activation of myelin basic protein kinase during echinoderm oocyte maturation and egg fertilization. Dev. Biol. **130:** 28-36.

34. RAY, L. B. & T. W. STURGILL. 1987. Rapid stimulation by insulin of a serine/threonine kinase in 3T3-L1 adipocytes that phosphorylates microtubule-associated protein 2 in vitro. Proc. Natl. Acad. Sci. USA **84:** 1502-1506.

35. HOSHI, M., E. NISHIDA & H. SAKAI. 1988. Activation of a Ca^{2+}-inhibited protein kinase that phosphorylates microtubule-associated protein 2 in vitro by growth factors, phorbol esters, and serum in quiescent cultured human fibroblasts. J. Biol. Chem. **263:** 5396-5401.

36. ROSSOMANDO, A. J., D. M. PAYNE, M. J. WEBER & T. W. STURGILL. 1989. Evidence that pp42, a major tyrosine kinase target protein, is a mitogen-activated serine/threonine protein kinase. Proc. Natl. Acad. Sci. USA **86:** 6940-6943.
37. GOTOH, Y., E. NISHIDA, T. YAMASHITA, M. HOSHI, M. KAWAKAMI & H. SAKAI. 1990. Microtubule-associated-protein (MAP) kinase activated by nerve growth factor and epidermal growth factor in PC12 cells. Identity with the mitogen-activated MAP kinase of fibroblastic cells. Eur. J. Biochem. **193:** 661-669.
38. GOTOH, Y., K. MORIYAMA, S. MATSUDA, E. OKUMURA, T. KISHIMOTO, H. KAWASAKI, K. SUZUKI, I. YAHARA, H. SAKAI & E. NISHIDA. 1991. Xenopus M phase MAP kinase: Isolation of its cDNA and activation by MPF. EMBO J. **10:** 2661-2668.
39. GOTOH, Y., E. NISHIDA, S. MATSUDA, N. SHIINA, H. KOSAKO, K. SHIOKAWA, T. AKIYAMA, K. OHTA & H. SAKAI. 1991. In vitro effects on microtubule dynamics of purified Xenopus M phase-activated MAP kinase. Nature **349:** 251-254.
40. LIN, L.-L., M. WARTMAN, A. Y. LIN, J. L. KNOPF, A. SETH & R. J. DAVIS. 1992. cPLA2 is phosphorylated and activated by MAP kinase. Cell **72:** 269-278.
41. NAKAMURA, M., Z. HONDA, T. IZUMI, C. SAKANAKA, H. MUTOH, M. MINAMI, H. BITO, Y. SEYAMA, T. MATSUMOTO, M. NOMA & T. SHIMIZU. 1991. Molecular cloning and expression of platelet-activating factor receptor from human leukocytes. J. Biol. Chem. **266:** 20400-20405.
42. HWANG, S.-B., M.-H. LAM & S.-S. PONG. 1986. Ionic and GTP regulation of platelet-activating factor to receptors and platelet-activating factor-induced activation of GTPase in rabbit platelet membranes. J. Biol. Chem. **261:** 532-537.
43. HWANG, S.-B. 1988. Identification of a second putative receptor of platelet-activating factor from human polymorphonuclear leukocytes. J. Biol. Chem. **263:** 3225-3233.
44. SHUKLA, S. D. 1991. Inositol phospholipid turnover in PAF transmembrane signalling. Lipids **26:** 1028-1033.
45. BITO, H., M. NAKAMURA, Z. HONDA, T. IZUMI, T. IWATSUBO, Y. SEYAMA, A. OGURA, Y. KUDO & T. SHIMIZU. 1992. Platelet-activating factor (PAF) receptor in rat brain: PAF mobilizes intracellular Ca^{2+} in hippocampal neurons. Neuron **9:** 285-294.
46. BELL, G. I. & T. REISINE. 1993. Molecular biology of somatostatin receptors. Trends Neurosci. **16:** 34-38.
47. BRUNO, J. F., Y. XU, J. SONG & M. BERELOWITZ. 1992. Molecular cloning and functional expression of a brain specific somatostatin receptor. Proc. Natl. Acad. Sci. USA **89:** 11151-11155.
48. KLUXEN, F. W., C. BRUNS & H. LUBBERT. 1992. Expression cloning of a rat brain somatostatin receptor cDNA. Proc. Natl. Acad. Sci. USA **89:** 4618-4622.
49. MEYERHOF, W., H.-J. PAUST, C. SCHÖNROCK & D. RICHTER. 1991. Cloning of cDNA encoding a novel putative G-protein-coupled receptor expressed in specific rat brain regions. DNA Cell Biol. **10:** 689-694.
50. MEYERHOF, W., I. WULFSEN, C. SCHÖNROCK, S. FEHR & D. RICHTER. 1992. Molecular cloning of a somatostatin-28 receptor and comparison of its expression pattern with that of a somatostatin-14 receptor in rat brain. Proc. Natl. Acad. Sci. USA **89:** 10267-10271.
51. O'CARROLL, A. M., S. J. LOLAIT, M. KONIG & L. C. MAHAN. 1992. Molecular cloning and expression of a pituitary somatostatin receptor with preferential affinity for somatostatin-28. Mol. Pharmacol. **42:** 939-946.
52. YAMADA, Y., S. R. POST, K. WANG, H. S. TAGER, G. I. BELL & S. SEINO. 1992. Cloning and functional characterization of a family of human and mouse somatostatin receptors expressed in brain, gastrointestinal tract, and kidney. Proc. Natl. Acad. Sci. USA **89:** 251-255.

53. YASUDA, K., D. S. RENS, C. D. BREDER, S. F. LAW, C. B. SAPER, T. REISINE & G. I. BELL. 1992. Cloning of a novel somatostatin receptor, SSTR3, coupled to adenylylcyclase. J. Biol. Chem. **267:** 20422-20428.
54. KAUPMANN, K., C. BRUNS, D. HOYER, K. SEUWEN & H. LÜBBERT. 1993. Distribution and second messenger coupling of four somatostatin receptor subtypes expressed in brain. FEBS Lett. **331:** 53-59.
55. RENS-DOMIANO, S., S. F. LAW, Y. YAMADA, S. SEINO, G. I. BELL & T. REISINE. 1992. Pharmacological properties of two cloned somatostatin receptors. Mol. Pharmacol. **42:** 28-34.
56. LAW, S. F., K. YASUDA, G. I. BELL & T. REISINE. 1993. Gi alpha 3 and G(o) alpha selectively associate with the cloned somatostatin receptor subtype SSTR2. J. Biol. Chem. **268:** 10721-10727.
57. BITO, H., M. MORI, C. SAKANAKA, T. TAKANO, Y. GOTOH, E. NISHIDA, Z. HONDA & T. SHIMIZU. 1994. Functional coupling of SSTR4, a major hippocampal somatostatin receptor, to adenylate cyclase inhibition, arachidonate release, and activation of the mitogen-activated protein kinase cascade. J. Biol. Chem. **269:** 12722-12730.
58. HONDA, Z., T. TAKANO, Y. GOTOH, E. NISHIDA, K. ITO & T. SHIMIZU. 1994. Transfected platelet-activating factor receptor activates mitogen-activated protein (MAP) kinase and MAP kinase kinase in Chinese hamster ovary cells. J. Biol. Chem. **269:** 2307-2315.
59. BLENIS, J. 1993. Signal transduction via the MAP kinases: Proceed at your own RSK. Proc. Natl. Acad. Sci. USA **90:** 5889-5892.
60. DAVIS, R. J. 1993. The mitogen-activated protein kinase signal transduction pathway. J. Biol. Chem. **268:** 14553-14556.
61. NISHIDA, E. & Y. GOTOH. 1993. The MAP kinase cascade is essential for diverse signal transduction pathways. Trends Biochem. Sci. **18:** 128-131.
62. MARIN, P., J. C. DELUMEAU, M. TENCE, J. CORDIER, J. GLOWINSKI & J. PREMONT. 1991. Somatostatin potentiates the alpha 1-adrenergic activation of phospholipase C in striatal astrocytes through a mechanism involving arachidonic acid and glutamate. Proc. Natl. Acad. Sci. USA **88:** 9016-9020.
63. GARDNER, A. M., R. R. VAILLANCOURT & G. L. JOHNSON. 1993. Activation of mitogen-activated protein kinase/extracellular signal-regulated kinase kinase by G protein and tyrosine kinase oncoproteins. J. Biol. Chem. **268:** 17896-17901.
64. GALLEGO, C., S. K. GUPTA, L. E. HEASLEY, N. X. QIAN & G. L. JOHNSON. 1992. Mitogen-activated protein kinase activation resulting from selective oncogene expression in NIH 3T3 and rat 1a cells. Proc. Natl. Acad. Sci. USA **89:** 7355-7359.
65. GUPTA, S. K., C. GALLEGO, G. L. JOHNSON & L. E. HEASLEY. 1992. MAP kinase is constitutively activated in gip2 and src transformed rat 1a fibroblasts. J. Biol. Chem. **267:** 7987-7990.
66. KAHAN, C., K. SEUWEN, S. MELOCHE & J. POUYSSEGUR. 1992. Coordinate, biphasic activation of p44 mitogen-activated protein kinase and S6 kinase by growth factors in hamster fibroblasts. Evidence for thrombin-induced signals different from phosphoinositide turnover and adenylylcyclase inhibition. J. Biol. Chem. **267:** 13369-13375.
67. ROBBINS, D. J., M. CHENG, E. ZHEN, C. A. VANDERBILT, L. A. FEIG & M. H. COBB. 1992. Evidence for a Ras-dependent extracellular signal-regulated protein kinase (ERK) cascade. Proc. Natl. Acad. Sci. USA **89:** 6924-6928.
68. SATOH, T., M. NAKAFUKU & Y. KAZIRO. 1992. Function of Ras as a molecular switch in signal transduction. J. Biol. Chem. **267:** 24149-24154.
69. FRANKLIN, R. A., B. MAZER, H. SAWAMI, G. B. MILLS, N. TERADA, J. J. LUCAS & E. W. GELFAND. 1993. Platelet-activating factor triggers the phosphorylation and activation of MAP-2 kinase and S6 peptide kinase activity in human B cell lines. J. Immunol. **151:** 1802-1810.

70. SAMIEI, M., J. S. SANGHERA & S. L. PELECH. 1993. Activation of myelin basic protein and S6 peptide kinases in phorbol ester- and PAF-treated sheep platelets. Biochim. Biophys. Acta **1176:** 287-298.
71. THOMAS, K. L. & S. P. HUNT. 1993. The regional distribution of extracellulary regulated kinase-1 and -2 messenger RNA in the adult rat central nervous system. Neuroscience **56:** 741-757.
72. FIORE, R. S., V. E. BAYER, S. L. PELECH, J. POSADA, J. A. COOPER & J. M. BARABAN. 1993. p42 Mitogen-activated protein kinase in brain: Prominent localization in neuronal cell bodies and dendrites. Neuroscience **55:** 463-472.
73. BADING, H. & M. E. GREENBERG. 1991. Stimulation of protein tyrosine phosphorylation by NMDA receptor activation. Science **253:** 912-914.
74. BARABAN, J. M., R. S. FIORE, J. S. SANGHERA, H. B. PADDON & S. L. PELECH. 1993. Identification of p42 mitogen-activated protein kinase as a tyrosine kinase substrate activated by maximal electroconvulsive shock in hippocampus. J. Neurochem. **60:** 330-336.
75. CAMPOS-GONZALEZ, R. & M. S. KINDY. 1992. Tyrosine phosphorylation of microtubule-associated protein kinase after transient ischemia in the gerbil brain. J. Neurochem. **59:** 1955-1958.
76. STRATTON, K. R., P. F. WORLEY, J. S. LITZ, S. J. PARSONS, R. L. HUGANIR & J. M. BARABAN. 1991. Electroconvulsive treatment induces a rapid and transient increase in tyrosine phosphorylation of a 40-kilodalton protein associated with microtubule-associated protein 2 kinase activity. J. Neurochem. **56:** 147-152.
77. STRATTON, K. R., P. F. WORLEY, R. L. HUGANIR & J. M. BARABAN. 1989. Muscarinic agonists and phorbol esters increase tyrosine phosphorylation of 40-kilodalton protein in hippocampal slices. Proc. Natl. Acad. Sci. USA **86:** 2498-2501.
78. HASLAM, R. J. & M. VANDERWEL. 1982. Inhibition of platelet adenylate cyclase by 1-*O*-Alkyl-2-acetyl-*sn*-blyceryl-3-phosphorylcholine (platelet-activating factor). J. Biol. Chem. **257:** 6879-6885.
79. SCHETTINI, G., T. FLORIO, O. MEUCCI, E. LANDOLFI, M. GRIMALDI, C. VENTRA & A. MARINO. 1989. Somatostatin inhibition of adenylate cyclase activity in different brain areas. Brain Res. **492:** 65-71.
80. CLARK, J. D., L.-L. LIN, R. W. KRIZ, C. S. RAMESHA, L. A. SULTZMAN, A. Y. LIN, N. MINOLA & J. L. KNOPF. 1991. A novel arachidonic acid-selective cytosolic PLA2 contains a Ca^{2+}-dependent translocation domain with homology to PKC and GAP. Cell **65:** 1043-1051.
81. TOBE, K., T. KADOWAKI, H. TAMEMOTO, K. UEKI, K. HARA, O. KOSHIO, K. MOMOMURA, Y. GOTOH, E. NISHIDA, Y. AKANUMA, Y. YAZAKI & M. KASUGA. 1991. Insulin and 12-*O*-tetradecanoylphorbol-13-acetate activation of two immunologically distinct myelin basic protein/microtubule-associated protein 2 (MBP/MAP2) kinases via de novo phosphorylation of threonine and tyrosine residues. J. Biol. Chem. **266:** 24793-24803.
82. KOSAKO, H., E. NISHIDA & Y. GOTOH. 1993. cDNA cloning of MAP kinase kinase reveals kinase cascade pathways in yeasts to vertebrates. EMBO J. **12:** 787-794.
83. MATSUDA, S., Y. GOTOH & E. NISHIDA. 1993. Phosphorylation of Xenopus mitogen-activated protein (MAP) kinase kinase by MAP kinase kinase kinase and MAP kinase. J. Biol. Chem. **268:** 3277-3281.
84. WINITZ, S., M. RUSSELL, N.-X. QIAN, A. GARDNER, L. DWYER & G. L. JOHNSON. 1993. Involvement of Ras and Raf in the Gi-coupled acetylcholine muscarinic m2 receptor activation of mitogen-activated protein (MAP) kinase kinase and MAP kinase. J. Biol. Chem. **268:** 19196-19199.
85. HOWE, L. R. & C. J. MARSHALL. 1993. Lysophosphatidic acid stimulates mitogen-activated protein kinase activation via a G-protein-coupled pathway requiring p21ras and p74raf-1. J. Biol. Chem. **268:** 20717-20720.

86. VAN CORVEN, E. J., P. L. HORDIJK, R. H. MEDEMA & J. L. BOS. 1993. Pertussis toxin-sensitive activation of p21ras by G protein-coupled receptor agonists in fibroblasts. Proc. Natl. Acad. Sci. USA **90:** 1257-1261.
87. MATSUOKA, K., M. SHIBATA, A. YAMAKAWA & T. TAKENAWA. 1992. Cloning of ASH, a ubiquitous protein composed of one Src homology region (SH) 2 and two SH3 domains, from human and rat cDNA libraries. Proc. Natl. Acad. Sci. USA **89:** 9015-9019.
88. LOWENSTEIN, E. J., R. J. DALY, A. G. BATZER, W. LI, B. MARGOLIS, R. LAMMERS, A. ULLRICH, E. Y. SKOLNIK, S. D. BAR & J. SCHLESSINGER. 1992. The SH2 and SH3 domain-containing protein GRB2 links receptor tyrosine kinases to ras signaling. Cell **70:** 431-442.
89. CLARK, S. G., M. J. STERN & H. R. HORVITZ. 1992. *C. elegans* cell-signalling gene sem-5 encodes a protein with SH2 and SH3 domains. Nature **356:** 340-344.
90. SIMON, M. A., D. D. BOWTELL, G. S. DODSON, T. R. LAVERTY & G. M. RUBIN. 1991. Ras1 and a putative guanine nucleotide exchange factor perform crucial steps in signaling by the sevenless protein tyrosine kinase. Cell **67:** 701-716.
91. BOWTELL, D., P. FU, M. SIMON & P. SENIOR. 1992. Identification of murine homologues of the *Drosophila* son of sevenless gene: Potential activators of rats. Proc. Natl. Acad. Sci. USA **89:** 6511-6515.
92. EGAN, S. E., B. W. GIDDINGS, M. W. BROOKS, L. BUDAY, A. M. SIZELAND & R. A. WEINBERG. 1993. Association of Sos Ras exchange protein with Grb2 is implicated in tyrosine kinase signal transduction and transformation. Nature **363:** 45-51.
93. ROZAKIS-ADCOCK, M., R. FERNLEY, J. WADE, T. PAWSON & D. BOWTELL. 1993. The SH2 and SH3 domains of mammalian Grb2 couple the EGF receptor to the Ras activator mSos1. Nature **363:** 83-85.
94. LI, N., A. BATZER, R. DALY, V. YAJNIK, E. SKOLNIK, P. CHARDIN, S. D. BAR, B. MARGOLIS & J. SCHLESSINGER. 1993. Guanine-nucleotide-releasing factor hSos1 binds to Grb2 and links receptor tyrosine kinases to Ras signalling. Nature **363:** 85-88.
95. GALE, N. W., S. KAPLAN, E. J. LOWENSTEIN, J. SCHLESSINGER & S. D. BAR. 1993. Grb2 mediates the EGF-dependent activation of guanine nucleotide exchange on Ras. Nature **363:** 88-92.
96. FEIG, L. A. 1993. The many roads that lead to Ras. Science **260:** 767-768.
97. THOMAS, S. M., M. DEMARCO, G. D'ARCANGELO, S. HALEGOUA & J. S. BRUGGE. 1992. Ras is essential for nerve growth factor- and phorbol ester-induced tyrosine phosphorylation of MAP kinases. Cell **68:** 1031-1040.
98. WOOD, K. W., C. SARNECKI, T. M. ROBERTS & J. BLENIS. 1992. Ras mediates nerve growth factor receptor modulation of three signal-transducing protein kinases: MAP kinase, Raf-1, and RSK. Cell **68:** 1041-1050.
99. DE VRIES-SMITH, A. M. M., B. M. T. BURGERING, S. J. LEEVERS, C. J. MARSHALL & J. L. BOS. 1992. Involvement of p21ras in activation of extracellular signal-regulated kinase 2. Nature **357:** 602-604.
100. HOWE, L. R., S. J. LEEVERS, N. GOMEZ, S. NAKIELNY, P. COHEN & C. J. MARSHALL. 1992. Activation of the MAP kinase pathway by the protein kinase raf. Cell **71:** 335-342.
101. KYRIAKIS, J. M., H. APP, X. F. ZHANG, P. BANERJEE, D. L. BRAUTIGAN, U. R. RAPP & J. AVRUCH. 1992. Raf-1 activates MAP kinase-kinase. Nature **358:** 417-421.
102. LANGE-CARTER, C. A., C. M. PLEIMAN, A. M. GARDNER, K. J. BLUMER & G. L. JOHNSON. 1993. A divergence in the MAP kinase regulatory network defined by MEK kinase and Raf. Science **260:** 315-319.
103. WANG, Y., M. S. SIMONSON, J. POUYSSEGUR & M. J. DUNN. 1992. Endothelin rapidly stimulates mitogen-activated protein kinase activity in rat mesangial cells. Biochem. J. **287:** 589-594.

104. KOLCH, W., G. HEIDECKER, G. KOCHS, R. HUMMEL, H. VAHIDI, H. MISCHAK, G. FINKENZELLER, D. MARMÉ & U. R. RAPP. 1993. Protein kinase C alpha activates RAF-1 by direct phosphorylation. Nature **364:** 249-252.
105. BAKER, S. J., T. K. KERPPOLA, D. LUK, M. T. VANDENBERG, D. R. MARSHAK, T. CURRAN & C. ABATE. 1992. Jun is phosphorylated by several protein kinases at the same sites that are modified in serum-stimulated fibroblasts. Mol. Cell. Biol. **12:** 4694-4705.
106. SETH, A., F. A. GONZALEZ, S. GUPTA, D. L. RADEN & R. J. DAVIS. 1992. Signal transduction within the mitogen-activated protein kinase. J. Biol. Chem. **267:** 24796-24804.
107. GILLE, H., A. D. SHARROCKS & P. E. SHAW. 1992. Phosphorylation of transcription factor p62TCF by MAP kinase stimulates ternary complex formation at c-fos promotor. Nature **358:** 414-417.
108. SHARP, J. D., D. L. WHITE, X. GRACE-CHIOU, T. GOODSON, G. C. GAMBOA, D. MACLURE, S. BURGETT, J. HOSKINS, P. L. SKATRUD, J. R. SPORTSMAN, G. W. BECKER, L. H. KANG, E. F. ROBERTS & R. M. KRAMER. 1991. Molecular cloning and expression of human Ca^{2+}-sensitive cytosolic phospholipase A2. J. Biol. Chem. **266:** 14850-14856.
109. SAMUELSSON, B., S. E. DAHLEN, J. A. LINDGREN, C. A. ROUZER & C. N. SERHAN. 1987. Leukotrienes and lipoxins: Structure, biosynthesis and biological effects. Science **237:** 1171-1176.
110. BITO, H., Z. HONDA, M. NAKAMURA & T. SHIMIZU. 1994. Cloning, expression and tissue distribution of rat platelet-activating-factor-receptor cDNA. Eur. J. Biochem. **221:** 211-218.

Prostanoid Receptors

Structure, Function, and Distribution[a]

SHUH NARUMIYA

Department of Pharmacology
Kyoto University Faculty of Medicine
Yoshida, Sakyo-ku
Kyoto 606, Japan

Prostanoids including prostaglandins (PGs) and thromboxanes (TXs) exhibit a wide variety of actions in various cells and tissues to maintain local homeostasis in the body. Characterization of the prostanoid receptors mediating these actions was carried out first pharmacologically by comparing effects and potencies of various types of PGs in different bioassay systems and by examining the effects of various PG analogues in these systems. Through these studies it was suggested that each prostanoid has a specific receptor and that there may be several receptor subtypes for the PGs of the E series.[1] PG receptors were characterized also biochemically as ligand-binding activities of radiolabeled PGs and their analogues, and by analyzing PG-induced signal transduction in cells. These analyses suggested that the PG receptor is a receptor coupling via a G protein to effectors such as adenylate cyclase and phospholipase C. However, none of the receptors was purified or cloned until we purified the TXA_2/PGH_2 receptor from human blood platelets[2] and cloned its cDNA.[3] These studies demonstrated that the TXA_2 receptor is indeed a G protein-coupled rhodopsin-type receptor with seven transmembrane domains and that it showed a very limited homology to other members of the rhodopsin-type receptors such as adrenergic and muscarinic receptors. These results suggest that the TXA_2 receptor belongs to a new subfamily of the rhodopsin superfamily of receptors and that the prostanoid receptors having homologous structures as a whole constitute this subfamily. Based on this assumption, we performed homology screening in various cDNA libraries and cloned several types and subtypes of the prostanoid receptors. These include the mouse TXA_2 receptor,[4] the mouse PGE receptor EP1, EP2, and EP3 subtypes,[5–7] the mouse PGF receptor,[8] and the mouse PGI receptor.[9] We also found that the EP3 receptor mRNA undergoes alternative splicing to produce several isoforms.[10–13] We also identified the human gene of the TXA_2 receptor, and revealed its organization.[14] These studies not only elucidated molecular structures of the prostanoid receptors, but also enabled us to study their properties in an isolated system. They also provided information on tissue and cell distribution of their expression. Based on such distribution, novel prostanoid actions have been examined and revealed.[15]

[a]This work was supported in part by grants-in-aid for scientific research from the Ministry of Education, Culture, and Science of Japan, the Mitsubishi Foundation, the Takeda Foundation, the Senri Life Science Foundation, and the Japanese Foundation on Metabolism and Diseases.

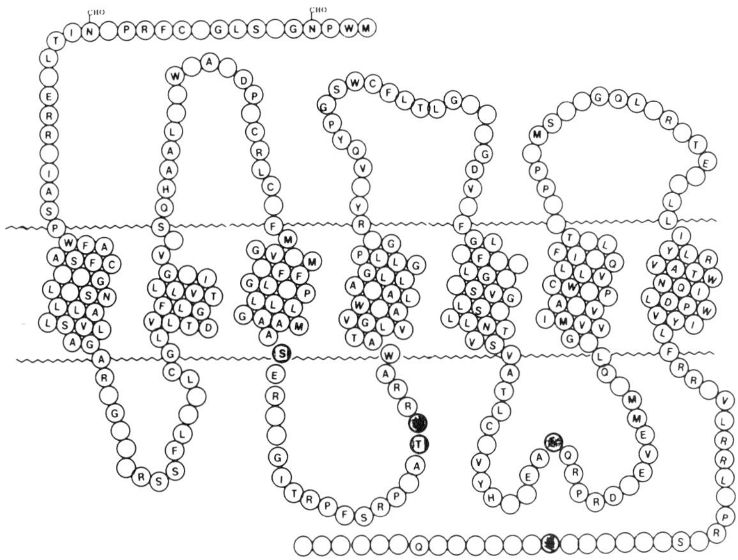

FIGURE 1. Membrane topology model of the TXA_2 receptor. The model was constructed on the amino acid sequence of the mouse TXA_2 receptor, and residues identical in the mouse and human receptors are shown in single-letter code. Potential phosphorylation sites by protein kinase C are indicated by *shadow*.

Here I review the results obtained by these studies and discuss implications of the findings.

TXA_2/PGH_2 RECEPTOR

Using a TXA_2 receptor antagonist, S-145, as an affinity ligand, we purified the TXA_2/PGH_2 receptor about 9,000-fold to apparent homogeneity from the solubilized membrane of human blood platelets.[2] Based on the partial amino acid sequences of the purified protein, an oligonucleotide probe was synthesized, and cDNA cloning was performed in the libraries of MEG-01 human megakaryocytic leukemia cells and human placenta. By this procedure, a cDNA for the human TXA_2 receptor was isolated.[3] A mouse homologue was then isolated by PCR and hybridization screening.[4] FIGURE 1 shows the membrane topology model of the TXA_2 receptor. The human and mouse TXA_2 receptors consist of 343 and 341 amino acids, respectively, and are 76% identical in the overall sequences. The hydrophobicity analysis and homology to other proteins suggested that they are rhodopsin-type proteins with seven putative transmembrane domains as shown in this figure. The receptors have N-terminal extracellular portions of less than 30 amino acids, and the 4th and 16th Asn in this region are glycosylated, as evidenced by peptide sequencing of the purified protein. Transmembrane segments consist mainly of hydrophobic amino acids, and the three-dimensional structure model based on the sequence suggests that these hydrophobic amino acids form a hydrophobic pocket for the hydrophobic ring structure of its ligand.[15,16] The cytoplasmic loops are short, and there are two conserved potential c-kinase phosphorylation sites in the second cytoplasmic loop, which together with several serine and threonine residues in the C-terminal may be involved in phosphorylation-mediated receptor desensitization. Desensitization of the TXA_2 receptor has already been reported.[17] The carboxy terminal cytoplasmic tail is also short, and there is no cysteine residue in this region, suggesting that the C-terminal of these receptors is not membrane attached by myristoylation but present freely in the cytoplasm. Ligand binding properties of the cloned receptors were examined by expression in cultured COS cells. The human and mouse receptors showed specific binding of the selective TXA_2 receptor ligand [^{3}H]S-145, with a K_d of 1.2 and 3.3 nM, respectively, and this binding was displaced selectively by various TXA_2/PGH_2 analogues in the order of S-145 > ONO-3708 > STA_2 > U-46619. Other PGs such as PGD_2 also displaced the binding, but their potencies were at least two orders of magnitude less than those of the above compounds. Signal transduction following the ligand binding was examined by expression of the cloned receptors in *Xenopus* oocytes. In these oocytes application of TXA_2 agonists evoked an inward current of an amplitude of several hundred nA. This response was blocked by prior treatment with the TXA_2 receptor antagonists and became much weaker upon repetitive application of the agonists, suggesting that they underwent homologous desensitization. There was no response found with other PGs and TXB_2 up to 1 μM concentration, confirming specificity of the cloned receptor. Because the observed inward current was induced by the rise in free calcium ion in oocytes, these results suggested that the cloned receptor evoked PI turnover and subsequent calcium release upon agonist stimulation, as suggested in cells and membranes containing the native TXA_2 receptor.[18]

There has been a controversy over the presence of receptor subtypes for TXA_2 receptor.[19] To clarify this issue, we have cloned the gene for this receptor from the human genomic library and characterized it. The gene is present as a single copy per haploid, spans over 15 kb, and contains 3 exons divided by two introns. The first intron, 6.3 kb long, exists in the 5′-noncoding region, 83 base pairs upstream from the ATG start site. Intron 2, 4.3 kb long, is located at the end of the sixth putative transmembrane domain and separates it from the downstream coding region. When RT-PCR analysis using primers flanking this intron was carried out on poly(A)$^+$RNA from MEG-01, human placenta, and mesangial cells, it amplified only a single DNA fragment of an expected size of 173 bp, suggesting that no alternative splicing of the coding region occurs and that only one type of TXA_2 receptor protein is formed from this gene in these cells. Although these findings cannot exclude the presence of a less homologous receptor, they favor the presence of one identical type of the TXA_2 receptor in platelets and smooth muscle cells. The chromosomal localization of this receptor gene was assigned to 19p13.3 of human chromosome.

Using the mouse TXA_2 receptor cDNA as a probe, tissue distribution of the receptor expression was examined by Northern blot analysis in various mouse tissues.[4] This revealed that the TXA_2 receptor is most abundantly expressed in thymus, followed by spleen, lung, kidney, heart, and uterus, in this order. Little, but significant, expression was observed in brain. In the similar analysis in human tissues,[3] placenta and cultured megakaryocytic leukemia cells showed higher expression than lung. Thus, thymus is one of the richest tissues in the receptor expression as placenta and platelet precursor cells. We used [^{3}H]S-145 as a specific radioligand for the TXA_2 receptor and examined a thymic cell population expressing the receptor.[20] We found that the thymic lymphocytes express the receptor at a density comparable to that found in human platelets. It is richest in immature thymocytes such as $CD4^-8^-$ and $CD4^+8^+$ cells and decreases in number during T cell development. We further found that the addition of STA_2, a TXA_2 agonist, induced apoptotic cell death of these immature thymocytes, and this action was antagonized by a TXA_2 receptor antagonist. These results, together with the finding that TX synthase is rich in dendritic cells and macrophages in thymus,[21,22] suggest that TXA_2 is released from these stromal cells and acts on thymocytes under some physiological conditions and that this TXA_2/TXA_2 receptor system plays a role in thymocyte differentiation and development.

PGE RECEPTOR EP1, EP2, AND EP3 SUBTYPES

Using the mouse TXA_2 receptor cDNA as a probe, two independent clones were isolated by cross-hybridization from the mouse lung cDNA library. Using one of the two cDNAs as a probe, a third clone was isolated. These clones encode different proteins, but all of them specifically bind PGE_2 and each generates a different signal on this binding, indicating that they all are PGE receptors. They are named EP1, EP2, and EP3 based on their similarities (see below) to the pharmacologically defined PGE receptor subtypes. The amino acid sequences of these receptors are shown in FIGURE 2 together with those of the mouse TXA_2 receptor, PGI, and PGF receptors. The EP1, EP2, and EP3 receptors consist of 405, 513 and 365 amino acids, respectively. The hydrophobicity analysis showed that all of them contain seven hydrophobic segments corresponding to the putative transmembrane domains, suggesting that they belong to the rhodopsin-type receptors. Although the overall homology among the three PGE receptors is not high, the homology is scattered throughout the entire sequences, suggesting that these receptors are derived from different genes that have evolved from a common ancestor. Strong conservations are found in several particular regions. The most conserved region is the seventh transmembrane domain, where the sequence of R-X-A-S-X-N-Q(P)-I-L-D-P-W-V(I)-Y-I(L)-L is shared. Another conserved region is the second extracellular loop, where the common sequence of G-R-Y(S)-X-X-Q-Y(W)-P-G-T-W-C-F- is seen. It is clear from FIGURE 2 that these sequences are very much conserved not only in the three PGE receptors but also in the receptors for TXA_2, PGF, and PGI of mouse as well as other species. These conserved sequences are presumed to reflect structural requirements as the prostanoid receptors. Ligand binding and recognition by the rhodopsin-type receptor occur in the outside half of the seven transmembrane segments.[22] It is interesting in this respect that the arginine residue in the seventh transmembrane domain is conserved in all the prostanoid receptors. This arginine is located at the position analogous to lysine^{296} of the bovine rhodopsin molecule. In rhodopsin, this lysine was identified as making a Shiff base with the aldehyde group of its ligand, all-*trans*-retinal. By this analogy and by the fact that the carboxyl group is essential for biological activities of most PGs, we presume that this arginine serves as the binding site for the α-carboxyl group of prostanoid molecules. On the other hand, there is no aspartic acid in the third transmembrane domain of the prostanoid receptors that is conserved in all of

```
                                            I__________________________                      II______
FP               MSMNSSKQPVSPAAGLIANTTCQTENRLSVFFSIIFMTVGILSNSLAIAILMKAYQRF-RQKSKA--SFLLLASG-   72
TP               MWPNGTSLGACFRPVNIT--LQERRAIASPWFAASFCALGLGSNLLALSVLAGA-----RPGAGPRSSFLALLCG-   68
EP1         MSPCGLNLSLADEAATCATPRLPNTSVVLPTGDNGTSPALPIFSMTLGAVSNVLALALLAQVAGRMRRRRSAA--TFLLFVAS-   81
EP3              MASMWAPEHSAEAHSNLS-STTDDCGSVSVAFPITMMVTGFVGNALAMLLVSRSY---RRRESKRKKSFL--LCIG   70
EP2    MAEVGGTIPRSNRELQRCVLLTTTIMSIPGVNASFSSTPERLNSPVTIPAVMFIFGVVGNLVAIVVL------CKSRKEQKETTFYTLVCG-   85
IP  M  KMMASDGHPGPPSVTPGSPLSAGGREWQGMAGGCWNITYVQDSVGPATSTLMFVAGVVGNGLALGIL------GARRRSH-PSAFAVLVTG-   85

       ___________                            III_______________________                 IV_________
FP     -LVITDFFGHLINGGIAVFVYASDKDWIRFDQSNILCSIFGISMVFSGLCPLFLGSAMAIERCIGVTNPIFHSTKITSKH-VKMILSGVCMF  162
TP     -LVLTDFLGLLVTGAIVASQHAALLDWRATDPSCRLCYFMGVAMVFFGLCPLLLGAAMASERFVGITRPFSRPTATSR-R-AWATVGLVWVA  157
EP1    -LLAIDLAGHVIPGALVLRLYTAGRA-PAGGA----CHFLGGCMVFFGLCPLLLGCGMAVERCVGVTQPLIHAARVSVAR-ARLALAVLAAM  166
EP3    WLALTDLVGQLLTSPVVILVYLSQRRWEQLDPSGRLCTFFGLTMTVFGLSSLLVASAMAVERALAIRAPHWYASHMKT-R-ATPVLLGVWLS  160
EP2    -LAVTDLLGTLLVSPVTIATYMKG-QWPGDQA---LCDYSTFILLFFGLSGLSIICAMSIERYLAINHAYFYSHYVDK-RLAGLTLFAIYAS  171
IP     -LAVTDLLGTCFLSPAVFVAY--G-LAHGGTM---LCDTFDFAMTFFDLASTLILFAMAVERCLALSHPYLYAQLDGP-PCARFALPSIYAF  176

       _____________                             V ________________________
FP     AVFVAVLPILGHRDYQIQASRTWCFYNTEHIE------DWE-DRFYLLFFSFLGLLALGVSFSCNAVTGVTLLRV-KFRSQ-----------  235
TP     AGALGLLPLLGLGRYSVQYPGSWCFLTLGT---------QRGDVVFGLIFALLGSASVGLSLLLNTVSV-ATL------------------  220
EP1    ALAVALLPLVHVGRYELQYPGTWCFISLGPRG------GWR-QALLAGLFAGLGLAALLAALVCNTLSGLALLRA-RWRRRRSRRFRKTAGP  250
EP3    VLAFALLPVLGVGRYSVQWPGTWCFISTGPAGNETDPAREPGSVAFASAFACLGLLALVVTFACNLATIKALVSR-----------------  235
EP2    NVLFCALPNMGLGRSERQYPGTWCFIDWTT--NVT------AYAAFSYMYAGFSSFLILATVLCNVLVCGALLRMHRQFMRRTSLGTEQHHA  255
IP     CCLFCSLPLLGLGEHQQYCPGSWCFI-RMR--SAP------GGCAFSLAYASLMALLVTSIFFCNGSVTLSLYHMYRQQRRHHGSFVP----  246

                                                      VI______________________
FP     ------------------------------------QHRQGRSHHLEMIIQLLAIMCVSCVCWSPFLV--TMANIAINGN-NSPVT--C-ET  285
TP     ----------CRVYHT----------------R--EATQRPRDCEVEMMVQLVGIMVVATVCWMPLLVFIMQTLLQTPPVMSFSGQLLRATE  284
EP1    DDRRRWGSRGPRLASASSASSITSATATLRSSRGGGSARRVHAHDVEMVGQLVGIMVVSCICWSPLLV--LVVLAIGGWN-SNSLQ--R-PL  336
EP3    ----------CRAKAAVSQS-----------------SAQWGRI-TTETAIQLMGIMCVLSVCWSPLLIMMLKMIFNQMSVEQCKTQMGKEKE  300
EP2    AAAAAVASVACRGHAGASPALQRLSDF--RRRR---SFRRIAGAEIQMVILLIATSLVVLICSIPLVVRVFINQLYQPNVVK---DISRNP-  339
IP     -------------------------------------TSRAREDEVYHLILLALMTVIMAVCSLPLMIRGFTQAIA-PDSRE---MG-----  302

        VII ____________________
FP     ---TLFALRMATWNQILDPWVYILLRKAVLRNLYKLASRCCGVNIISLHIWELSSIKNSLKVAAISESPAAEKESQQASSEAGL         366
TP     -HQLLIYLRVATWNQILDPWVYILFRRSVLRRLHPRFSSQLQAVSLRRPPAQAMLSGP                                   341
EP1    ---FL-AVRLASWNQILDPWVYILLRQAMLRQLLRLLPLRVSAKGGPTELGLTKSAWEASSLRSSRHSGFSHL                    405
EP3    CNSFLIAVRLASLNQILDPWVYLLLRKILLRKFCQIRDHTNYASSSTSLPCPGSSALMWSDQLER                            365
EP2    ---DLQAIRIASVNPILDPWIYILLRKTVLSKAIEKIKCLFCRIGGSGRDSSAQHCSESRRTSSAMSGHSRSFLARELKEISSTSQTL(90) 513
IP     ---DLLAFRFNAFNPILDPWVFILFRKAVFQRLKFWLCCLCARSVHGDLQAPLSRPAS(60)                               416
```

FIGURE 2. Amino acid sequence alignment of the mouse prostanoid receptors. The amino acid sequences of the mouse PGF receptor (FP), TXA_2 receptor (TP), EP1 receptor (EP1), EP3 receptor (EP3), EP2 receptor (EP2), and prostacyclin receptor (IP) are aligned to obtain the optimum homology. The approximate positions of the putative transmembrane regions are indicated by *horizontal lines* above the sequences. Amino acids conserved are **bold faced**.

TABLE 1. Ligand Binding Properties of the Cloned Prostanoid Receptors

Type and Subtype		K_d (nM); Radioligand	Rank Order of Ligands
TP		1.2; [^{3}H]S-145	S-145>ONO-3708 > STA_2 >> PGD_2
EP			PGE_2 $PGF_{2\alpha}$
	EP1	21; [^{3}H]PGE_2	PGE_2 > iloprost > PGE_1 >> $PGF_{2\alpha}$ > PGD_2 17-PhenylE_2 > sulprostone > M&B >>> AH-6809 . . .butaprost
	EP2	11; [^{3}H]PGE_2	PGE_2 = E_1 >>> PGD_2, $PGF_{2\alpha}$, iloprost Miso. >> M&B >>> Sulprostone SC-19220 butaprost
	EP3	3; [^{3}H]PGE_2	PGE_2 = E_1 >> iloprost > PGD_2 > $PGF_{2\alpha}$ M&B >>> butaprost, SC-19220 = 0
FP		1.3; [^{3}H]$PGF_{2\alpha}$	$PGF_{2\alpha}$ = 9α, 11β-PGF > $PGF_{1\alpha}$ > PGD_2 > STA_2 > PGE_2 > iloprost
IP		4.5; [^{3}H]iloprost	Cicaprost > iloprost > PGE_1 > carbacyclin >> PGD_2 $PGF_{2\alpha}$ PGE_2 STA_2

NOTE: Cloned cDNAs for the prostanoid receptors were expressed in either cultured COS or CHO cells, and ligand binding properties were examined using the respective radioligand. The affinities of the receptors to various PGs and analogues were determined by displacement experiments and shown in "Rank Order of Ligands." Under EP, the upper line shows the rank order of various prostanoids and the lower that of various PGE analogues. TP, the TXA_2 receptor; EP, the PGE receptor; FP, the PGF receptor; IP, the prostacyclin (PGI) receptor.

the monoamine receptors and is suggested to be the binding site for the amine of their ligands. In addition, the transmembrane domains of the prostanoid receptors are more hydrophobic than those of the monoamine receptors, which may reflect binding to the cyclopentane ring and aliphatic side chains of prostanoid molecules.

Ligand Binding Properties

Ligand binding specificities of the cloned PGE receptors were examined by their expression in cultured cells (TABLE 1). EP1, EP2, and EP3 show specific [^{3}H]PGE_2 binding, with K_d values of 21, 11, and 3 nM, respectively. EP1 receptor shows about 10 times higher affinity to PGE_2 than PGE_1, while about the same affinity to PGE_2 and PGE_1 is seen in EP2 and EP3. It is also noteworthy that EP1 has a high affinity to iloprost, while the other two receptors have an affinity to this compound more than two orders of magnitudes less than PGE_2. Among the PGE analogues tested,

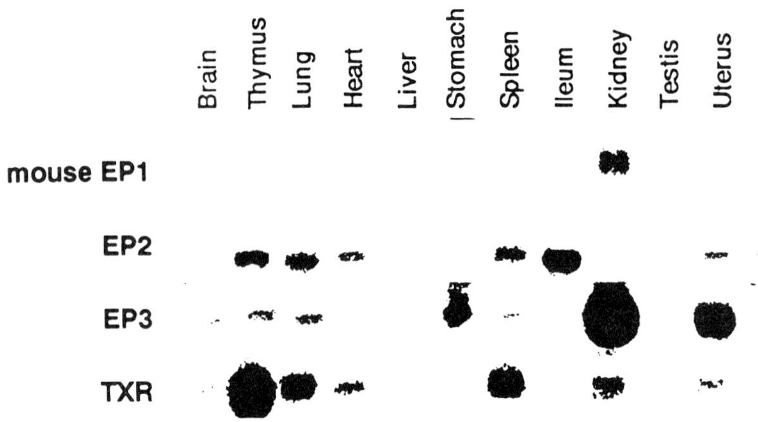

FIGURE 3. Northern blot analyses of the mouse TXA_2 receptor (TXR) and the three subtypes of PGE receptor, EP1, EP2, and EP3. Northern blot analysis was carried out using the receptor cDNAs as probes on poly(A)$^+$RNAs isolated from various mouse tissues. Only the regions showing positive hybridization bands are shown.

17-phenyl-PGE_2 showed the highest affinity to EP1, while misoprostol and M&B-28767 are the best ligands to EP2 and EP3, respectively. These results generally agree well with those proposed for these pharmacologically defined "subtypes."[1] However, there are two apparent discrepancies. One is the inability of AH-6809, the so-called EP1/DP antagonist, to displace efficiently [^{3}H]PGE_2 binding to EP1. The other is the failure of butaprost, the well-known EP2 agonist, to bind to the cloned receptor. The reason for these discrepancies are not known at present. It may be due to species difference of the receptors or may imply that there are still other molecular subtypes of PGE receptors.

Signal Transduction

The expressed receptor systems were also used for examination of signal transduction of these PGE receptors. Stimulation of EP1 by PGE_2 induced the rise in intracellular Ca^{2+} concentration. The response consisted of two phases: a peak of about 30-s duration, followed by a low but sustained rise for more than 3 min. The PI response evoked by this receptor was weak (about only 120% of the control) and occurred slowly, making it difficult to assess the contribution of the PI response to the rapid transient rise in free calcium ion concentration in the cells. Changes in the intracellular cAMP level was negligible in the EP1-expressing cells. In contrast, PGE_2 increased the intracellular cAMP level in the EP2-expressing cells and decreased it in the EP3-expressing cells.

Tissue Distribution

Tissue distribution of each subtype of PGE receptor was examined by Northern blot analysis of its mRNA expression (FIGURE 3). These studies have shown that each subtype has a specific expression pattern. Expression of EP1 is low compared to other subtypes. Moderate expression was found in kidney. Little, but significant, expression was noted in lung, but expression in other organs was not detected. On the other hand, EP2 is expressed most highly in ileum, followed by thymus, lung, spleen, heart, and uterus, in that order. EP3 mRNA is expressed very highly in kidney and uterus, and expression was seen also in stomach, thymus, spleen, lung, and brain. We also noted that in the same organ the three subtypes are localized differently. For example, in kidney EP3 is mainly localized in the medulla, EP1 in papilla, and EP2 in the cortex (Sugimoto *et al.*, unpublished observation).

EP3 RECEPTOR ISOFORMS

Previous pharmacological studies suggest that the receptors with the EP3-like ligand binding property mediate a variety of PGE_2 actions. Some of these actions can be understood by the mechanism described above for the cloned EP3–that is, inhibition of adenylate cyclase. For example, EP3 mediates PGE_2-induced inhibition of vasopressin-induced water reabsorption in kidney and of histamine-induced acid secretion in stomach. As suggested by the previous works,[23,24] both inhibitions by EP3 appear elicited by inhibition of adenylate cyclase, because vasopressin and histamine induce the above actions by increasing cAMP in the respective tissue. However, EP3-like receptors also mediate PGE_2-induced contraction of uterine and ileum smooth muscles.[25,26] These actions are difficult to explain by adenylate cyclase inhibition, and several people have claimed that EP3 evokes calcium mobilization in these tissues.[26] These findings suggest two possibilities. One is that a single form of EP3 can couple to different G proteins to induce different signal transduction pathways in different tissues. The other possibility is that there are more than one EP3 receptors with similar ligand-binding properties. In order to clarify these issues, we carried out cDNA cloning of EP3 receptor in the cDNA library of bovine adrenal gland, because we found that an apparently single class of EP3 receptor couples to more than one signaling pathway in this tissue (Namba *et al.*, unpublished observations). This cloning revealed that the EP3 mRNA undergoes alternative splicing in this tissue to produce at least four EP3 isoforms (EP3A, B, C, and D).[10] These isoforms are identical from the amino terminus to the 10 amino acids after the seventh transmembrane domain and are different only in the end part of the carboxyl terminal tails (FIGURE 4). Because they share the same transmembrane and extracellular structures, they show almost identical ligand binding specificity. Yet, these isoforms couple to different G proteins to induce different signaling in the cells: EP3A couples to Gi to induce inhibition of adenylate cyclase; EP3B and EP3C couple to Gs to increase cAMP; and EP3D couples to Gp, in addition to Gi and Gs, to induce pertussis toxin-insensitive PI turnover and calcium mobilization in the cells. We do not yet know if one isoform is preferentially produced in some tissue or not. However, such isoform production fits well with the functional diversity of EP3 receptors. The presence of such EP3 isoforms is not limited to bovine tissues. We have already identified three isoforms in mouse EP3. They show alternative splicing at the identical position of the C-terminal tail. The first and second isoforms (EP3α and β) utilize the same signal transduction–that is, inhibition of adenylate cyclase–but differ in

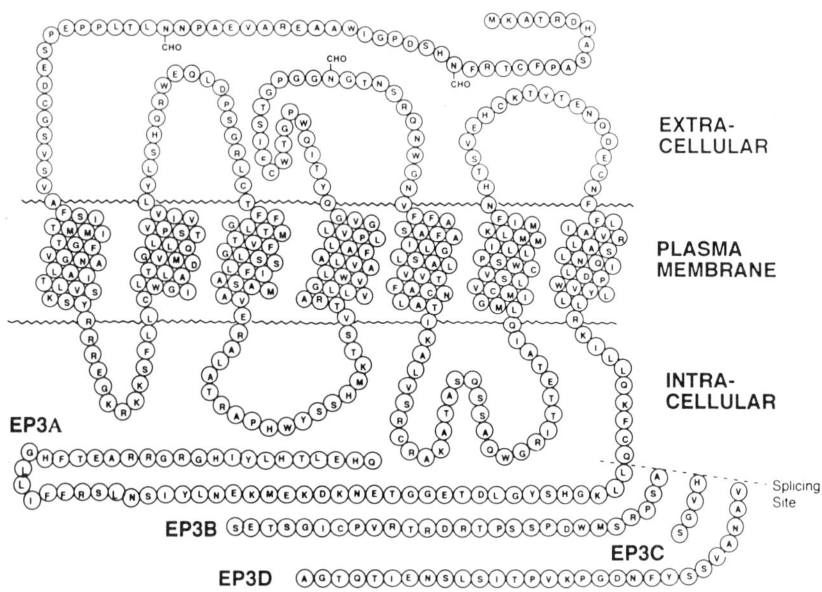

FIGURE 4. Alternative splicing of the bovine EP3 receptor. Structures of four isoforms of the bovine EP3 receptor, EP3A, EP3B, EP3C, and EP3D are shown.

coupling properties with Gi protein and sensitivity in agonist-induced desensitization.[11,12] On the other hand, the third isoform (EP3) shows an additional signal transduction: it can couple to Gs in addition to Gi.[13]

PGF RECEPTOR

PGF receptor is known to be rich in corpus luteum. PCR was performed with degenerate primers corresponding to the conserved sequences of the putative second extracellular loop and the seventh transmembrane domain (see above) on the cDNA synthesized from bovine corpora luteum total RNA. By this procedure a 390-bp fragment was obtained that showed about 50% sequence homology to the corresponding region of human TXA_2 receptor. Cross-hybridization screening was carried out with this fragment of DNA as a probe, and positive clones were isolated from the cDNA library of pregnant mouse ovary. The representative clone showed a 1,098-bp open-reading frame that encodes a protein of 366 amino acids with a calculated molecular weight of 40,077. It is again a rhodopsin-type protein with seven hydrophobic segments and shares sequence similarity to other prostanoid receptors (FIGURE 3). High homology to the mouse TXA_2 receptor and the EP1 subtype of PGE receptor is observed: 39.3 and 39.8% in total sequences, respectively. When expressed in cultured COS cells, it bound [^{3}H]$PGF_{2\alpha}$ with a K_d of 1.32 nM, and this [^{3}H]$PGF_{2\alpha}$ binding was displaced by unlabeled PGs in the order of $PGF_{2\alpha}$ = 9α, 11β-PGF_2 > $PGF_{1\alpha}$ > PGD_2 > STA_2 > PGD_2 > iloprost. The affinity and specificity of this binding are in good agreement with the $PGF_{2\alpha}$ receptor characterized previously in corpus luteum membranes.[27] $PGF_{2\alpha}$ added to the transfected COS cells evoked the formation of inositol-1,4,5-trisphosphate in a concentration-dependent fashion from 1 nM. The response reached a plateau of 180% of the control at 1 μM. These results indicated that the expressed receptor couples to activation of phospholipase C and the consequent calcium mobilization. This was confirmed also by the electrophysiologic response of *Xenopus laevis* oocyte microinjected with this mRNA. $PGF_{2\alpha}$ elicited the inward current, indicating that calcium mobilization occurred in response. RNA blot analysis showed two major hybridization bands at 2.3 and 6 kilobases. The highest expression was observed in the pregnant ovary. Expression was also prominent in kidney and significant in lung, heart, and stomach. *In situ* hybridization analysis in ovary showed that the mRNA expression was observed exclusively in the large luteal cells of corpus luteum, while no labeling was found in any stages of ovarian follicle. This distribution of the receptor expression appears consistent with the known $PGF_{2\alpha}$ actions in these organs, such as induction of luteolysis,[28] contraction of mesangial cells in kidney glomerulus,[29] and bronchoconstriction.[30]

PGI (PROSTACYCLIN) RECEPTOR

PGI receptor cDNA was cloned by PCR-based homology screening from the cDNA libraries of mouse mastocytoma P-815 cells that express the high amount of the receptor.[10] It is a protein of 417 amino acids with a calculated molecular weight of 44,722. The amino acid sequence in the transmembrane domains shows the highest homology, 39%, to the EP2 subtype of the PGE receptor; the homology to other receptors are 32, 28, and 32 to EP3, EP1, and TXA_2 receptor, respectively (FIGURE 3). When expressed in cultured CHO cells, it binds [^{3}H]iloprost, a specific radioligand for PGI receptor, with a K_d of 4.5 nM. This binding was displaced by various PGI analogues and prostanoids in the order of cicaprost > iloprost > PGE_1 > carbacyclin > PGD_2, STA_2, PGE_2 > $PGF_{2\alpha}$. The affinity to iloprost and the specificity of the binding agree well with the previous characterization of the receptor in platelets and other sources.[31–33] As suggested in these systems, the expressed receptor stimulates adenylate cyclase upon the iloprost binding with an EC_{50} of 0.1 nM. In addition, it evoked phosphatidylinositol turnover at the higher iloprost concentrations, with an EC_{50} of 100 nM; at 1 μM iloprost, IP3 formation was observed at 30 s after the agonist addition and reached a maximum of about 3-fold over the control at 2 min. It has been reported that prostacyclin or its analogues induces responses other than cAMP generation, such as contraction of rabbit stomach and guinea pig ileum and [Ca^{++}] increase in several cultured cell lines.[34–37] The above finding suggests the possibility that these responses are mediated, indeed, by the prostacyclin receptor coupling to the signaling other than cAMP elevation. Northern blot analysis using the PGI receptor cDNA revealed that its mRNA is most abundantly expressed in thymus, followed by spleen, heart and lung, in that order. In the thymus, this receptor is expressed almost exclusively in the medulla. While this is in contrast to the TXA_2 receptor, which is expressed much more strongly in the cortex than the medulla, these findings again suggest an unexpected immunomodulatory role for a PG the major role of which has been focused only in the circulatory system. There has been some concern regarding the existence of subtypes or isoforms of the PGI receptor. However, we have not obtained any evidence for the presence of the homologous molecules or alternatively spliced variants of the cloned receptor.

CONCLUDING REMARKS

In this review I summarize our work of cDNA cloning of various types, subtypes, and isoforms of the prostanoid receptors. Through this work it is now firmly established that the diverse actions of prostanoids are exerted by a family of homologous receptors. The presence of the receptors corresponding to each prostanoid and of multiple receptors to PGE provides a good model for analysis of the ligand binding structure of this group of receptors. Clarification of such structures will make possible the development of more specific drugs to control and mimic prostanoid actions. cDNA cloning has already made possible screening and evaluation of various compounds in an isolated receptor system using a series of human receptors. In addition to such practical development, it also stimulates and revives basic prostanoid research. As seen in the distribution study of the TXA_2 receptor, these studies will give us a hint of a new prostanoid action. Prostanoid research has been mainly in smooth muscles and other systems where PG actions can be evaluated easily, and it is very likely that many important actions have escaped notice. Distribution analysis of cloned receptors will certainly be of help to discover these actions. The cloning of the prostanoid receptors reported here has also opened a way to make gene-targeted animals devoid of each receptor. Such experiments will clarify the contribution of each prostanoid and its receptor system to various physiological regulations and pathophysiological derangements in health and disease. Molecular biology of the prostanoid receptors will thus open new fields in prostanoid biology.

ACKNOWLEDGMENTS

The author thanks the following people for the studies presented in this review: F. Ushikubi, M. Hirata, Y. Hayashi, T. Namba, H. Oida, K. Nakamura, and A. Kakizuka of the Department of Pharmacology, Kyoto University Faculty of Medicine; Y. Sugimoto, M. Negishi, A. Honda, A. Watabe, S. Irie, and A. Ichikawa of the Department of Physiological Chemistry, Kyoto University Faculty of Pharmaceutical Sciences; and Y. Aiba, O. Mazda, and Y. Katsura of the Department of Immunology, Chest Disease Research Institute, Kyoto University.

REFERENCES

1. Coleman, R. A., I. Kennedy, P. P. A. Humphrey, K. Bunce & P. Lumley. 1990. Prostanoids and their receptors. *In* Comprehensive Medicinal Chemistry, Vol. 3. Membranes and Receptors. J. C. Emmett, Ed.: pp. 643-714. Pergamon Press. Oxford.
2. Ushikubi, F., M. Nakajima, M. Hirata, M. Okuma, M. Fujiwara & S. Narumiya. 1989. J. Biol. Chem. **264:** 16496-16501.
3. Hirata, M., Y. Hayashi, F. Ushikubi, Y. Yokota, R. Kageyama, S. Nakanishi & S. Narumiya. 1991. Nature **349:** 617-620.

4. NAMBA, T., Y. SUGIMOTO, M. HIRATA, Y. HAYASHI, A. HONDA, A. WATABE, M. NEGISHI, A. ICHIKAWA & S. NARUMIYA. 1992. Biochem. Biophys. Res. Commun. **184:** 1197-1203.
5. SUGIMOTO, Y., T. NAMBA, A. HONDA, Y. HAYASHI, M. NEGISHI, A. ICHIKAWA & S. NARUMIYA. 1992. J. Biol. Chem. **267:** 6463-6466.
6. HONDA, A., Y. SUGIMOTO, T. NAMBA, A. WATABE, A. IRIE, M. NEGISHI, S. NARUMIYA & A. ICHIKAWA. 1993. J. Biol. Chem. **268:** 7759-7762.
7. WATABE, A., Y. SUGIMOTO, A. HONDA, A. IRIE, T. NAMBA, M. NEGISHI, S. ITO, S. NARUMIYA & A. ICHIKAWA. 1993. J. Biol. Chem. **268:** 20175-20178.
8. SUGIMOTO, Y., K. HASUMOTO, T. NAMBA, A. IRIE, M. KATSUYAMA, M. NEGISHI, A. KAKIZUKA, S. NARUMIYA & A. ICHIKAWA. 1994. J. Biol. Chem. **269:** 1356-1360.
9. NAMBA, T., H. OIDA, Y. SUGIMOTO, A. KAKIZUKA, M. NEGISHI, A. ICHIKAWA & S. NARUMIYA. 1994. J. Biol. Chem. **269:** 9986-9992.
10. NAMBA, T., Y. SUGIMOTO, M. NEGISHI, A. IRIE, F. USHIKUBI, A. KAKIZUKA, S. ITO, A. ICHIKAWA & S. NARUMIYA. 1993. Nature **365:** 166-170.
11. SUGIMITO, Y., M. NEGISHI, Y. HAYASHI, T. NAMBA, A. HONDA, A. WATABE, M. HIRATA, S. NARUMIYA & A. ICHIKAWA. 1993. J. Biol. Chem. **268:** 2712-2718.
12. NEGISHI, M., Y. SUGIMTO, A. IRIE, S. NARUMIYA, & A. ICHIKAWA. 1993. J. Biol. Chem. **268:** 9517-9521.
13. IRIE, A., Y. SUGIMOTO, T. NAMBA, A. HARAZONO, A. HONDA, A. WATABE, M. NEGISHI, S. NARUMIYA & A. ICHIKAEA. 1993. Eur. J. Biochem. **217:** 313-318.
14. NÜSING, R. M., M. HIRATA, A. KAKIZUKA, T. EKI, K. OZAWA & S. NARUMIYA. 1993. J. Biol. Chem. **268:** 25253-25259.
15. NARUMIYA, S., M. HIRATA, T. NAMBA, Y. HAYASHI, F. USHIKUBI, Y. SUGIMOTO, M. NEGISHI & A. ICHIKAWA. 1993. J. Lipid Mediators **6:** 155-161.
16. YAMAMOTO, Y., K. KAMIYA & S. TERAO. 1993. J. Med. Chem. **36:** 820-825.
17. MURRAY, R. & G. A. FITZGERALD. 1989. Proc. Natl. Acad. Sci. USA **86:** 124-128.
18. ARITA, H., T. NAKANO & K. HANASAKI. 1989. Prog. Lipid Res. **28:** 273-301.
19. HALUSHKA, P. V., D. E. MAIS, P. R. MAYEUX & T. A. MORIMELLI. 1989. Annu. Rev. Pharmacol. Toxicol. **10:** 213-239.
20. USHIKUBI, F., Y. AIBA, K. NAKAMURA, T. NAMBA, M. HIRATA, O. MAZDA, Y. KATSURA & S. NARUMIYA. 1993. J. Exp. Med. **178:** 1825-1830.
21. NÜSING, R., R. LESCH & V. ULLRICH. 1990. Eicosanoids **3:** 53-58.
22. HOMO-DELARCHE, F., D. DUVAL & M. PAPIERNIK. 1985. J. Immunol. **135:** 506-512.
23. SAVARESE, T. M. & C. M. FRASER. 1992. Biochem. J. **283:** 1-19.
24. GARCIA-PEREZ, A. & W. L. SMITH. 1984. J. Clin. Invest. **74:** 63-74.
25. CHEN, M. C., D. A. AMIRIAN, M. TOOMEY, M. J. SANDERS & A. H. SOLL. 1988. Gastroenterology **94:** 1121-1129.
26. COLEMAN, R. A., I. KENNEDY & R. L. G. SHELDRICK. 1987. Br. J. Pharmacol. **91:** 406P.
27. GOUREAU, O., Z. TANFIN, S. MARC & S. HARBON. 1992. Am. J. Physiol. **263:** C257-C265.
28. POWELL, W. S., S. HAMMARSTRÖM & B. SAMUELSSON. 1974. Eur. J. Biochem. **41:** 103-107.
29. HORTON, E. W. & N. L. POYSER. 1976. Physiol. Rev. **56:** 595-651.
30. MENÉ, P., G. R. DUBYAK, A. SCARPA & M. J. DUNN. 1987. Biochem. Biophys. Res. Commun. **142:** 579-586.
31. WASSERMAN, M. A. 1975. Eur. J. Pharmacol. **32:** 146-155.
32. LEIGH, P. J., W. A. CRAMP & J. MACDERMOT. J. Biol. Chem. **259:** 12431-12436.
33. ARMSTRONG, R. A., R. A. LAWRENCE, R. L. JONES, N. H. WILSON & A. COLLIER. 1989. Br. J. Pharmacol. **97:** 657-668.
34. HASHIMOTO, H., M. NEGISHI & A. ICHIKAWA. 1990. Prostaglandins **40:** 491-505.

35. Whittle, B. J. R., K. G. Mugridge & S. Moncada. 1979. Eur. J. Pharmacol. **53:** 167-172.
36. Lawrence, R. A., R. L. Jones & N. H. Wilson. 1992. Br. J. Pharmacol. **105:** 271-278.
37. Watanabe, T., Y. Yatomi, S. Sunaga, I. Miki, A. Ishii, A. Nakao, M. Higashihara, Y. Seyama, M. Ogura, H. Saito, K. Kurokawa & T. Shimizu. 1991. Blood **78:** 2328-2336.
38. Vassaux, G., D. Gaillard, G. Ailhaud & R. Negrel. 1992. J. Biol. Chem. **267:** 11092-11097.

Cyclooxygenase-Dependent Formation of the Isoprostane 8-EPI Prostaglandin $F_{2\alpha}$[a]

DOMENICO PRATICO, JOHN A. LAWSON, AND GARRET A. FITZGERALD[b]

Centre for Cardiovascular Science
Department of Medicine and Experimental Therapeutics
University College
Dublin, Ireland

INTRODUCTION

Isoprostanes are a family of free radical catalyzed prostaglandin isomers.[1] Urinary excretion of one of these compounds, 8-epi prostaglandin (PG) $F_{2\alpha}$, is being investigated as a potential noninvasive marker of free radical generation *in vivo,*[2,3] and isoprostane formation may contribute to free radical catalyzed membrane injury.[4]

Two isoprostanes, 8-epi $PGF_{2\alpha}$ and 8-epi PGE_2, have been reported to have biological effects, vasoconstriction, and induction of the platelet shape change, but not aggregation, which are inhibited by antagonists of the thromboxane receptor.[5–7] 8-Epi $PGF_{2\alpha}$ will also activate the recombinant thromboxane receptor.[8,9] Indirect evidence has been provided that suggests that 8-epi $PGF_{2\alpha}$ may act through a distinct but related receptor.[8] Although a single thromboxane receptor gene has been cloned,[10] it is also possible that posttranslational modifications of the thromboxane receptor might modulate its affinity for isoprostanes;[11] in this regard, it is of note that one can discriminate a form of the thromboxane receptor that mediates platelet shape change and vasoconstriction from one which mediates platelet aggregation.[12]

Intuitively, the concept of discrete cellular receptors for by-products of lipid peroxidation is somewhat surprising, although given the structural approximation of 8-epi $PGF_{2\alpha}$ to the PGG/H synthase (cyclooxygenase) product $PGF_{2\alpha}$, it would be unsurprising if it activated recombinant receptors[13–15] for this eicosanoid. However, it apparently does not do this (A. Ford Hutchinson, personal communication, 1994). Nonetheless, thromboxane antagonists have been reported to blunt responses to $PGF_{2\alpha}$ in several biological systems,[16] so a similar incidental activation of thromboxane receptors may characterize certain F-isoprostane isomers.

[a] This work was supported by a Program Grant from the Wellcome Trust and by the Center for Cardiovascular Science. Dr. Pratico was the recipient of a fellowship in cardiovascular science from Bristol Myers Squibb.

[b] Address for correspondence: Dr. G. A. FitzGerald, Center for Experimental Therapeutics, 909 Biomedical Research Building, University of Pennsylvania, Philadelphia, Pennsylvania 19104.

Given the biological effects of 8-epi $PGF_{2\alpha}$ on platelet function and evidence of increased urinary excretion of the compound in some syndromes associated with platelet activation,[17] we were interested in characterizing its formation in human platelets.

METHODS

8-Epi $PGF_{2\alpha}$, thromboxane (Tx) B_2, and 12-hydroxyeicosatetraenoic acid (HETE) were measured using stable isotope dilution/gas chromatography/mass spectrometry methods. Briefly, internal standards labeled with stable isotopes ($[^{18}O_2]$8-epi $PGF_{2\alpha}$, $[^2H_4]TxB_2$, $[^2H_8]$12-HETE) were added to samples before extraction, purification, and derivatization. Pentafluorobenzyl esters were prepared, due to their excellent fragmentation patterns in electron capture negative ion chemical ionization mass spectrometry. TxB_2 and 12-HETE were derivatized as trimethylsilyl (TMS) ethers, and 8-epi $PGF_{2\alpha}$ was derivatized as the *tert*-butyldimethylsilyl ether.

Platelet aggregation was measured using the photometric method. Platelets were washed; calcium studies measuring changes in fluorescence of FURA-2-loaded platelets were performed as previously described.[18]

RESULTS

Aggregation of washed human platelets with arachidonic acid resulted in a time-dependent increase in 8-epi $PGF_{2\alpha}$ formation, which coincided with the onset of aggregation and roughly corresponded to the time course of formation of both TxB_2 and 12-HETE (FIG. 1). Similar results were obtained when platelets were aggregated with thrombin or collagen. Treatment of platelets with aspirin (100 μM) inhibited aggregation and prevented formation of both 8-epi $PGF_{2\alpha}$ and TxB_2. A minor inhibitory effect on 12-HETE formation was also evident. Pretreatment of platelets with vitamin E (1 mM) had no effect on generation of any of the eicosanoids.

8-Epi $PGF_{2\alpha}$ formation from arachidonic acid was also achieved using the commercially available PG endoperoxide synthase (Cayman Chemical Co., Ann Arbor, MI) *in vitro*. The identity of the product with synthetic 8-epi $PGF_{2\alpha}$ was confirmed by a total electron impact mass spectrum (FIG. 2). Importantly, $[^3H]$8-epi $PGF_{2\alpha}$ was synthesized and used as a tracer to confirm that the endogenous material that coeluted with the peak corresponding to the internal standard in the quantitative selective ion monitoring (SIM) assay actually coeluted with it over four highly resolving HPLC purification procedures, as previously described.[19]

We confirmed that 8-epi $PGF_{2\alpha}$ induced platelet shape change and behaved as a competitive inhibitor for platelet aggregation induced by thromboxane mimetics.[20] Coincident with the shape change response 8-epi $PGF_{2\alpha}$ evokes an increase in platelet intracellular calcium (FIG. 3).

DISCUSSION

F_2 isoprostanes are formed from arachidonic acid in a free radical catalyzed manner *in vitro* and levels are increased in plasma and urine of animal models of

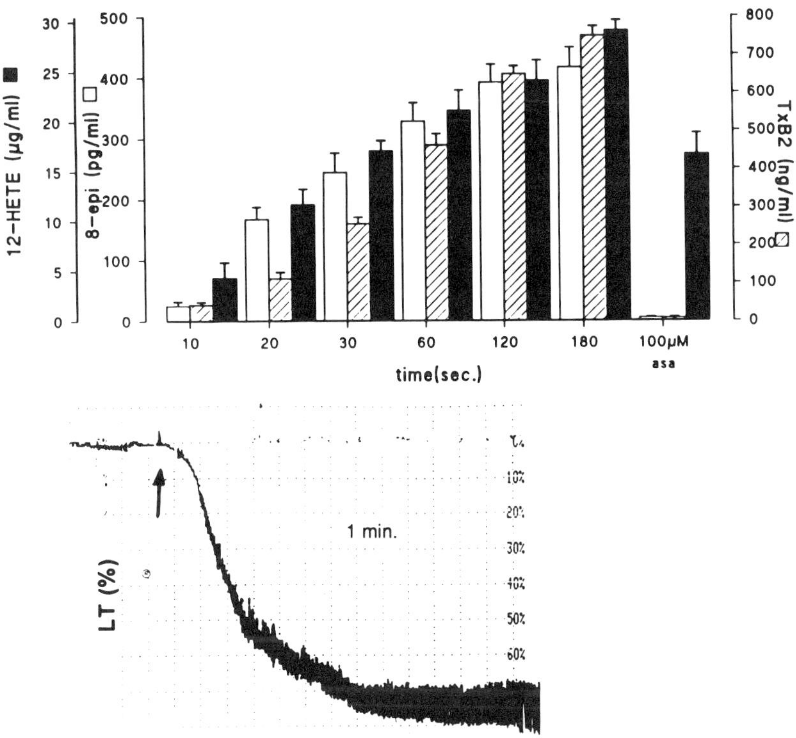

FIGURE 1. Time-dependent formation of 8-epi $PGF_{2\alpha}$, 12-HETE, and TxB_2 in washed human platelets stimulated with arachidonic acid (100 µM). The time course of the corresponding platelet aggregation is shown as a progressive decrease in light transmission (LT) in the **lower panel.** Pretreatment of platelets with aspirin (100 µM) abolished aggregation and both 8-epi $PGF_{2\alpha}$ and TxB_2 formation at 180 seconds after addition of arachidonic acid. Suppression of 12-HETE from corresponding levels in the absence of aspirin attained significance ($p < 0.01$).

free radical-induced organ injury.[2] Similar increments in urinary F_2 isoprostanes are observed in human syndromes putatively associated with oxidant damage.[3,17] Some of these settings, such as reperfusion of thrombosed coronary arteries with thrombolytic drugs and chronic cigarette smoking are also associated with platelet activation, as reflected by increased excretion of urinary metabolites of the platelet cyclooxygenase product TxB_2.[21,22] Given these observations and our particular interest in the effects of 8-epi $PGF_{2\alpha}$ on platelets and other cells relevant to thrombotic vascular occlusion, we explored the formation of this compound by human platelets.

Although 8-epi $PGF_{2\alpha}$ is formed in much smaller amounts than TxB_2, it is generated coincident with aggregation in a cyclooxygenase-dependent manner. Thus, inhibition of aggregation and TxB_2 production with aspirin or indomethacin prevented 8-epi $PGF_{2\alpha}$ formation. Furthermore, when aggregation was driven in a PGG/H synthase-independent manner by high doses of collagen or thrombin, formation of 8-epi $PGF_{2\alpha}$ did not occur, indicating its dependence on enzyme turnover rather than aggregation.

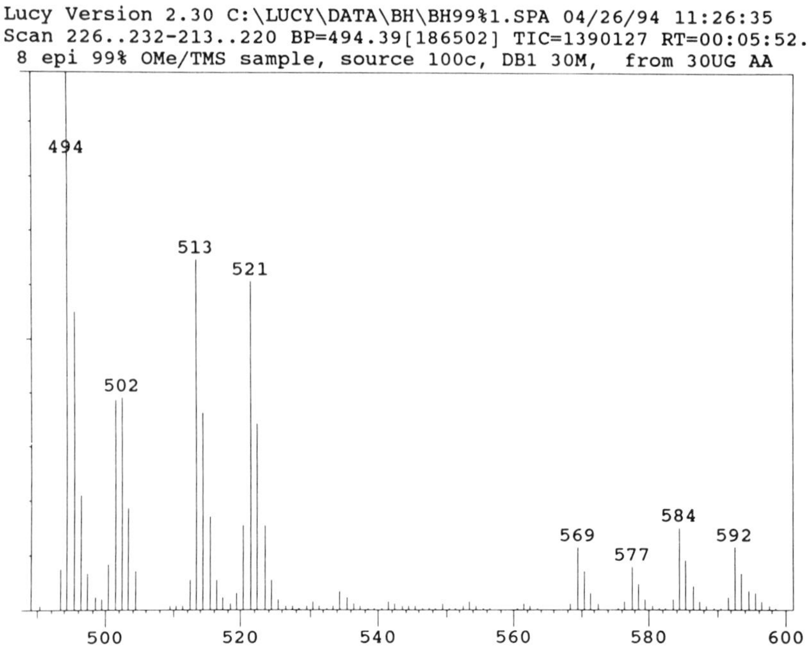

FIGURE 2. Electron impact mass spectrum (detail) of a mixture of authentic 8-epi $PGF_{2\alpha}$ and the product of an incubation of $[^2H_8]$arachidonic acid with platelet-rich plasma. Both are derivatized as the methyl ester TMS ether. Note the pairs of ions that differ by 8 mass units. The lower region of the spectrum becomes less straightforward to interpret, since the losses often involve one or more deuterons; e.g., loss of TMS-OH from the molecular ion of the unlabeled species (*m/z* 584) yields the ion at *m/z* 494, but the analogous loss from the $^2H_8$ species can occur as TMS-OH or TMS-O^2H, yielding the pair of ions at *m/z* 501/502.

8-Epi $PGF_{2\alpha}$ appears to be formed directly, rather than via free radicals generated during enzyme turnover, in that pretreatment of platelets with vitamin E and other oxygen scavengers did not influence 8-epi $PGF_{2\alpha}$ formation following activation.

The assumption that 8-epi $PGF_{2\alpha}$ was being formed under these conditions is based on the selective increase of a peak that comigrates with $^{18}O_2$-labeled authentic 8-epi $PGF_{2\alpha}$ in a SIM assay, using GC/MS. Peaks corresponding to other isoprostanes were not altered in response to platelet activation. Morrow *et al.*[19] recently employed a four-step HPLC purification procedure to support the interpretation that the comigrating peak actually corresponded to endogenous 8-epi $PGF_{2\alpha}$. The same approach confirmed the capacity of PGG/H synthase to generate 8-epi $PGF_{2\alpha}$ directly.

The biological implications of these observations remain to be established. We have demonstrated in a controlled clinical study that aspirin administration does not depress urinary 8-epi $PGF_{2\alpha}$ excretion in healthy volunteers, despite profound suppression of urinary Tx metabolites.[23] However, 8-epi $PGF_{2\alpha}$ is formed in serum, and this is suppressed *ex vivo* when volunteers take aspirin (Ref. 23 and J. Maclouf, personal communication, 1994). These observations suggest that platelets contribute

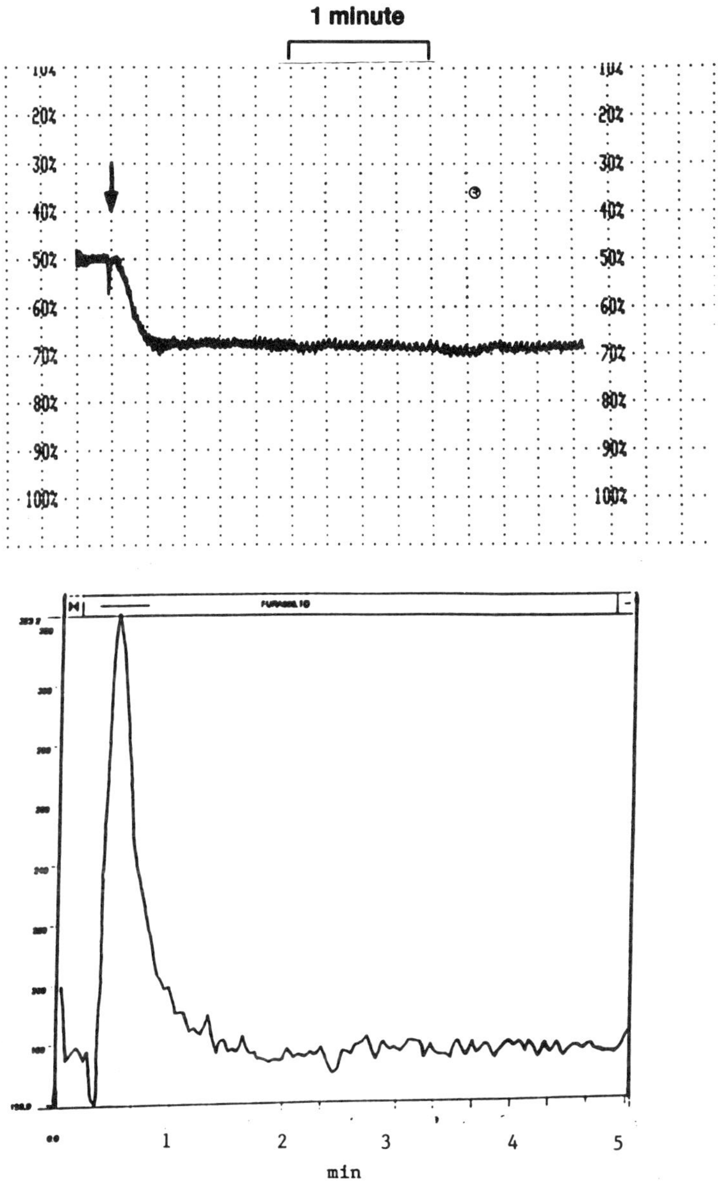

FIGURE 3. Platelet shape change induced by 8-epi $PGF_{2\alpha}$ (2 μM) indicated as an increase in light transmission (LT) in the **upper panel.** The corresponding change in platelet intracellular calcium is shown in the **lower panel.**

to a minor degree, if at all, to urinary 8-epi $PGF_{2\alpha}$ under physiological circumstances. Further investigation will clarify if this mechanism of formation is reflected in urinary levels of 8-epi $PGF_{2\alpha}$ in the setting of platelet activation.

Urinary 8-epi $PGF_{2\alpha}$ may be a useful guide to selecting clinical syndromes for evaluation of putative antioxidant drugs. However, this particular F_2-isoprostane can be formed in a PGG/H synthase-dependent manner, at least in human platelets. Characterization of the consequences of its formation may provide novel insights into platelet biology.

REFERENCES

1. MORROW, J. D., T. M. HARRIS & L. JACKSON ROBERTS II. 1990. Noncyclooxygenase oxidative formation of a series of novel prostaglandins: Analytical ramifications for measurements of eicosanoids. Anal. Biochem. **184:** 1-10.
2. MORROW, J. D., K. E. HILL, R. F. BURK, T. M. NAMMOUR, K. F. BADR & L. JACKSON ROBERTS II. 1990. A series of prostaglandin $F_{2\alpha}$-like compounds are produced in vivo in humans by a non-cyclooxygenase, free radical-catalyzed mechanism. Proc. Natl. Acad. Sci. USA **87:** 9383-9387.
3. LYNCH, S. M., J. D. MORROW, L. JACKSON ROBERTS II & B. FREI. 1994. Formation of non-cyclooxygenase-derived prostanoids ($F_{2\alpha}$-isoprostanes) in plasma and low density lipoprotein exposed to oxidative stress in vitro. J. Clin. Invest. **93:** 998-1004.
4. MORROW, J. D., J. A. AWAD, H. J. BOSS, I. A. BLAIR & J. J. ROBERTS. 1992. Non-cyclooxygenase-derived prostanoids ($F_{2\alpha}$-isoprostanes) are formed in situ on phospholipids. Proc. Natl. Acad. Sci. USA **89:** 10721-10725.
5. TAKAHASHI, K., T. M. NAMMOUR, M. FUKUNAGA, J. EBERT, J. D. MORROW, L. JACKSON ROBERTS II, R. L. HOOVER & K. F. BADR. 1992. Glomerular actions of a free radical-generated novel prostaglandin, 8-epi-prostaglandin $F_{2\alpha}$, in the rat. J. Clin. Invest. **90:** 136-141.
6. MORROW, J. D., T. A. MINTON & L. J. ROBERTS II. 1992. The F_2-isoprostane, 8-epi-$PGF_{2\alpha}$, a potent agonist of the vascular thromboxane/endoperoxide receptor, is a platelet thromboxane/endoperoxide receptor antagonist. Prostaglandins **44:** 155-163.
7. MORROW, J. D., T. A. MINTON, C. R. MUKUNDAN, M. D. CAMPBELL, W. E. ZACKERT, V. C. DANIEL, K. F. BADR, I. A. BLAIR & L. J. ROBERTS II. 1994. Free radical-induced generation of isoprostanes in vivo. **6:** 4317-4326.
8. FUKUNAGA, M., N. MAKITA, L. J. ROBERTS II, J. D. MORROW, K. TAKAHASHI & K. F. BADR. 1993. Evidence for the existence of F_2-isoprostane receptors on rat vascular smooth muscle cells. J. Am. Physiol. C1619-C1624.
9. KINSELLA, B. T., D. J. O'MAHONY, J. A. LAWSON, D. PRATICO & G. A. FITZGERALD. 1994. Cellular activation of thromboxane receptors. *In* Platelet-Dependent Vascular Occlusion. Annals of the New York Academy of Science, Volume 714. G. A. FitzGerald, L. Jennings & C. Patrono, Eds.: 270-278. The New York Academy of Sciences. New York, NY.
10. NUSING, R. M., M. HIRATA, A. KAKIZUKA, T. EKI, K. OZAWA & S. NARUMIYA. 1993. Characterization and chromosomal mapping of the human thromboxane A_2 receptor gene. J. Biol. Chem. **33:** 25253-25259.
11. KINSELLA, T., D. O'MAHONY & G. A. FITZGERALD. 1993. Regulation of the thromboxane receptor by phosphorylation in vivo. Blood **82:** 610 (Abstr.).
12. TAKAHARA, K., R. MURRAY, G. A. FITZGERALD & D. J. FITZGERALD. 1990. The response to thromboxane A2 in human platelets: Discrimination of two binding sites linked to distinct effector systems. J. Biol. Chem. **256:** 6836-6844.

13. ABRAMOWITZ, M., Y. BOIE, T. NGUYEN, T. H. RUSHMORE, M. A. BAYNES, K. M. METTERS, D. M. SLIPETZ & R. GRYGORCZK. 1994. Cloning and expression of a cDNA for the human prostanoid FP receptor. J. Biol. Chem. **269:** 2632-2636.
14. SAKAMOTO, K., T. EZASHI, K. MIWA, E. OKUDA-ASHITAKA, T. HOUTANI, T. M. SUGIMOTO, S. ITO & O. HAYAISHI. 1994. Molecular cloning and expression of a cDNA of the bovine prostaglandin $F_{2\alpha}$ receptor. J. Biol. Chem. **5:** 3881-3886.
15. SUGIMOTO, Y., K. HASUMOTO, T. NAMBA, A. IRIE, M. KATSUYAMA, M. NEGISHI, A. KAKIZUKA, S. NARUMIYA & A. ICHIKAWA. 1994. Cloning and expression of a cDNA for mouse prostaglandin F receptor. J. Biol. Chem. **2:** 1356-1360.
16. DARIUS, H., J. B. SMITH & A. M. LEFER. 1985. Beneficial effects of a new potent and specific thromboxane receptor antagonist (SQ-29,548) in vitro and in vivo. J. Pharm. Exper. Ther. **235:** 274-281.
17. DELANTY, N., M. REILLY, J. A. LAWSON, D. J. FITZGERALD & G. A. FITZGERALD. 1994. Urinary 8-epi-prostaglandin $F_{2\alpha}$ excretion as a non-invasive marker of free radical injury in humans. AFCR (Baltimore). Clin. Res. **42:** 182A (Abstr.).
18. PRATICO, D., L. IULIANO, S. M. PULCINELLI, M. S. BONAVITA, P. P. GAZZANIGA & F. VIOLI. 1992. Hydrogen peroxide triggers activation of human platelets selectively exposed to non aggregating concentrations of arachidonic acid and collagen. J. Lab. Med. **119:** 364-370.
19. MORROW, J. D., T. A. MINTON, K. F. BADR & L. J. ROBERTS II. 1994. Evidence that the F_2-isoprostane, 8-epi-prostaglandin $F_{2\alpha}$ is formed in vivo. Biochim. Biophys. Acta **1210:** 244-248.
20. MORROW, J. D., T. A. MINTON & L. J. ROBERTS II. 1992. The F_2-isoprostane, 8-epi $PGF_{2\alpha}$ a potent agonist of the vascular thromboxane/endoperoxide receptor, is a platelet thromboxane/endoperoxide receptor antagonist. Prostaglandins **44:** 155-163.
21. NOWAK, J., J. J. MURRAY, J. A. OATES & G. A. FITZGERALD. 1987. Biochemical evidence of a chronic abnormality in platelet and vascular function in apparently healthy chronic cigarette smokers. Circulation **76:** 6-14.
22. FITZGERALD, D. J., F. CATELLA, L. FOY & G. A. FITZGERALD. 1988. Marked platelet activation in vivo after intravenous streptokinase in patients with acute myocardial infarction. Circulation **77:** 142-150.
23. CATELLA, F., M. P. REILLY, F. MEAGHER, N. MORAN, N. DELANTY, J. A. LAWSON & G. A. FITZGERALD. 1994. Physiological formation of 8-epi-$PGF_{2\alpha}$ in vivo is not affected by cyclooxygenase inhibition. Adv. Prostaglandin Thromboxane Leukotriene Res. In press.

Diverse Mechanisms of Specificity of Human Receptors for Eicosanoids[a]

EDWARD J. GOETZL,[b] JINHONG YANG, MENGHANG XIA, LI ZENG, SUE WON SO, AND SONGZHU AN

Departments of Medicine and Microbiology-Immunology
University of California Medical Center
San Francisco, California 94143-0711

INTRODUCTION

Many of the products of oxygenation of arachidonic acid are potent mediators of physiological responses and may contribute to disorders of development, cellular function, tissue repair, and host defense in a variety of diseases.[1,2] The enzymes of mobilization and oxygenation of arachidonic acid that produce eicosanoid mediators are distributed widely in most mammalian tissues.[3] Regional differences in the levels of generation and degradation of each type of eisosanoid are one determinant of their relative biological activity. However, recently emerging evidence suggests that expression of distinct sets of receptors (Rs) is the predominant mechanism of regulation of tissue responses to eicosanoid mediators. This section will describe selected aspects of the structure, tissue distribution, and signal transduction pathways of human subtypes of cellular Rs for cyclooxygenase and lipoxygenase-derived mediators. Each of multiple transcripts of a subtype of eicosanoid R is assigned here to a functional class, based on the pattern of use of secondary mediators of signal transduction, as a provisional system of classification until characterization of the fundamental biological roles is completed. A new antisense genetic approach will be discussed for elucidating the complex roles of families of guanine nucleotide-binding (G) proteins in the transductional functions, and possibly cellular expression, of eicosanoid Rs. The tentative conceptual framework of multiple levels of specificity of eicosanoid Rs that is proposed now may permit a more meaningful integration of the accumulating characteristics of these Rs.

HUMAN CELLULAR RECEPTORS FOR PROSTAGLANDIN E2

Prostaglandin E2 (PGE2) is a potent mediator of many cardiovascular, pulmonary, renal, endocrine, gastrointestinal, neural, and immune responses.[4] Cellular recognition

[a] This work was supported by National Institutes of Health Grant HL31809.

[b] Address for correspondence: Edward J. Goetzl, M.D., UB8B, Box 0711, University of California, 533 Parnassus, San Francisco, California 94143-0711.

and effects of PGE2 are mediated by three subtypes of high-affinity Rs, termed EP1, EP2, and EP3. The initial definition of the three subtypes of PGE2Rs was based on the different specificities of their combining sites, as assessed by the rank order of affinities for PGE2, PGE1, and other natural prostanoids, as well as the affinities for a series of synthetic analogs of PGE2.[4] Recent cloning and expression of mouse and human PGE2Rs has demonstrated structural differences between the subtypes, confirmed the distinctive binding specificities with combinant Rs, and begun to examine the divergent pathways of signal transduction.[5–7]

A human lung tissue cDNA library was screened with a 500-bp cDNA probe generated by PCR amplification of a sequence in first-strand cDNAs of human blood polymorphonuclear leukocytes by using two sets of degenerate primers representing conserved sequences in the fourth to fifth transmembrane domains and the seventh transmembrane domain, respectively, of mouse kidney EP3 subtype of PGE2R and human thromboxane A2 (TxA2)R.[7] Ten clones were isolated that had overlapping sequences, of which one had a 3.0-kb insert with a 1464-bp open-reading frame, encoding a protein of 488 amino acids with seven transmembrane domains typical of G protein-associated Rs.[7,8] This human EP2 subtype of PGE2R exhibited amino acid sequence homologies of 88% with mouse EP2 subtype of PGE2R and 30% and 27%, respectively, with mouse EP3 subtype of PGE2R and human TxA2R (TABLE 1A). Expression of the human EP2 R in COS-7 and K293 cells revealed a single K_d for PGE2 of 2.2 nM, with the expected order of mean IC_{50} values of 0.7, 0.8, 110, and 976 nM for PGE2, PGE1, PGF2α, and PGD2, respectively.[7] The application of synthetic analogs of PGE2, with defined preferences for subtypes of PGE2Rs, confirmed the specificity of the combining site of human EP2Rs (TABLE 1B). Both M&B 28767 and misoprostol competitively inhibited the binding of PGE2, as has been observed previously for mouse EP2Rs,[5] whereas neither sulprostone nor SC-19220 interacted detectably with the human EP2R. The human EP2R mediated an increase in the intracellular concentration of cyclic AMP ([cAMP]i) (TABLE 1A), but failed to transduce a detectable increase in the intracellular concentration of Ca^{++} ([Ca^{++}]i) in either COS-7 or K293 cell transfectants, as assessed by the Fura-2 dye technique with several internal positive controls. The failure of PGE2 to evoke significant increases in turnover of phosphatidylinositol 4,5-bisphosphate in K293 cells expressing EP2Rs, as contrasted with a mean increase of 3.2-fold in the turnover of phosphatidylinositol, was consistent with the lack of effect on [Ca^{++}]i.

A similar PCR-based cloning strategy, using first-strand cDNAs from human cutaneous mast cells and two primers representing conserved sequences in the fourth transmembrane domain and the seventh transmembrane domain, respectively, of mouse kidney EP3R and human TxA2R led to the isolation of a cDNA encoding a human EP3-subtype of PGE2R.[9] This human EP3R has 85% amino acid sequence homology with the mouse kidney-derived EP3R and a similarly short third cytoplasmic loop.[6] The human recombinant EP3R in COS-7 cells bound PGE2 with a K_d of 2.6 nM and a high order of specificity revealed by the respective mean IC_{50} values of 0.7, 1.4, 140, and > 1,000 nM for PGE2, PGE1, PGF2α, and PGD2. The specificity of this human EP3R was confirmed by the similarly high affinities of M&B 28767 and sulprostone, the intermediate affinity of misoprostol, and the lack of interaction of SC-19220 (TABLE 1B). The human EP3R transduced a decrease in the forskolin-elevated [cAMP]i of COS-7 cell transfectants, as expected, and a modest

TABLE 1A. Human Receptors for Prostaglandin E2

Subtype	Number of Amino Acids (Size in kDa)	Receptor Structural Homologies	PGE2 Binding (K_d)	Responses to PGE2: cAMP	Responses to PGE2: $[Ca^{++}]i$
EP2	488 (53)	MoEP2 = 88% MoEP3(a) = 30% HuTxA2R = 27%	2.2 nM	EC_{50} = 4.3 nM (increase)	None
EP3, type I	367 (41)	MoEP3(a) = 85% HuTxA2R = 32%	2.6 nM	EC_{50} = 11 nM (decrease)	Mean 30% increase by 0.1 μM PGE2

TABLE 1B. Inhibition of Binding by Synthetic Analogues of PGE2 (Mean IC_{50}, nM; n = 3)

Specificity	Misoprostol (EP2/EP3)	M&B 28767 (EP2/EP3)	Sulprostone (EP3/EP1)	SC-19220 (EP1)
EP2	4,500	150	>10,000	>10,000
EP3, type1	155	0.74	2.13	>1,000

increase in $[Ca^{++}]i$. Four additional cDNAs, encoding different bovine EP3 subtype PGE2Rs, have recently been identified and partially characterized.[10] Their structural variations result from alternative splicing of mRNA, which has been detected as well in multiple human EP3Rs.

Five different cDNAs that encode structurally distinct EP3Rs that share a common amino-terminal and seven transmembrane-domain structure up to the first ten amino acids of the cytoplasmic tail and differ in the remainder of the tail sequence have been isolated from a human uterine library.[11] Two human uterine EP3R transcripts are identical to the human kidney EP3R identified first,[9] but differ in 3′-untranslated sequence, one resembles the bovine EP3R designated "d," and the other two are novel in structure. This additional level of diversity among Rs of the same PGE2-binding affinity and specificity effectively couples one combining site to different signal transduction pathways. As many alternative transcripts are being identified for some subtypes of eicosanoid Rs, it is recommended that each be assigned first to a functional class based on the patterns of use of secondary mediators of signal transduction. For example, most EP3Rs thus would fall into three functional classes: 1 transduces inhibition of adenylyl cyclase (AC) and modest increases in $[Ca^{++}]i$; 2 transduces inhibition of AC and at least threefold increases in $[Ca^{++}]i$; and 3 transduces increases in AC without a change in $[Ca^{++}]i$. The number of functional classes will presumably be limited by the possible associations of any one eicosanoid R with G proteins, will reflect critical G protein-coupling sequences, and may relate to biological role more closely than the total sequence.

Northern blot analyses of human tissue-derived mRNA have demonstrated extreme differences in the distribution of PGE2Rs. The single EP2R is most strongly represented in lungs, kidneys, pancreas, small intestine, immune organs, and blood leukocytes.[7] In contrast, the human EP3R analogous to the mouse kidney-derived EP3R is expressed predominantly in kidney, with a weak signal in pancreas.

Human cellular Rs for mediators derived from the lipoxygenation of arachidonic acid have been solubilized and purified for structural studies.[12,13] These are expected to share the same general topological structure of seven transmembrane domains as other G protein-associated Rs and to exhibit substantial amino acid sequence homology with Rs for prostaglandins, TxA2, and related lipid mediators. Signature amino acid sequences appear in most of the Rs for lipid mediators (TABLE 2). Although of no known functional significance, these sequences in the sixth and seventh membrane-spanning domains are not observed consistently in G protein-associated Rs with binding specificity for other ligands.

ROLES OF G PROTEINS COUPLED TO EICOSANOID RECEPTORS

Analyses of the effects of G protein-selective toxins on high-affinity binding of eicosanoid mediators and their cellular signaling activities in intact cells and of similar effects of guanine nucleotides in membrane preparations suggested initially a functional interaction of LTB4Rs with pertussis toxin-sensitive G proteins.[14] This was confirmed by the chromatographic demonstration of a trimolecular complex of LTB4, LTB4Rs, and a G protein, the assembly and stability of which was dependent on the presence of LTB4.[15]

TABLE 2. Distinctive Structural Features of Human Receptors for Eicosanoids

Receptor	Seven Transmembrane Domains	Short Third Intracellular Loop[a]	Sixth TMD[b] Signature Sequence	Seventh TMD Signature Sequence
Hu EP2	+	+	-**C**SI**P**-	-IL**D**PW-
Mo EP2	+	+	-**C**SI**P**	-IL**D**PW-
Hu EP3(I)	+	++	-**C**WS**P**-	-IL**D**PW-
Mo EP3(a)	+	++	-**C**WS**P**-	-IL**D**PW-
Hu PAF	+	++	-**C**FV**P**-	-VL**D**PV-
Hu TxA2	+	++	-**C**WL**P**-	-IL**D**PW-

[a] + = 60 amino acids or shorter, ++ = 30 amino acids or shorter.
[b] TMD = transmembrane domain.

To identify the specific G protein(s) associated with LTB4Rs and examine their functional contributions, an antisense genetic approach was used to suppress the cellular levels of alpha-chains of the pertussis toxin-sensitive Go, Gi1, Gi2, and Gi3 proteins.[16] Alpha-chains of individual G proteins were suppressed in separate cultures of human differentiated HL-60 cells by transfection with the high-level expression vector pcDNA I, containing a cDNA encoding the full-length antisense message for the alpha-chains of Gi1, Gi2, Gi3, or Go, or the vector alone as a control.[16] The extent of suppression of G protein alpha-chains by antisense mRNA was quantified by sodium dodecylsulfate-polyacrylamide gel electrophoresis (SDS-PAGE) and Western blot analysis of extracts of an equal number of each transfectant. Although Go alpha-chain usually is not detected in HL-60 cells induced to differentiate by exposure to one stimulus, the application of transforming growth factor-beta (TGF-β) and dihydroxy-vitamin D3 together elicited detectable levels of Go. The degree of suppression of cellular content of Gi2 and Go, respectively, was 80% to 95% and 80% at 7 days. The anti-G protein antibodies used to detect Gi2 and Go cross-react with Gi1 and Gi3, respectively (Calbiochem). The residual alpha-chain immunoreactivity apparently attributable to unsuppressed Go thus may reflect the unaffected content of Gi3. Development of two additional Western blots with monospecific rabbit antiserum to bovine brain Go alpha-chain confirmed the mean 90% decrease in Go alpha-chain in Go antisense transfectants. In contrast, the much lower level of Gi1 expected from prior Northern blot quantification of message is not reflected in analyses of Gi2. Immunoquantification thus may underestimate the suppression Go, which like Gi2 would be reduced consistently to less than 15% of normal. In three studies, the selectivity of the antisense effect was demonstrated by the lack of suppression of Go in Gi2 antisense transfectants and of Gi2 in Go antisense transfectants. Western blot analyses of the course of suppression of G protein alpha-chains showed maximal effects of antisense mRNA at 6-7 days, with less than half of the maximum at 3-4 days ($n = 3$) and loss of more than half of the maximum effect after 14 days ($n = 2$).

Chemotactic responses of Go-deficient transfectants to LTB4 and fMLP, each of which is recognized by a distinct set of high-affinity G protein-coupled receptors, were significantly less than those of the sham and Gi2-deficient transfectants. The response of Go-deficient transfectants to the highest concentration of LTB4 was indistinguishable from that of sham transfectants. The lysosomal degranulation of HL-60 leukocytes through G protein-coupled receptors for LTB4 and fMLP, as reflected in release of beta-glucuronidase, was significantly less for Go-deficient, but not Gi2-deficient, transfectants than for sham transfectants. The possibility that the deficient functional responses of Go-deficient transfectants to LTB4 were attributable to abnormalities in binding of LTB4 was excluded by the results of three studies of the characteristics of binding of [^{3}H]-LTB4. The mean percentage displacement of identical levels of maximum binding of [^{3}H]-LTB4 by 3, 10, and 30 nM nonradioactive LTB4 were 10, 41 and 89, respectively, for sham transfectants; 20, 59, and 88 for Go-deficient transfectants; and 13, 57, and 74 for Gi2-deficient transfectants. The respective K_d values for sham, Go-deficient, and Gi2-deficient antisense transfectants were 12, 7.6, and 8.2 nM, which are insignificantly different and the same as published values for untransfected HL-60 leukocytes.[17] Binding of [^{3}H]-fMLP to the HL-60 leukocyte transfectants in two studies also was not changed significantly by the reduction in individual levels of Go or Gi2 or by the combined deficiencies of Go and Gi1-3 attained by cotransfections.

A disturbance of one or more signal transduction pathways in Go-deficient transfectants was suggested by depressed functional responses to normal binding of LTB4. The deficiency in Go alpha-chain prevented the expected increase in [Ca^{++}]i elicited by optimal concentrations of either LTB4 or fMLP. In contrast, normal increases in [Ca^{++}]i were observed after LTB4 or fMLP was added to combined transfectants deficient in alpha-chains of Gi1, Gi2, and Gi3. The inhibition of calcium responses in Go-deficient HL-60 leukocyte transfectants was not altered substantially by the combined cotransfection of antisense messages directed to Gi1-3. In three additional studies, only Go deficiency suppressed significantly the [Ca^{++}]i responses of HL-60 leukocyte transfectants to LTB4 and fMLP. The biochemical consequences of Gi2 deficiency in HL-60 leukocyte transfectants was examined in relation to the known inhibition of adenylyl cyclase activity mediated by Gi proteins. LTB4 inhibited by 75% or more the forskolin-induced increase in concentration of adenosine 3′,5′-monophosphate (cAMP) in sham and Go-deficient transfectants, but inhibited by a mean of only 40% the increase observed in Gi2-deficient HL-60 leukocytes.

The application of antisense mRNA approaches to suppress selectively the levels of individual G proteins in HL-60 leukocytes bearing receptors for LTB4 and fMLP after differentiation allows assignment of the specific roles of some G proteins in signal transduction. Go appears to mediate increases in [Ca^{++}]i that are coupled to both chemotactic and degranulation responses, whereas the inhibition of adenylyl cyclase attributable to Gi2 had no detectable involvement in either functional response. The maintenance of the high-affinity state typical of 3-4% of the receptors for LTB4 and fMLP is dependent on their association with Gi or Go proteins, but depletion of over 80% of Go, Gi2, or Gi1-3 together had no significant effect on binding of [^{3}H]-LTB4 or [^{3}H]-fMLP by the HL-60 leukocyte transfectants. The leukocyte level of G proteins required to mediate each receptor-dependent function is presumed to have a distinctive threshold, which appears to be much lower for high-affinity binding than for transduction of biochemical or functional responses.

TABLE 3. Mechanisms of Specificity of Eicosanoid Receptors

1. Multiplicity of eicosanoid binding sites
2. Differences in cellular and tissue distribution
3. Diversity of signal transduction pathways
4. Distinctive regulation of cellular expression

Current evidence from analyses of other G protein-coupled receptors suggests associations with more than one type of G protein, which may transmit distinct signals to different effectors, as well as additive or opposing signals to the same effector. The physical coupling of two or more different G proteins to one receptor was shown by immunoprecipitation of solubilized receptors with antibodies specific for G proteins[18] and by anti-G protein antibody-mediated uncoupling of receptor-G protein complexes.[19] The functional interactions of one receptor with multiple G proteins have been demonstrated by introduction into intact cells of anti-G protein antibodies by patch-clamp pipettes[20,21] and of antisense oligonucleotides,[22] application of anti-G protein antibodies in cell membrane preparations,[23] and specific competition by a novel peptide antagonist for G protein binding to recombinant receptors in phospholipid vesicles.[24] Now the present data from studies of a nonpharmacological and nonantibody technique for uncoupling LTB4Rs from one or more G proteins confirms functionally relevant associations with two or more G proteins. Further studies of short-term and sustained G protein-deficient transfectants may facilitate identification of the specific links of individual G proteins to other cellular functions.

MULTIPLE CELLULAR DETERMINANTS OF THE SPECIFICITY OF EICOSANOID RECEPTORS

Although many of the critical molecular and cellular characteristics of eicosanoid Rs are incompletely defined, it is clear that an extraordinary spectrum of mechanisms contribute to the multiple levels of their specificity (TABLE 3). Differences in tissue content, cellular expression, ligand-binding affinity, and signal transduction pathways are subject to separate regulatory factors and determine the distinctive biological role(s) of each eicosanoid R. The primary variable that defines an eicosanoid R is the principal ligand, as assigned by affinity of binding relative to other eicosanoids. Many of the basic sets of such Rs can be further divided into subsets by differences in recognition of synthetic analogs of the principal ligand. The binding specificity of the combining site of an eicosanoid R is an incompletely understood function of amino acid sequence and of the resultant chemical nature and configuration of complementarity-determining domains in cellular membranes. These limited structures of the combining site thus are not reflected in the degree of amino acid sequence homology observed between the full-length structures of highly related Rs, such as EP2 and EP3. Similarly, the factors governing prevalence of the mRNA and protein of any one eicosanoid R in different tissues have not been elucidated. Tissue and cellular distribution are often subset dependent and quantitatively prominent, as with the finding of human EP3Rs predominantly in kidneys but human EP2 receptors being more widespread.

The finding of a high-density of cell surface Rs specific for an eicosanoid in one or a few tissues is clearly the primary determinant of a major biological effect of the eicosanoid. The nature and duration of the response elicited, however, are dependent on many additional effector mechanisms and subject to multiple levels of regulation. Some of these latter variables have been identified only recently and have been shown to have complex relationships with G proteins. Different forms of mouse and human EP3Rs have been recognized that result from alternative splicing of mRNA encoding the carboxyl-terminal tail and possibly other substituents (TABLE 1).[11,25] The capacity of each form of EP3R to transduce a unique biochemical response to PGE2 is presumed, but not proven, to result from association with one or more distinctive G proteins. An analogous set of third intracellular loop splice variants of the type I rat R for pituitary adenylyl cyclase-activating polypeptide (PACAP) has been found, each member of which transduces a different pattern of responses in adenylyl cyclase and phospholipase C.[26]

A complex pattern of dependence of LTB4R signal transduction on G proteins has been found by the application of antisense genetic techniques. The LTB4Rs expressed on HL-60 cells induced to differentiate by dihydroxy-vitamin D3 and TGF-β require Go alpha-chains for a response of $[Ca^{++}]i$ and of chemotaxis and degranulation to LTB4.[16] In contrast, the coupling of LTB4Rs of HL-60 cells to Gi2 alpha-chains permits LTB4 to inhibit forskolin-evoked rises in [cAMP]i without apparent functional consequences. Therefore, an eicosanoid R may associate with more than one G protein and evoke divergent biochemical and functional responses. Structural variations in the domains of eicosanoid Rs that interact with G proteins, as in the case of the EP3Rs, may expand by another order the range of possible transductional and functional outcomes of eicosanoid recognition by the R.

Heterotrimeric G proteins are known to affect the membrane localization and traffic of Rs and other surface proteins of cells.[27] One or more such mechanisms may account for the differences in sensitivity of two mouse EP3 isoforms to PGE2-induced desensitization.[25] The desensitization-susceptible isoform, which has one carboxy-terminal structure, is sequestered and reduced in total number by exposure to PGE2, whereas no changes were elicited by PGE2 in the resistant isoform with another carboxy-terminal structure. At this point in our understanding of the bases for biological specificity of eicosanoid Rs, G protein interactions seem capable of governing the pathways of biochemical responses, coupling such responses to productive functions, responsiveness to repetitive stimulation, and cellular distribution of the Rs. Each of the G protein-dependent mechanisms of regulation of eicosanoid Rs represents a potential level of biological specificity beyond those endowed by ligand binding and the tissue content of each R.

REFERENCES

1. SAMUELSSON, B., M. GOLDYNE, E. GRANSTROM, M. HAMBERG, S. HAMMARSTROM & C. MALMSTEN. 1978. Annu. Rev. Biochem. **47:** 997-1029.
2. SAMUELSSON, B. 1983. Science **220:** 568-575.

3. NEEDLEMAN, P., J. TURK, B. A. JAKSCHIK, A. R. MORRISON & J. B. LOFKOWITH. 1986. Annu. Rev. Biochem. **55:** 69-103.
4. COLEMAN, R. A., I. KENNEDY, P. P. A. HUMPHREY, K. BUNCE & P. LUMLEY. 1990. *In* Comprehensive Medicinal Chemistry: Membranes and Receptors, Vol. 3. J. C. Emmett, Ed.: 643-714. Pergamon Press. New York, NY.
5. HONDA, A., Y. SUGIMOTO, T. NAMBA, A. WATABE, A. IRIE, M. NIGISHE, S. NARUMIYA & A. ICHIKAWA. 1993. J. Biol. Chem. **268:** 7759-7762.
6. SUGIMOTO, Y., T. NAMBA, A. HONDA, Y. HAYASHI, M. NEGISHI, A. ICHIKAWA & S. NARUMIYA. 1992. J. Biol. Chem. **267:** 6463-6466.
7. AN, S., J. YANG, M. XIA & E. J. GOETZL. 1993. Biochem. Biophys. Res. Commun. **197:** 263-270.
8. DOHLMAN, H. G., J. THORNER, M. G. CARON & R. J. LEFKOWITZ. 1991. Annu. Rev. Biochem. **60:** 653-688.
9. YANG, J., M. XIA & E. J. GOETZL. 1994. Biochem. Biophys. Res. Commun. **198:** 999-1006.
10. NAMBA, T., Y. SUGIMOTO, M. NEGISHI, A. IRIE, F. USHIKUBI, A. KAKIZUKA, S. ITO, A. ICHIKAWA & S. NARUMIYA. 1993. Nature **365:** 166-170.
11. AN, S., J. YANG, S. W. SO, M. XIA & E. J. GOETZL. 1994. FASEB J. **8:** A412.
12. HALUSHKA, P. V., D. E. MAIS, P. R. MAYEUX & T. A. MORINELLI. 1989. Annu. Rev. Pharmacol. Toxicol. **10:** 213-239.
13. GOLDMAN, D. W., L. A. GIFFORD, R. N. YOUNG, T. MAROTTI, M. K. L. CHEUNG & E. J. GOETZL. 1991. J. Immunol. **146:** 2671-2677.
14. GOLDMAN, D. W., F.-H. CHANG, L. A. GIFFORD, E. J. GOETZL & H. R. BOURNE. 1985. J. Exp. Med. **162:** 145-156.
15. SHERMAN, J. W., M. A. MENDELSON, J. M. BOGGS, C. H. KOO & E. J. GOETZL. 1992. J. Cell. Biochem. **48:** 367-372.
16. GOETZL, E. J., R. S. SHAMES, J. YANG, F. W. BIRKE, Y.-F. LIU, P. R. ALBERT & S. AN. 1994. J. Biol. Chem. **269:** 809-812.
17. GOLDMAN, D. W., D. M. OLSON, D. G. PAYAN, L. A. GIFFORD & E. J. GOETZL. 1986. J. Immunol. **136:** 4631-4636.
18. MURRAY-WHELAN, R. & W. SCHLEGEL. 1992. J. Biol. Chem. **267:** 2960-2965.
19. LAW, S. F., K. YASUDA, G. I. BELL & T. REISINE. 1993. J. Biol. Chem. **268:** 10721-10727.
20. LLEDO, P. M., V. HOMBURGER, J. BOCKAERT & J. D. VINCENT. 1992. Neuron **8:** 455-463.
21. KLEUSS, C., H. SCHERUBL, J. HESCHELER, G. SCHULTZ & B. WITTIG. 1992. Nature **358:** 424-426.
22. EASON, M. G., H. KUROSE, B. D. HOLT, J. R. RAYMOND & S. B. LIGGETT. 1992. J. Biol. Chem. **267:** 15795-15801.
23. ARAGAY, A. M., A. KATZ & M. I., SIMON. 1992. J. Biol. Chem. **267:** 24983-24988.
24. MUKAI, H., E. MUNEKATA, & T. HIGASHIJIMA. 1992. J. Biol. Chem. **267:** 16237-16243.
25. NEGISHI, M., Y. SUGIMOTO, A. IRIE, S. NARUMIYA & A. ICHIKAWA. 1993. J. Biol. Chem. **268:** 9517-9521.
26. SPENGLER, D., C. WAEBER, C. PANTOLONI, F. HOLSBOER, J. BOCKAERT, P. H. SEEBURG & L. JOURNOT. 1993. Nature **365:** 170-175.
27. BOMSEL, M. & K. MOSTOV. 1992. Mol. Biol. Cell **3:** 1317-1328.

Leukotriene D_4–Induced Signal Transduction[a]

ANITA SJÖLANDER AND EVA GRÖNROOS

Department of Cell Biology
University of Linköping
S-581 85 Linköping, Sweden

INTRODUCTION

The inflammatory reaction is an important part of the body defense against invading microorganisms. Studies of the inflammatory response have revealed that a large number of different cell types are involved. Consequently, cell-cell interactions between different types of cells are likely to represent important regulatory events of the inflammatory process. For a number of years noninflammatory cells in this context were looked upon only as target cells, but lately more and more attention has been focused on these cells as important regulators of the inflammatory process. A condition for noninflammatory cells to participate in the regulation of the inflammatory reaction is a direct and/or indirect cross talk between these cells and inflammatory cells. One important mechanism for such a cellular cross talk is the release and recognition of soluble inflammatory mediators, such as leukotrienes.

Leukotrienes belong to a family of biologically active conjugated trienes formed from arachidonic acid by the 5-lipoxygenase pathway[1,2] and are among the most important mediators of the inflammatory reaction. Leukotrienes have also been indicated as crucial mediators of several pathological processes, such as asthma and other hypersensitivity reactions.[1,3,4] In addition, the so-called slow-reacting substances of anaphylaxis has been shown to consist of leukotrienes C_4, D_4, and E_4 (LTC_4, LTD_4 and LTE_4).[5] These leukotrienes are extremely potent and possess a large number of functional effects, such as increased vascular permeability and contraction of smooth muscles. In addition, LTD_4 has been suggested to induce mucus secretion.[6] Consequently, it is important to study and reveal the immediate cellular effects of these leukotrienes, not only on inflammatory cells, but also on different types of noninflammatory cells.

Leukotriene-induced effects are generally agreed to be mediated via interaction with specific plasma membrane receptors.[7] Specific leukotriene receptor binding sites have been identified with the help of radioactive ligands in a large number of different tissues and cells.[8–10] In terms of the LTD_4 receptor a fairly recent report, in which target-size analyses were performed, estimated the receptor to have a molecular mass of 65 kDa and to be a glycosylated protein.[10] In another study, however, the LTD_4 receptor was suggested to be at least a dimer, and the molecular mass of one of the

[a]This work was supported by grants from the Swedish Medical Research Council (04X-10356), the King Gustaf's Memorial Foundation, and the Swedish Society of Medicine.

subunits was calculated to be around 70 kDa.[11] Consequently, although there is evidence for the existence of a specific LTD_4 receptor, no detailed information is available about its molecular structure. Furthermore, despite the fact that the signal transduction pathways of different groups of plasma membrane receptors have been extensively studied over the last 20 years,[12] little is known today about the transmembrane signaling properties of the LTD_4 receptor and the second messengers that it might generate.

RESULTS AND DISCUSSION

The concept that noninflammatory cells also might participate in the regulation of the inflammatory reaction includes the ability of these cells to recognize inflammatory mediators, generate intracellular signals, and respond to these mediators. To further clarify the basis for noninflammatory cells in the regulation of the inflammatory reaction, we have for the last couple of years studied the ability of a human epithelial cell line, Intestine 407, to respond to the inflammatory lipid mediator LTD_4. In particular, we have focused our work on the signal transduction properties of LTD_4 in these noninflammatory cells.

LTD_4 Activates at Least Two Different G Proteins

Several years ago it was demonstrated that high-affinity stereoselective specific binding sites for $^3H\text{-}LTD_4$ were present in guinea pig lung membranes.[8] This specific binding of LTD_4 to plasma membrane fractions was found to be affected by preincubation with a GTP analog,[8] suggesting that the LTD_4 receptor is able to interact with and activate one or several different types of G proteins. Other groups have subsequently published data in support of this concept.[11] Further studies of the signaling mechanism of the LTD_4 receptor have revealed that this receptor also causes an increased hydrolysis of phosphatidylinositols.[7,13] Consequently, it seems as if the binding of LTD_4 to its receptor activates a G protein-regulated phospholipase C, which is the enzyme directly responsible for the hydrolysis of phosphatidylinositols.[13] The hydrolytic products of phospholipase C action are inositol trisphosphate and diacylglycerol, which are both plausible second messengers for LTD_4.[7] We and others have also shown that cells stimulated with LTD_4 respond with a prompt and transient increase in their cytosolic free Ca^{2+} concentration, already in the nM concentration range.[14,15] This LTD_4-triggered cytosolic free Ca^{2+} transient is generated by both an intracellular mobilization of Ca^{2+} and an influx of Ca^{2+} over the plasma membrane. Preincubation of epithelial cells with pertussis toxin revealed that the Ca^{2+}-signaling mechanism of LTD_4 appears to consist of dual G protein-regulated pathways. The reason for this assumption is that the mobilization of Ca^{2+} is regulated by a pertussis toxin-insensitive G protein, whereas the influx is regulated by a pertussis toxin-sensitive G protein.[15] These data can be summarized in a schematic model as outlined in FIGURE 1. It is worth pointing out that the pertussis toxin-sensitive G protein involved in the regulation of the Ca^{2+} influx across the plasma membrane is most likely a heterotrimeric G protein, whereas the G protein involved in the mobilization of intracellular Ca^{2+} is a pertussis toxin-insensitive G protein of unknown structure.

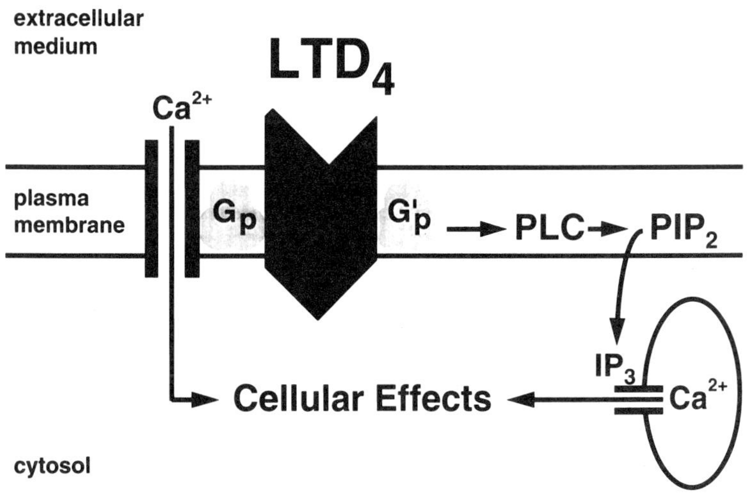

FIGURE 1. Model of LTD_4-induced signal transduction. Gp, G protein; G′p, a different G protein; PLC, phospholipase C; PIP_2, phosphatidylinositol bisphosphate; IP_3, inositol trisphosphate.

Effects of a Calcium Signal in Epithelial Cells

In order to investigate the importance of a calcium signal in epithelial cells for the regulation of an inflammatory response we mimicked the LTD_4-induced calcium signal by incubating the cells with a calcium ionophore (5 μM of A23187). This artificially induced calcium signal led to an activation of the 5-lipoxygenase pathway as revealed by increased production of LTB_4 and 5-hydroxyeicosatetraenoic acid (5-HETE).[16] These data support the concept that epithelial cells can sense and respond to an inflammatory mediator, and by doing so play an active role in the regulation of the inflammatory reaction.

The LTD_4-Induced Calcium Influx is cAMP-Dependent

Ca^{2+} channels in different tissues differ to some extent in their primary structure and/or in their subunit composition. Therefore, it is not unexpected that the regulatory mechanisms also differ among the various types of Ca^{2+} channels found in cells. It is well established that β-adrenoreceptor stimulation causes an increase of the Ca^{2+} current largely by a cAMP-dependent phosphorylation. However, recent data suggest that in addition this channel may also be regulated directly by a G protein.[17] This is not surprising since many ion channels found in plasma membranes are directly regulated by G proteins.

In order to further investigate the LTD_4-induced calcium influx in epithelial cells, we decided to explore a possible dual regulation by on one hand a heterotrimeric G protein and on the other a cAMP-dependent protein kinase. One fundamental condition

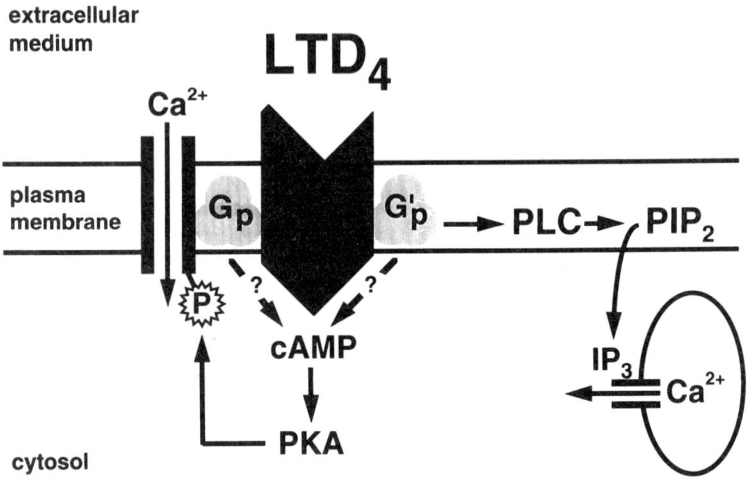

FIGURE 2. Possible model of how the LTD_4-induced calcium influx is regulated by cAMP. Activation of protein kinase A positively modulates the opening of the calcium channel, most likely via a direct phosphorylation of the channel. Gp, G protein; G′p, a different G protein; PLC, phospholipase C; PIP_2, phosphatidylinositol bisphosphate; IP_3, inositol trisphosphate; cAMP, cyclic adenosine monophosphate; PKA, protein kinase A.

for such a dual regulatory mechanism is that the LTD_4 stimulation causes a cAMP signal in epithelial cells. Unpublished data from our laboratory show that LTD_4 can, at least in Intestine 407 cells, induce a rapid increase in the cellular level of cAMP. In agreement with our findings Giembycz and co-workers have shown that LTD_4 stimulation of guinea pig lung parenchyma cells causes a pronounced increase (350%) in the cellular level of cAMP.[18] However, conflicting data exist in the literature regarding the ability of LTD_4 to affect the cellular level of cAMP.[19,20] Not only LTD_4-induced increases but also cellular decreases of the cAMP level have been reported from studies performed on different cell types.[21] The most likely explanation for these conflicting data is the use of different cell types, although differences in the experimental conditions cannot be totally excluded at this point.

Data from our laboratory also indicate that the LTD_4-induced increase in cAMP is directly involved in the regulation of the LTD_4-triggered calcium influx. Cells preincubated with the cAMP inhibitor Rp-cAMPS caused a total block of the LTD_4-induced influx of calcium, whereas it did not reduce the LTD_4-induced intracellular mobilization of calcium in epithelial cells. These data suggest a direct regulatory role of cAMP in the LTD_4-induced influx mechanism of Ca^{2+}. The mechanism whereby cAMP controls the influx of Ca^{2+} could be an activation of a protein kinase A and a subsequent phosphorylation of the Ca^{2+} channels in the plasma membrane. A schematic model of such a cAMP-dependent influx is outlined in FIGURE 2. Howard and co-workers have presented data that indicate that the LTD_4 receptor interacts both with the adenylate cyclase system and with the PIP_2 pathway.[20] These findings lend additional support for our model (FIG. 2) of a cAMP-dependent step in the LTD_4-induced influx mechanism of Ca^{2+}.

SUMMARY

In a human epithelial cell line LTD_4 induces a calcium signal that is dependent on both intracellular mobilization and influx of calcium. This calcium signal is generated via the activation of dual G protein pathways. Whereas the intracellular mobilization of calcium is regulated by a pertussis toxin-insensitive G protein, the subsequent influx of calcium is regulated by a pertussis toxin-sensitive G protein. Furthermore, a LTD_4-induced cellular elevation of cAMP also participates in the regulation of this calcium signal. The increase in cAMP is directly related to the LTD_4-induced influx of calcium, perhaps by an activation of protein kinase A and a subsequent phosphorylation of a plasma membrane channel. This model of the LTD_4-induced signaling pathway in epithelial cells is outlined in FIGURE 2.

ACKNOWLEDGMENTS

The authors are indebted to Åsa Schippert for excellent technical assistance and Birgitta Bohman for linguistic revision of the manuscript.

REFERENCES

1. SAMUELSSON, B. 1983. Science **220:** 568-575.
2. SAMUELSSON, B., S.-E. DAHLEN, J. Å. LINDGREN, C. A. ROUZER & C. N. SERHAN. 1987. Science **237:** 1171-1176.
3. GOETZL, E. J., D. G. PAYAN & D. W. GOLDMAN. 1984. J. Clin. Immunol. **4:** 79-84.
4. FORD-HUTCHINSON, A. W. 1985. Fed. Proc. **44:** 25-29.
5. SAMUELSSON, B. & S. HAMMARSTRÖM. 1981. Immunol. Today **2:** 3-6.
6. MAROM, Z., J. H. SHELHAMER, M. K. BACH, D. R. MORTON & M. KALINER. 1981. Am. Rev. Respir. Dis. **126:** 449-451.
7. SARAU, H. M., S. MONG, J. J. FOLEY, H.-L. WU & S. T. CROOKE. 1987. J. Biol. Chem. **262:** 4034-4041.
8. MONG, S., H.-L. WU, G. K. HOGABOOM, M. A. CLARK, J. M. STADEL & S. T. CROOKE. 1985. Eur. J. Pharmacol. **106:** 241-253.
9. SIMONSON, M. S., P. MENÉ, G. R. DUBYAK & M. J. DUNN. 1988. Am. J. Physiol. **255**(24): C771-C780.
10. ROCHETTE, C., D. W. NICHOLSON & K. M. METTERS. 1993. Biochim. Biophys. Acta **1177:** 283-290.
11. WATANABE, T., T. SHIMIZU, I. MIKI, C. SAKANAKA, Z.-I. HONDA, Y. SEYAMA, T. TERAMOTO, T. MATSUSHIMA, M. UI & K. KUROKAWA. 1990. J. Biol. Chem. **265:** 21237-21241.
12. BERRIDGE, M. J. 1993. Nature **361:** 315-325.
13. VEGESNA, R. V. K., S. MONG & S. T. CROOKE. 1988. Eur. J. Pharmacol. **147:** 387-396.
14. CROOKE, S. T., M. MATTERN, H. M. SARAU, J. D. WINKLER, J. BALCAREK, A. WONG & C. F. BENNETT. 1989. Trends Pharmacol. **3:** 103-107.
15. SJÖLANDER, A., E. GRÖNROOS, S. HAMMARSTRÖM & T. ANDERSSON. 1990. J. Biol. Chem. **265:** 20976-20981.
16. SJÖLANDER, A., Å. SCHIPPERT & S. HAMMARSTRÖM. 1993. Prostaglandins **45:** 85-96.
17. HARTZELL, H. C. & R. FISCHMEISTER. 1992. Trends Pharmacol. Sci. **13:** 380-385.
18. GIEMBYCZ, M. A., J. DIAMOND & I. W. RODGER. 1993. Br. J. Pharmacol. **108:** 824-832.

19. Torphy, T. J., M. Burman, L. W. Schwartz & M. A. Wasserman. 1986. J. Pharmacol. Exp. Ther. **237:** 332-340.
20. Howard, S., M. Chan-Yeung, L. Martin, S. Phaneuf & H. Salari. 1992. Eur. J. Pharmacol. **227:** 123-129.
21. Andersson, R., L. Gustafsson, S. Hedman, P. Hedqvist & B. Samuelsson. 1982. Acta Physiol. Scand. **116:** 97-99.

Prostaglandin J2 and Related Compounds

Mode of Action in G1 Arrest and Preclinical Results

MASANORI FUKUSHIMA,[a,b] HIROSHI SASAKI,[c] AND SHOJI FUKUSHIMA[d]

[b]*Aichi Cancer Center*
Chikusa-ku
Nagoya 464, Japan

[c]*Jikei University School of Medicine*
Minato-ku
Tokyo 105, Japan

[d]*Kobe-Gakuin University*
Nishi-ku
Kobe 651-21, Japan

Prostaglandin D2 undergoes dehydration in human plasma and forms Δ^{12}-PGJ2, which is one of the final metabolites in the human body.[1] In plasma or serum we can identify Δ^{12}-PGJ2; however, in serum-free media or aqueous solution PGJ2 is the primary compound derived from PGD2. Thus the step to PGJ2 is nonenzymatic, and Δ^{12}-PGJ2 is formed via an enzymatic process in which albumin is involved. Although the passage to Δ^{12}-PGJ2 is not major PGD2 metabolism, the biological or pharmacological activities of Δ^{12}-PGJ2 are specific and significant. Presently, at least four different activities are described.[2] Antiproliferative activity against cell and virus are well known and have been studied extensively by many researchers. Accumulated experimental results strongly suggest that the effects on cells and also probably on viruses are universal; however, cellular responses to such PGs seem different between normal cells and malignant cells. For example, in malignant cells Δ^{12}-PGJ2 suppresses *myc* gene family expression prior to G1 arrest, whereas in normal cells, it does not.

As to antiviral effect, interestingly Δ^{12}-PGJ2 inhibits retrovirus including HTLV-1 and HIV production. PG effect on virus production is unique, since it seems to affect both DNA and RNA viruses. In the herpes simplex II virus system the primary action of Δ^{12}-PGJ2 and also its analogue, Δ^{7}-PGA1, lies in the early step of transcription of the viral genome.[3] Stimulation of osteogenesis is a new research field that was opened by Koshihara *et al.*[4] Recently her group developed a very active analogue *in vivo* that may be a candidate for clinical trial. Another activity reported

[a]To whom correspondence should be addressed.

is immunostimulation. Δ^{12}-PGJ2 at very low doses apparently augments various immunological parameters.

As is well known, series of PGA and PGJ are actively incorporated into cells and transferred into the nuclei.[1] In this process the fatty acid-binding protein family is suggested to play a role of carrier protein.[5] When such PGs transfer to the nuclei, expression of the *myc* gene family is suppressed, and, inversely, genes for heat shock protein are stimulated.[2] This cellular process seems to be a common stress response to various stressors, including heat shock, hypoxia, and arsenite.[2] This cellular response currently attracts many researchers' interest in relation to apoptosis. In 1993 Korean investigators reported an apoptotic effect of Δ^{12}-PGJ2 using L1210 mouse leukemia cells.[6] However, before discussions of the apoptotic effect of such PGs, one should consider dose and time effect of PGs and cell kinetics.[1] The effect of PGA or PGJ on cell growth or cell survival strictly depends on the dose and time of exposure. Incorporation of such PGs into the cell is largely affected by the existence of serum. Serum-free media at a dose of 10 μg/ml of Δ^{12}-PGJ2 with 10-min incubation is enough to observe cell cycle arrest at G1 phase, whereas in the presence of 5 to 10% serum it requires more than 4 hours. This data suggests that the influx of a certain number of PG molecules into the cytoplasm, possibly into the nucleus, leads to cellular responses in changing gene expressions. The number of PGs leading G1 arrest in cells was roughly calculated as 600 pmoles/10^6 cells. Of course, the exact number must be much smaller than this, since in this experiment cell cycle was not synchronized, and the sensitivity for the detection of G1 phase cell accumulation was low. The IC_{50} concentration is critical to analyze the cellular effect of PGA and PGJ. When the dose of PG exceeds the IC_{50} level, it leads to cell death. Increasing the dose of PG causes the number of cells to decrease faster. When we look at the cell cycle population every 12 h until 48 h after the addition of Δ^{12}-PGJ2 into the culture medium, it is apparent that at IC_{50}, which is defined as 50% inhibition of the cell growth at day 4 in culture, cells accumulated in G1 phase, while the number of G2/M-phase cells decreased. In the presence of Δ^{12}-PGJ2 at IC_{90}, defined at day 4 of incubation, the number of G2/M-phase cells doubled at 24 h, while slight accumulation of G1 phase cells was observed. In this condition, the number of cells decreased significantly, thereby implying that at the IC_{90} level Δ^{12}-PGJ2 leads to cell death.

The question arose as to whether or not this PG effect requires a natural configuration. In order to answer this question, we separated four stereoisomers of Δ^{7}-PGA1. Dose-response curves for cell growth inhibition were almost identical for these four stereoisomers. Furthermore, cell cycle analysis at IC_{50} for four stereoisomers again showed the same pattern.[7] All isomers arrested cells at G1 phase at a similar level at IC_{50} dose of Δ^{12}-PGJ2.

To summarize the PG effect on cell cycle depends strictly on the concentration (TABLE 1). At the IC_{50} level cells arrest at G1 phase. This effect is partially reversible, since when the exposure time to PG is prolonged, the time-dose effect occurs. At the IC_{90} level cells arrest at G2/M and then irreversibly begin cell death mechanisms. This effect is independent of PG configuration. This fact supports the hypothesis that the chemical reaction responsible for the above PG effect in the cell is the reaction of enone with the -SH group(s) of certain proteins that play a key role(s) in cellular mechanisms for the protection of cells.[8]

TABLE 1. Effect of Prostaglandin on Cell Cycle

PG Concentration	Effect	
IC_{50}[a]	G1 arrest	Reversible
More than IC_{50}	G2/M arrest and cell death	Irreversible

[a] The concentration required for 50% inhibition of cell growth.

Identification of the molecule(s) responsible for G1 arrest may help to elucidate the mechanism of G1 regulation. However, from the clinical viewpoint we should know whether such PGs can be used for patients as an anticancer drug. During the past decade we screened more than 200 PG analogues using *in vitro* and also *in vivo* experimental systems. Because of its ready availability in large amounts at low cost and its chemical stability, we have chosen Δ^7-PGA1, an analogue of Δ^{12}-PGJ2, as a candidate for clinical development. However, Δ^7-PGA1 is oily, water insoluble. Integration of this compound into lipid microspheres was the method used to overcome this obstacle to clinical use.

In order to integrate PG into lipid microspheres efficiently, Δ^7-PGA1 methylester was synthesized. In this form, the cytotoxicity of the compound was somewhat strengthened.[9] When Δ^7-PGA1-Me with a mixture of soybean oil, lecithin, glycerin, and water passes through a microfluidizer apparatus, it forms uniform particles of 180-nm diameter. Δ^7-PGA1-Me is integrated into the inner space of the lecithin vesicle with soybean oil.[9] The particle is easily soluble in water and is very stable over the period of a year at 4°, even at 37°. Although in 5% glucose or in distilled water the particle size did not change, in saline the diameter of the lipid microsphere increased up to 400 nm during 2-h incubation at 37°; at 4° the size increased very slowly. Δ^7-PGA1 in lipid microspheres was perfectly stable at 4° over a period of a year. At 37° decomposition of Δ^7-PGA1 was evident in 2 weeks or more. These pharmaceutical data guarantee ease in future clinical use.

As a first step in preclinical drug evaluation, we examined the acute toxicity of lipo-Δ^7-PGA1-Me with a single shot using rats and mice. Female rats could tolerate a larger dose than male rats. LD_{10} for the male rats was 18.6 mg and for female rats

TABLE 2. Acute Toxicity of Lipo-Δ^7-PGA1 by Single Shot

Species	Sex	Injection Route	LD_{10} (mg/kg)	Estimated MTD in Human (mg/kg)
Mouse	Male	Intravenous	40.5	3.4
Rat	Male	Intravenous	18.6	2.6
	Female	Intravenous	35.8	9.6
	Male	Intraperitoneal	71.8	4.9
	Female	Intraperitoneal	57.5-100.6	7.0-17.2

NOTE: $n = 10$. MTD = maximum tolerated dose.

35.8 mg. When the drug was given intraperitoneally, the LD_{10} was 71.8 mg for male rats. In female rats, LD_{10} of ip injection could not be determined clearly. It ranged from 57.5 to 100.8 mg/kg. From these data the maximum tolerable dose was estimated for humans: it was estimated at about 2.6 mg/kg for males and 9.6 mg/ for females in iv use.[9]

According to the results of acute toxicity, consecutive injections were given over 28 days. Rats were divided into four groups consisting of 10 rats in each group, and 2.5, 5.0, and 10 mg/kg lipo-Δ^7-PGA1 and PG free lipid microsphere solution were injected intravenously. In the 10 mg/kg group half or more of the rats showed some signs of toxicity, such as decreased locomotion and tachypnea; however, the effects were transient, disappearing within 20 min. In this experiment the observed responses to the drug were very similar between male and female rats. During the 28 days of the consecutive injection period, rats gained their body weight normally. Only a slight suppression of weight increase was observed in the 10 mg/kg group; however, there was no significant difference between control and this group. This subchronic toxicology data suggests that 10 mg/kg is the maximum tolerable dose for consecutive administration in rats. *In vivo* antitumor activity of lipo-Δ^7-PGA1-Me was studied at this dose, using human cancer xenograft in nude mice.

In past experiments we have shown a significant antitumor effect of lipo-Δ^7-PGA1-Me using human ovarian cancer models.[10,11] Here we describe the effect of this compound on another type of tumor derived from human cancer. When a subcutaneously inoculated 2-mm^3 fragment of the tumor formed a palpable mass, lipo-Δ^7-PGA1-Me was injected iv once a day at 10 mg/kg for 10 days. In this therapeutic experimental system, even conventional clinically useful anticancer drugs failed to show activities. Lipo-Δ^7-PGA1-Me failed to show any activity against HT-29 human colon cancer and A-375 and G361 human melanoma. However, Hela S3 human cervical cancer responded well to 10 days of consecutive injection of 10 mg/kg of the drug. In this experiment, 10 mg/kg of 5-fluorouracil did not show any antitumor effect in the same dosing schedule. Lovo human colon cancer also showed some response to this dose and schedule of the drug. In the case of Hela S3 at day 16 T/C was 33.0%, and at day 27 it had reached 56.3%. Any additive effect with 5-fluorouracil was not obtained in these experimental conditions.

In order to determine the optimal dosing schedule of lipo-Δ^7-PGA1-Me we examined the stability of Δ^7-PGA1-Me and Δ^7-PGA1 in serum from rat and human. Unexpectedly, Δ^7-PGA1 decayed rapidly in the serum, with a half-life of 87 min in human serum and 31.6 min in rat serum. Interestingly, the half-life of methylester of Δ^7-PGA1 was shorter than that of Δ^7-PGA1 itself: in rat serum it disappeared within a min, whereas in human it decayed with a half-life of 13.9 min. However, Δ^{12}-PGJ2 was very stable in the serum over hours in the same conditions. Serum ultrafiltrate, cut off at a molecular mass of 5 kDa, lost catalytic activity for Δ^7-PGA1 and its methylester. Five percent human serum albumin restored the catalytic activity partially. In analbuminemia rat serum, the decay of Δ^7-PGA1 slowed down to about one-third of the half-life of normal rat serum; however, analbuminemia rat serum still has the catalytic activity of Δ^7-PGA1 methylester. These data suggest that albumin is partly responsible for the metabolism of Δ^7-PGA1, which may not be the case for its methylester. HPLC analysis data suggests that a possible metabolite is

Δ^7-PGC1; the metabolic profile of Δ^7-PGA1 methylester to Δ^7-PGA1 and to Δ^7-PGC1 differs between human and rat serum.

In conclusion, first, lipo-Δ^7-PGA1-Me showed a significant *in vivo* antitumor effect against several human cancer xenografts in nude mice. Second, the maximum tolerable dose for humans was estimated from acute and subchronic toxicological results; however, the dose-limiting toxicity has not yet been determined. Third, Δ^7-PGA1-Me is metabolized possibly to Δ^7-PGC1, a type that is inactive against tumors. The half-life of Δ^7-PGA1-Me in human serum is 13 min. Based on the above-described results, we are planning to conduct a phase I clinical trial.

REFERENCES

1. FUKUSHIMA, M. 1990. Prostaglandin J2, anti-tumor and anti-viral activities and the mechanisms involved. Eicosanoids **3:** 189-199.
2. FUKUSHIMA, M. 1992. Biological activities and mechanisms of action of PGJ2 and related compounds. Update. Prostaglandins Leukotrienes Essent. Fatty Acids **47:** 1-12.
3. YAMAMOTO, N., T. TSURUMI, M. FUKUSHIMA, K. MAENO & Y. NISHIYAMA. 1987. Mechanism of inhibition of herpes simplex virus replication by Δ^7-prostaglandin A1 and Δ^{12}-prostaglandin J2. Biochem. Biophys. Res. Commun. **146:** 1425-1431.
4. KOSHIHARA, Y., R. TAKAMORI, K. NOMURA, S. SUGIURA & S. KUROZUMI. 1991. Enhancement of in vitro mineralization in human osteoblasts by a novel prostaglandin A1 derivative TEI-3313. J. Pharmacol. Exp. Ther. **258:** 1120-1126.
5. KHAN, S. H. & S. SOROF. 1990. Preferential binding of growth inhibitory prostaglandins by a target protein of a carcinogen. Proc. Natl. Acad. Sci. USA **87:** 9401-9405.
6. KIM, I-K., J-H. LEE, H-W. SOHN, H-S. KIM & S-H. KIM. 1993. Prostaglandin A2 and Δ^{12}-prostaglandin J2 induce apoptosis in L1210 cells. FEBS Lett. **321:** 209-214.
7. SASAKI, H., Y. TERASHIMA, M. FUKUSHIMA, A. HAZATO & S. KUROZUMI. 1993. Cell cycle arrest at G1 phase by antitumor prostaglandins The Cell Cycle 93. Regulation, Targets, and Clinical Applications. XIIIth International Spring Symposium. Washington, D.C. Abstr. 146.
8. NOYORI, R. & M. SUZUKI. 1993. Organic synthesis of prostaglandins: Advancing Biology. Science **259:** 44-45.
9. FUKUSHIMA, S., Y. TAKEUCHI, M. NAKANO, T. KITA, Y. KIKUCHI & M. FUKUSHIMA. 1993. Δ^7-PGA1 lipid microspheres—pharmaceutical properties and preclinical evaluations. *In* Advances in Polyunsaturated Fatty Acids Research. Y. Yasugi *et al.*, Eds.: 67-70. Elsevier. Tokyo.
10. SASAKI, H., K. TAKEDA, Y. TERASHIMA, H. EKIMOTO, K. TAKAHASHI, T. TSURUO & M. FUKUSHIMA. 1991. Human ovarian cancer cell lines resistant to cisplatin, doxorubicin and L-phenylalanine mustard are sensitive to Δ^7-prostaglandin A1 and Δ^{12}-prostaglandin J2. Gynecol. Oncol. **41:** 36-40.
11. KIKUCHI, Y., T. KITA, M. MIYAUCHI, J. HIRATA, H. SASA, I. NAGATA & M. FUKUSHIMA. 1992. Adjuvant effects of antineoplastic prostaglandins to cisplatin in nude mice bearing human ovarian cancer cells. J. Cancer Res. Clin. Oncol. **118:** 453-457.

Cell-Cell Interactions in Lipoxin Generation and Characterization of Lipoxin A_4 Receptors[a]

CHARLES N. SERHAN,[b] STEFANO FIORE, AND BRUCE D. LEVY

Hematology-Oncology Division
Department of Medicine
Brigham and Women's Hospital and
Harvard Medical School
Boston, Massachusetts 02115

Lipoxins are tetraene-containing eicosanoids that display potent biological actions.[1-3] They are generated in microgram quantities by a few individual cell types alone, such as activated macrophages of rainbow trout,[4] or in nanogram levels by transcellular biosynthetic routes documented during cell-cell interactions,[3] in particular, those observed during human neutrophil-platelet and eosinophil transcellular biosynthesis of eicosanoids.[3,5,6] Their formation in human cell types can be triggered by either initial oxygenation of arachidonic acid by 5-lipoxygenase followed by interactions with 12-lipoxygenase[5] or by 15-lipoxygenation and interaction with 5-lipoxygenase and subsequent reactions.[3,6] One member of this series, lipoxin A_4 (LXA_4), has been documented *in vivo* in bronchoalveolar lavage fluid from patients

[a] This work was supported in part by National Institutes of Health Grant GM 38765 and the Pew Scholars Program. C.N.S. is the recipient of an Established Investigator Award from the American Heart Association and the Clifford M. Clarke Science Award from the Arthritis Foundation. S.F. is the recipient of the 1992 Frederic C. McDuffie Fellowship from the Arthritis Foundation.

[b] Address for correspondence: Charles N. Serhan, Ph.D., Brigham and Women's Hospital, Longwood Medical Research Center 611, 221 Longwood Avenue, Boston, Massachusetts 02115.

ABBREVIATIONS:

5-HETE	=	5(S)-hydroxy-(6E,8Z,11Z,14Z)-eicosatetraenoic acid
12-HETE	=	12(S)-hydroxy-(5Z,8Z,10E,14Z)-eicosatetraenoic acid
15-HETE	=	15(S)-hydroxy-(5Z,8Z,11Z,13E)-eicosatetraenoic acid
5(S),6(R)-DiHETE	=	5(S),6(R)-dihydroxy-(7E,9E,11Z,14Z)-eicosatetraenoic acid
$iLXA_4$	=	immunoreactive LXA_4
LTA_4	=	5(S)-(E)-5,6-oxido-(7E,9E,11Z, 14Z)-eicosatetraenoic acid
LTB_4	=	5(S),12(R)-dihydroxy-(6Z,8E,10E,14Z)-eicosatetraenoic acid
LTC_4	=	5(S)-hydroxy-6(R)-S-glutathionyl-(7E,9E,11Z, 14Z)-eicosatetraenoic acid
LTD_4	=	5(S)-hydroxy-6(R)-S-cysteinyl-glycyl-(7E,9E, 11Z,14Z)-eicosatetraenoic acid
LTE_4	=	5(S)-hydroxy-6(R)-S-cysteinyl-(7E,9E,11Z,14Z)-eicosatetraenoic acid
LXA_4	=	5(S),6(R),15(S)-trihydroxy-(7E,9E,11Z,13E)-eicosatetraenoic acid
LXB_4	=	5(S),14(R),15(S)-trihydroxy-(6E,8Z,10E,12E)-eicosatetraenoic acid

with pulmonary diseases,[7] is generated by coronary artery components in patients undergoing balloon angioplasty,[8] and is detected following activation of human bone marrow *ex vivo*.[9]

Lipoxin A_4 is active with some cell types in the nanomolar range,[9–11] suggesting that it can play important regulatory roles. It both stimulates vasodilation and can counter the proinflammatory actions of leukotriene (LT)D_4[12] and LTB_4[13] observed in animal *in vivo* models. Lipoxin A_4 has also recently been shown to modulate LTC_4-induced airway obstruction after inhalation by asthmatic patients.[14] Human neutrophils (PMN) exposed (15 min, 37°C) to LXA_4 ($IC_{50} \approx 10^{-8}$ M) *in vitro* are less responsive when challenged with either the chemotactic peptide (fMLP) or LTB_4.[10,11] In addition, LXA_4 (10^{-10} M) stimulates colony formation in human marrow cells treated with suboptimal concentrations of granulocyte-macrophage colony-stimulating factor (GM-CSF).[9] Results of these studies suggest that lipoxins may represent regulatory molecules in cellular events ranging from those involved with human myelopoiesis to pulmonary inflammation.

As is the case with other bioactive eicosanoids, it is critical to have rapid and sensitive methods available for detection of LXA_4 to assess the role of lipoxins in both physiological and pathophysiological events. To this end, we developed a solid-phase, enzyme-linked immunosorbent assay that is capable of monitoring LXA_4 in both *in vivo*- and *in vitro*-derived fluids.[15] Human PMN exposed to LXA_4 rapidly remodel (within 5-15 s) their membrane lipids, an event that is pertussis toxin sensitive.[16] These cells possess specific recognition sites for LXA_4 (K_d = 0.5 nM) that may mediate its actions.[17] In this communication, we review evidence on the formation of LXA_4 in a cell-cell interaction model using PMN-platelet coincubations and induction of specific [11,12-^{3}H]LXA_4 binding sites that confers functional responses in HL-60 cells.[15,18]

MATERIALS AND METHODS

Materials

fMLP, A23187, nordihydroguaiaretic acid (NDGA) and natural human PDGF-AB were purchased from Sigma Chemical Co. (St. Louis, MO), and thrombin and LSM were from Enzyme Research Laboratories (South Bend, IN). GM-CSFrh was purchased from Boehringer Mannheim (Indianapolis, IN). Synthetic lipoxins were obtained from Cascade Biochem Limited (Reading, Berkshire, England). HPLC grade solvents were purchased from Doe and Ingalls (Medford, MA), methyl formate was from Eastman Kodak Co. (Rochester, NY), and Sep-Pak C_{18} cartridges were from Waters Associates (Milford, MA).

PLT:PMN Incubations

Freshly isolated PMN (as in Refs. 17,18) were incubated in Dulbecco's phosphate-buffered saline (PBS) containing both $CaCl_2$ (0.6 mM) and $MgCl_2$ (1.0 mM; pH 7.4) for 90 minutes at 37°C in the presence or absence of GM-CSFrh (200 pM). Aliquots (0.5 ml) of each cell suspension (10-30 × 10^6 PMN/ml) were combined with aliquots (0.5 ml) of isolated PLT suspensions with the cell ratio adjusted from 0:1 to 100:1 (PLT:PMN). PMN and PLT were warmed (5 min, 37°C) before exposure to soluble stimuli (20 min, 37°C). Incubations were stopped by the addition of 1 volume of 0.05 M Tris-HCl (pH 8.0) and vigorous mixing. Materials were kept at −30°C for at least 30 minutes prior to analysis. LXA_4 enzyme-linked immunosorbent assay (LXA_4-ELISA) was prepared and used as described.[15]

Cell Cultures and Differentiation

HL-60 cells were seeded in RPMI medium supplemented with 100 U/ml penicillin, 100 μg/ml streptomycin, and 10% fetal calf serum (Hyclone, Logan, UT), incubated at 37°C with 5% CO_2 atmosphere in 250-ml flasks. Individual flasks containing ~50 × 10^6 cells per flask next received dimethyl sulfoxide (DMSO, 1.12% v/v, 120 h), retinoic acid (RA, 1 μM, 120 h), or phorbol myristate acetate (PMA, 20 nM, 48 h). Before performing binding assays, cells were washed twice in PBS, Ca^{2+} and Mg^{2+} free, enumerated, and suspended at 20 × 10^6 cells/ml in Tris buffer (10 mM), pH 7.4. Nitro blue tetrazolium reduction was performed to monitor induction of polymorphonuclear phenotype as described,[19] and cell adherence was determined for induction of macrophage-like phenotype.[20] Human umbilical vein endothelial cells (HUVEC) maintained in culture third passage were obtained from Dr. M. Gimbrone (Brigham and Women's Hospital Department of Pathology).

Ligand Binding

^{3}H-LXA_4 and ^{3}H-LTB_4 binding were performed.[17,18] After suspending cells in Tris buffer 10 mM (pH 7.4, Ca^{2+} 2.5 mM, Mg^{2+} 1.2 mM), aliquots (0.5 ml) were incubated (20 min at 4°C) with ^{3}H-ligands alone (0.3 nM) or in the presence of increasing concentrations of homoligands or other compounds (3-300 nM). The incubations were rapidly centrifuged (60 s, 12,000 *g*) on silicon oil (d = 1.028), and cell-associated radioactivity was determined by liquid scintillation counting (Wallac 1409, Pharmacia, Piscataway, NJ). Binding experiments with HUVEC cells were performed in 12 well plates with 3.5 × 10^5 cells/well. After 10 min, wells were washed twice with PBS, and cell-associated label was recovered by adding glacial acetic acid (0.5 ml). Results obtained from these assays were submitted for further analysis using the Ligand program (Elsevier-Biosoft, Cambridge, UK).

RESULTS AND DISCUSSION

ELISA Sensitivity and Selectivity

After conditions for the enzyme immunoassay of LXA_4 were optimized, calibrated dose-response curves were derived using serial dilutions of competing unlabeled synthetic LXA_4.[15] Inhibition of 50% of the maximal binding (IC_{50}) observed with enzyme-labeled LXA_4 occurred with 0.45 pmol LXA_4/ml, and the minimum detectable amount (80% of maximal binding) was 90.8 fmol LXA_4/ml. This IC_{50} is comparable to those obtained by ELISA for other eicosanoids: 0.4 pmol[21] and 0.2 pmol/ml for LTC_4,[22] 0.49 pmol/ml for 6-keto-$PGF_{1\alpha}$,[23] 1 pmol/ml for 13,14-dihydro-15-keto-$PGF_{2\alpha}$,[24] and 0.54[25] and 0.4 pmol/ml for TxB_2.[22,25] Cross-reactivity of the antibody for related compounds was 0.1% for 15S-HETE, and < 0.1% for 5S-HETE, 12S-HETE, LTB_4, LTC_4, $LTD_{,4}$, LTE_4 and C20:4.[15] Further experiments on the stereoselectivity of antibody binding employed lipoxygenase products with structures identical to portions of native LXA_4, namely 5(S),6(R)-DiHETE (C_1-C_6) and LXB_4 (C_1-C_5 and C_{15}-C_{20}). Cross-reactivity for these closely related eicosanoids was < 5% for 5(S),6(R)-DiHETE and between 1-7.5% for LXB_4. Thus, the antiserum proved to be both sensitive and selective for LXA_4.

Coincubations of autologous human PLT and PMN in the presence of stimuli have been demonstrated by both HPLC and GC/MS to generate LXA_4 from endogenous sources.[5] For purposes of direct comparison to prior results and to evaluate the ELISA's use in the determination of LXA_4 generation by cellular sources, GM-CSFrh-primed PMN (15×10^6) were coincubated with PLT (150×10^6) in the presence of A23187 (10^{-7} M) and thrombin (1 U/ml). After the coincubations, serial dilutions of the materials generated were performed, and the quantities of LXA_4 present were determined using both ELISA and ED-UV RP-HPLC, which gave a correlation coefficient of 0.99.[15] In addition, the quantities of $iLXA_4$ identified are in agreement with those reported using HPLC-based quantitation[5,26,27] (reviewed in Ref. 3) and indicated that lower cell numbers than those required for analysis by HPLC and GC/MS can be used to evaluate LXA_4 in each incubation setting.

Characteristics of $iLXA_4$ Generation during PLT-PMN Interactions

We used the LXA_4-ELISA assay to examine factors regulating LXA_4 generation by suspensions of human PMN coincubated with increasing numbers of autologous PLT (FIG. 1). In previous experiments we found that platelets incubated alone without a second cell type do not generate lipoxins, but, when exposed to cells that could provide a source of LTA_4, they convert it to lipoxins.[5,26] When in contact with LTA_4-emitting cells, platelets become a substantial source of lipoxins. Bidirectional routes of transcellular leukotriene and lipoxin formation have been demonstrated using human platelets and neutrophils (PLT:PMN) as a model system. These cell types are readily accessible, and they each possess the enzymes of interest. Human platelets contain a 12-lipoxygenase, human leukocytes possess both 5- and 15-lipoxygenase, and human neutrophils in particular display substantial 5-lipoxygenase activity when exposed to stimuli. Platelets also contribute to LX formation by donating arachidonate to leukocytes that is converted to LTA_4 by the 5-LO and further to LX by platelets; or platelet-derived arachidonate can be converted to 15-H(P)ETE that is further transformed to lipoxins.[5,26] Thus, in view of previous results with labeled precursors, the platelet-PMN coincubation was a useful model to evaluate and extend the utility of the LXA_4-ELISA to screen additional agonists.

Since GM-CSFrh is known to increase arachidonic acid availability and LXA_4 generation,[26] the impact of adding this cytokine on LXA_4 generation was examined at different ratios of PLT:PMN. At each cell ratio examined (FIG. 1), PMN primed with GM-CSFrh (200 pM) generated higher levels of $iLXA_4$ when exposed (20 min,

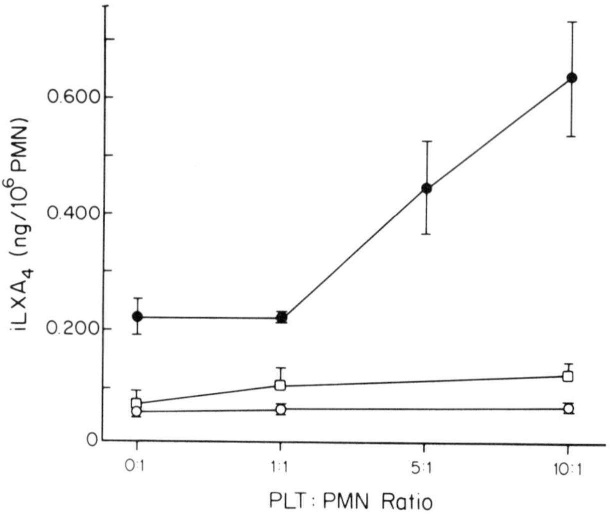

FIGURE 1. Generation of $iLXA_4$ from endogenous sources during coincubations of PLT-PMN: Effect of cell ratio (PLT:PMN) and GM-CSFh priming of PMN. PMN ($8-10 \times 10^6$) were incubated in the presence of GM-CSFrh (200 pM) for 90 min at 37°C, and then combined with autologous PLT. After warming (5 min, 37°C), cells were exposed to either A23187 (10^{-7} M) and thrombin (1 U/ml) (●) or vehicle control (○) for 20 minutes at 37°C. In parallel, PMN, not primed with GM-CSFrh, were coincubated with PLT and exposed to the soluble stimuli (*open squares*). Incubations were stopped with 1 volume of cold, 0.05 M Tris-HCl (pH 8.0). ELISA was performed without further purification of products (n = 3 separate donors, mean ± SEM). (Reprinted from Levy *et al.*[15] with the permission of American Oil Chemists' Society Press.)

37°C) to A23187 (10^{-7} M) and thrombin (1 U/ml). As the PLT:PMN ratio was increased to 10:1, GM-CSFrh-primed PMN led to generation of 5-6 times more LXA_4 than activated PMN that were not exposed to GM-CSF. At higher PLT number, homotypic PLT-PLT interactions led to decreased LXA_4 generation. Also, in the presence of stimuli (A23187 or fMLP plus thrombin), neither PMN incubated alone (without priming) nor isolated PLT (data not shown) generated $iLXA_4$ in statistically significant amounts greater than incubations performed in the absence of stimuli. Primed PMN in the presence of stimuli, but without deliberate addition of PLT, did generate nanomolar quantities (~3 nM $iLXA_4$/incubation) of $iLXA_4$. Because limitations in the isolation of PMN resulted in ~1-2 PLT per PMN contaminating these suspensions, we were unable to assess the amount of $iLXA_4$ produced by PMN alone without platelets present. Intentional addition of PLT dramatically increased formation of $iLXA_4$ (FIG. 1) to levels that can elicit a range of bioactions (reviewed in Ref. 3), including the stimulation of myelopoiesis.[9] These findings are consistent with production of leukotriene A_4 by activated PMN that is released and transformed to LXA_4 by platelet 12-lipoxygenase.[5,26] Additionally, stimulus-induced formation of $iLXA_4$ was maximal during coincubations that were held motionless; shaking the coincubations reduced $iLXA_4$ formation (>83%, data not shown). Thus, interference with PLT-PMN interactions impairs the transcellular exchanges of biosynthetic intermediates, such as LTA_4. Taken together these results, obtained using independent methodologies, both confirm and extend the findings that activated cytokine-primed PMN interacting with PLT generate biologically relevant quantities of LXA_4.[26]

PDGF promotes activation of PMN.[28,29] To determine if this cytokine also stimulates LXA_4 generation, GM-CSFrh-primed PMN were coincubated with autologous PLT in the presence of PDGF and the chemotactic peptide fMLP, which has been previously shown to activate PMN for transcellular biosynthetic generation of LXA_4.[5,26] Indeed, cells exposed to either fMLP (10^{-7} M) or PDGF-AB (10 ng/ml) generated more $iLXA_4$ from endogenous sources of substrates than unstimulated cells (FIG. 2). When exogenous arachidonic acid was added to these incubations, increases in $iLXA_4$ levels were obtained upon exposure to either fMLP (410%, $p < 0.02$) or PDGF (381%, $p < 0.09$). Agonist-induced formation of $iLXA_4$ from endogenous sources of substrate proved to be both concentration and time dependent (FIG. 2, upper and lower panels). PDGF (10 ng/ml) and fMLP (10^{-6} M) plus thrombin (1 U/ml) led to approximately equivalent levels of $iLXA_4$ (FIG. 2, upper panel). In addition, the divalent cation ionophore A23187 led to a concentration-dependent (10^{-8}-10^{-5} M) increase in $iLXA_4$ (0.004-2.340 ng $iLXA_4/10^6$ PMN). With each stimulus examined, $iLXA_4$ generation was rapid in onset and statistically significant ($p < 0.05$) within 60 seconds (FIG. 2, lower panel). Exposure to NDGA (100 μM), a commonly used nonselective LO inhibitor that blocks 5-, 12- and 15-LO, led to complete inhibition of $iLXA_4$ formation during coincubations (cells from two separate donors with four determinations). These results demonstrate that $iLXA_4$ is triggered by either fMLP or PDGF, indicating that naturally occurring ligands can stimulate the formation of lipoxins using $iLXA_4$ as an index to monitor lipoxin production from endogenous sources of arachidonic acid.

Induction of LXA_4 Receptors

After preparation of synthetic [11,12-^{3}H]LXA_4, we characterized its specific binding to PMN[17] and promyelocytic cells (HL-60).[18] In both series of experiments, direct comparisons were carried out with specific binding of [14,15-^{3}H]LTB_4. Human neutrophils do not metabolize LXA_4 at either 4° or 37°C.[17] This is unlike LTB_4, which is rapidly converted to ω-oxidation products by human PMN. When routine phenotypic markers were monitored,[18,20] untreated HL-60 cells displayed a low level of specific binding for both ^{3}H-LXA_4 and ^{3}H-LTB_4 ligands (FIG. 3). Differentiation induced by exposure to either DMSO (1.12%) or retinoic acid (1 μM) for 5 days (FIG. 3) was accompanied with a three- to fivefold increase in specific binding for both radioligands. PMA-treated cells (20 nM, 48h) displaying characteristics of macrophagic-like phenotype—that is, NBT-negative cells that adhere to plastic,[19,20] also led to the appearance of specific binding with both ^{3}H-LXA_4 and ^{3}H-LTB_4 (FIG. 3). Equilibrium binding with ^{3}H-LXA_4 at 4°C was reached at 10 min, which remained virtually unchanged for the next 20 min.[18] To assess whether induction of specific binding for both ^{3}H-ligands required *de novo* synthesis, actinomycin D (2 μg/ml) was added with PMA incubations. This agent blocked the PMA-induced increments in specific binding (FIG. 3) for both labeled eicosanoids, suggesting that inhibition

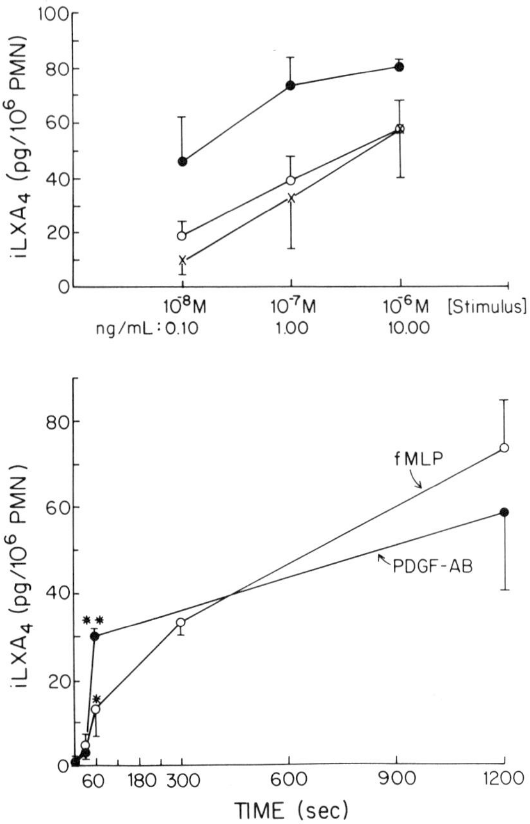

FIGURE 2. Generation of iLXA$_4$ from endogenous and exogenous arachidonic acid during coincubations of GM-CSFrh-primed PMN with autologous PLT. PMN (10×10^6) were exposed (90 min, 37°C) to 200 pM GM-CSFrh and then incubated (20 min, 37°C) with autologous PLT (100×10^6) in the presence of stimuli. Incubations were terminated and assayed as described in METHODS. Values are expressed in pg/10^6 PMN (mean ± SEM, $n = 3$) minus values obtained without stimuli present in parallel incubations. **Upper panel:** Concentration response. Incubations were carried out in the presence of PDGF-AB reported in ng/ml (x), fMLP alone (●) or fMLP and thrombin (○). **Lower panel:** Time course. Incubations were terminated at the indicated time points after exposure to either 10^{-7} M fMLP (○) or 10 ng/ml PDGF-AB (●) ($*p < 0.05$, $**p < 0.01$). (Reprinted from Levy *et al.*[15] with the permission of American Oil Chemists' Society Press).

of *de novo* protein biosynthesis also blocked appearance of specific binding sites. The impact of protease and glycosidase treatments was assessed with both differentiated HL-60 cells and human PMN for ^{3}H-LXA$_4$-specific binding. Protease treatment reduced specific binding and provided additional evidence in support of a protein component of LXA$_4$-specific binding sites.[18] Thus, specific binding of ^{3}H-LXA$_4$ was sensitive to protease and glycosidase treatment with both PMN[17] and HL-60 cells.[18]

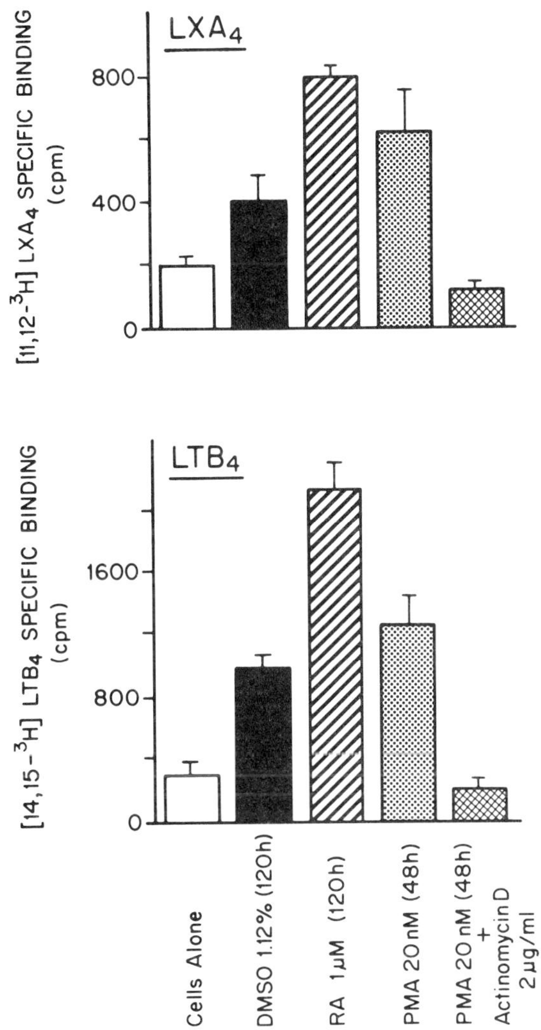

FIGURE 3. Induction of [11,12-^{3}H]LXA$_4$- and [14,15-^{3}H]LTB$_4$-specific binding to HL-60 cells. HL-60 cells were cultured in the presence or absence of DMSO (1.12%, 120 h), retinoic acid (1 μM, 120 h), or PMA (20 nM, 48 h). Binding assays were performed in Tris buffer, pH 7.4 with 10^7 cells (0.5 ml aliquots, 4°C for 20 min) with either [11,12-^{3}H]LXA$_4$ (0.3 nM) (**upper panel**), or [14,15-^{3}H]LTB$_4$ (**lower panel**). HL-60 cells exposed to PMA (20 nM, 48 h) and actinomycin D (2 μg/ml) were used in the *last column*. Results represent the mean ± SEM from three separate experiments. (Reprinted from Fiore *et al.*[18] with the permission of W. B. Saunders Co.)

TABLE 1. Cell Types Displaying [11,12-^{3}H]LXA$_4$-Specific Binding

Cell Type	Cell Number	K_d (nM)	B_{max} (M)	% Association of Added Label
HL-60	10^7/0.5 ml	0.58 ± 0.3	2.8×10^{-10}	8.8
HUVEC	3×10^5/0.5 ml	11.0 ± 2.6	2.5×10^{-10}	5.8
Platelets	2.5×10^8/0.5 ml	—	—	13.6
Red blood cells	10^7/0.5 ml	—	—	8.2
Raji	5×10^6/0.5 ml	—	—	6.4
Jurkat	5×10^6/0.5 ml	—	—	4.3

NOTE: Aliquots (0.5 ml) of cells suspended in Tris buffer were incubated for 20 min at 4°C with ^{3}H-LXA$_4$ (0.3 nM). HL-60 cells were differentiated by exposure to retinoic acid as in FIGURE 3. The label was competed with excess unlabeled LXA$_4$ (3-300 nM range) to determine specific binding. Results represent the mean ± SD of at least three separate experiments with duplicate determinations in each.

Results from isotherm binding assays (4°C, 10 min) with differentiated HL-60 cells and [11,12-^{3}H]LXA$_4$ (0.1-30 nM) showed that [11,12-^{3}H]LXA$_4$-specific binding sites gave a K_d = 0.6 ± 0.3 nM (FIG. 4). A nonlinear portion of the Scatchard plot was observed for concentrations < 0.2 nM. This type of nonlinearity from a Scatchard analysis observed with both HL-60 cells and PMN[17] could, at 37°C, implicate conversion of labeled LXA$_4$. However, PMN do not metabolize LXA$_4$ at either 4° or 37°C.[17] Our results obtained with LTB$_4$-specific binding and HL-60 cells, K_d = 0.12 nM, were essentially in agreement with values recently reported by Harada,[30] namely, K_d = 0.23 nM. Differentiated HL-60 cells convert LXA$_4$ to 15-oxo and dihydro derivatives as do human monocytes (37°C), but these transformations are negligible at 4°C and thus do not appear to alter binding results obtained at 4°C.[31]

Next, competition binding experiments were performed with differentiated HL-60 cells to further characterize the interactions of ^{3}H-LXA$_4$ with its specific binding sites. LXA$_4$, LXB$_4$, LTB$_4$, LTC$_4$, and the leukotriene receptor antagonists SKF 104353 (LTD$_4$ antagonist) and ONO-4057 (LTB$_4$ antagonist; see Ref. 18) were assessed as potential competing ligands. Neither LXB$_4$, LTB$_4$, nor trihydroxyheptanoic acid (methyl ester, 300 nM) displaced ^{3}H-LXA$_4$-specific binding with differentiated HL-60 cells; while LTC$_4$ caused ~30% decrease of specific binding when added in 3 log molar excess. The finding that LXA$_4$ (300 nM) was unable to compete ^{3}H-LTB$_4$ (0.3 nM) binding to differentiated HL-60 cells suggests that LXA$_4$ and LTB$_4$ interact with separate classes of specific binding sites.

Leukotriene receptor antagonists SKF 104353 and ONO-4057 did not displace ^{3}H-LXA$_4$ binding with differentiated HL-60 cells. In contrast, SKF 104353 and LTD$_4$ were effective in competing the specific ^{3}H-LXA$_4$ binding with HUVEC (FIG. 5). HUVEC displayed a K_d of 11.0 ± 2.6 nM and a B_{max} of 2.5×10^{-10} M for ^{3}H-LXA$_4$, and virtually identical values were calculated for LTD$_4$ competition. In the case of ^{3}H-LTB$_4$, HUVEC did not specifically bind LTB$_4$, but nonspecific cell association was evident with this ^{3}H-ligand (n = 3). In addition, specific binding of ^{3}H-LXA$_4$ was not evident with several other cell types surveyed, including washed platelets,

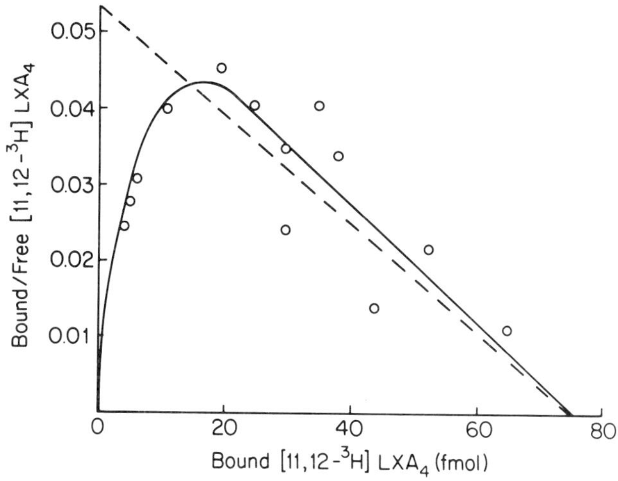

FIGURE 4. Scatchard analysis of [11,12-^{3}H]LXA$_4$-specific binding. Results from isotherm binding assays (4°C, 10 min) with RA (1 μM, 120 h)-differentiated HL-60 cells (5 × 10^6/0.5 ml) and [11,12-^{3}H]LXA$_4$ (0.1-30 nM) were analyzed with the program Ligand. For concentrations > 0.25 nM, the nonlinear analysis (least square) of ^{3}H-LXA$_4$ data indicated a high-affinity binding site with K_d = 0.6 ± 0.3 nM with B_{max} ~2.8 × 10^{-10}M. Statistical significance as given by the Ligand program (sum of squares = 970.8, mean square 138.7) was reached with <6 runs with random point distribution ($p > 0.5$). (Reprinted from Fiore *et al.*[18] with the permission of W. B. Saunders Co.)

RBCs, the B cell (Raji), and T cell (Jurkat) cultured cell lines (TABLE 1). Taken together, these results indicate that LXA$_4$ interacts with unique binding sites in differentiated HL-60 cells that are not sensitive to either leukotriene receptor antagonists (SKF 104353 or ONO-4057). In HUVEC, ^{3}H-LXA$_4$-specific binding was sensitive to both LTD$_4$ and SKF 104353 but not ONO-4057, suggesting that ^{3}H-LXA$_4$-specific binding in endothelial cells may reflect its interaction with putative LTD$_4$ receptors or a subtype of these receptors.

LXA$_4$ rapidly stimulates lipid remodeling and phosphatidic acid (PA) formation by human neutrophils.[16] To determine if ^{3}H-LXA$_4$ binding confers PLD activation, phosphatidylethanol (PEt) and PA were monitored in both PMN and HL-60 cells. Results in FIGURE 6 indicate that LXA$_4$ stimulates PLD activity in these cell types with similar temporal responses. PMN exposed to LXA$_4$ (10^{-10} M) rapidly generated the ethanol trapping product PEt[18] within 60 s, which declined to baseline levels by 5 min. In the absence of added EtOH, PEt was not formed in statistically significant amounts. A biphasic concentration dependence was obtained for PEt formation in both PMN (upper insert) and differentiated HL-60 cells (lower insert). An apparent maximal response was noted with the concentration range of 10^{-9}-10^{-10} M for LXA$_4$, and a second peak of activity was observed at 10^{-7} M LXA$_4$. Below 10^{-8} M, both the chemotactic peptide fMLP and LXA$_4$ gave results of similar magnitude, with fMLP appearing to be slightly more potent with PMN from some donors. The potential contribution of other biosynthetic pathways was also evaluated, and LXA$_4$-induced

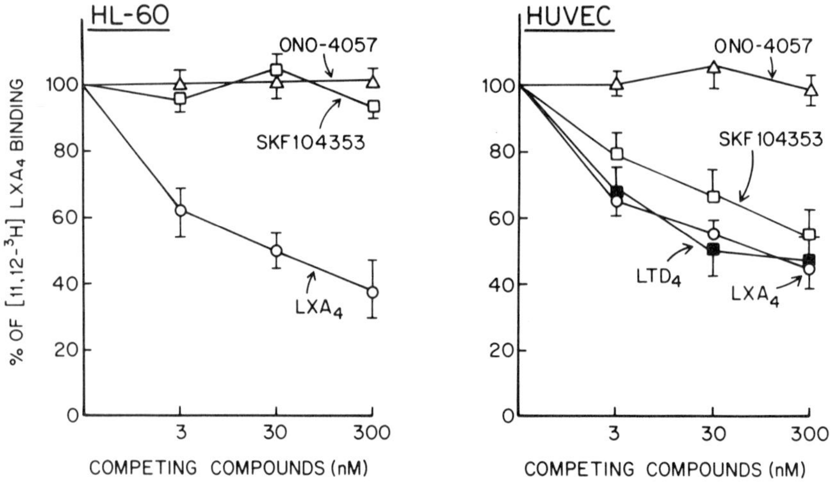

FIGURE 5. Competition of [11,12-^{3}H]LXA$_4$ binding by leukotriene receptor antagonists. RA-differentiated HL-60 cells (10^7/0.5 ml) or HUVEC (3.5×10^5/well) were incubated for 10 min at 4°C with ^{3}H-LXA$_4$ and increasing concentrations of either unlabeled LXA$_4$, ONO 4079, LTD$_4$, or SKF 104353 (each 3-300 nM). Results are the mean ± SD of three separate experiments with duplicate determinations in each. (Reprinted from Fiore *et al.*[18] with the permission of W. B. Saunders Co.)

PA formation was examined in both ^{32}PγATP and [1-^{14}C]-arachidonate-labeled PMN. ^{32}P-labeled PA was evident in statistically significant levels only after 30 min of exposure to LXA$_4$ (10^{-7} M). Similar results were obtained with ^{14}C-labeled PA formation derived from ^{14}C-arachidonate-labeled precursors.[18] These findings indicated that LXA$_4$ can also stimulate other routes of PA formation in PMN, but only after 30 min of exposure.

Only differentiated HL-60 cells (expressing specific binding sites for ^{3}H-LXA$_4$) incubated with LXA$_4$ rapidly generated PEt that was observed within 30 s. Undifferentiated HL-60 cells incubated with LXA$_4$ (10^{-9} M) did not rapidly generate PEt. The concentration dependence with these cells also gave a biphasic response with LXA$_4$ and gave an apparent maximum at 10^{-9} M. To evaluate possible signal transduction events involved in LXA$_4$-mediated PLD activation, PMN and HL-60 cells were exposed to either islet activating protein (IAP) or staurosporine. Results from these experiments indicated that the LXA$_4$-mediated PLD activity evoked within the lower concentration range (10^{-9}-10^{-10} M) was sensitive to IAP treatment in both cell types, and, similarly, the PLD activity stimulated at higher concentrations of LXA$_4$ (10^{-7} M) was inhibited by staurosporine.[18] Thus, in both PMN and HL-60 cells at concentrations below 10^{-8} M, LXA$_4$ rapidly interacts with specific binding sites that trigger PLD activity and hence confers a functional response, while within submicromolar concentrations of LXA$_4$, it may stimulate additional processes that can also lead to activation of PLD by different mechanisms.

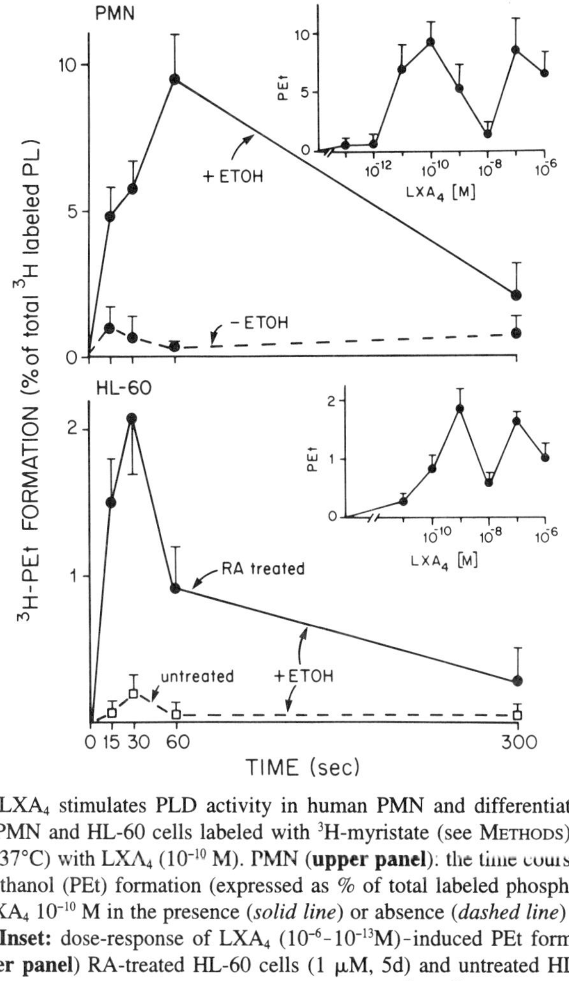

FIGURE 6. LXA_4 stimulates PLD activity in human PMN and differentiated HL-60 cells. Both human PMN and HL-60 cells labeled with ^{3}H-myristate (see METHODS) were incubated (10^7 cells/ml, 37°C) with LXA_4 (10^{-10} M). PMN (**upper panel**): the time course of ^{3}H-carrying phosphatidylethanol (PEt) formation (expressed as % of total labeled phospholipids) by cells exposed to LXA_4 10^{-10} M in the presence (*solid line*) or absence (*dashed line*) of ethanol 0.5% (v/v; n = 4). **Inset:** dose-response of LXA_4 (10^{-6}-10^{-13}M)-induced PEt formation (t = 60 s; n = 6). (**Lower panel**) RA-treated HL-60 cells (1 μM, 5d) and untreated HL-60 cells. **Inset** shows the dose-response of PEt formation with LXA_4 (10^{-6}-10^{-12}M, t = 60 s). In each panel, full-scale total labeled PL were equivalent to ~10,000 dpm. Results with HL-60 cells are the mean ± SD of three separate experiments. (Reprinted from Fiore *et al.*[18] with the permission of W. B. Saunders Co.)

CONCLUSIONS

To firmly establish a role for lipoxins in physiologic and pathophysiologic events, it is important to determine when and where they are generated as well as to delineate their modes of action. Here, we have reviewed recent evidence obtained with a new LXA_4-ELISA that demonstrates the formation of LXA_4 during cell-cell interactions triggered by natural ligands.[15] These results not only confirmed the formation of LX by PMN-platelet coincubation,[5,26,27] but extend our knowledge of known ligands that stimulate LX to include PDGF as well as GM-CSF-primed cells stimulated with either fMLP, thrombin, or the divalent cation ionophore. This new assay will also be useful in other models of eicosanoid transcellular metabolism and can be used to monitor LXA_4 in patient-derived fluids.[15] Its availability will permit further evaluation of the formation of LXA_4 *in vivo*. Aspirin-sensitive asthmatics generate LXA_4 after ingestion of aspirin. LXA_4 is detected in nasal lavages of these patients after taking a threshold dose of aspirin but not placebo.[15] Thus, in addition to angioplasty-triggered formation of LXA_4,[8] bronchoalveolar lavage fluids from pulmonary diseases,[7] and various cell types *in vitro*[4–6] (reviewed in Ref. 3), the present results give additional support to the notion that lipoxins are indeed generated *in vivo* by biosynthetic routes that can involve transcellular biosynthesis.

Once generated, LXA_4 can exert its actions in several organ systems.[3] LXA_4 has specific recognition sites on PMN[17] that are subject to G protein-type regulation. The sites are inducible in HL-60 cells, and binding with LXA_4 confers functional responses such as lipid remodeling.[18] The LXA_4 receptors induced in HL-60 cells appear to be specific for LXA_4. In contrast, those present on endothelial cells also compete with LTD_4, and thus may represent a subset or subtype of LTD_4/LXA_4 receptors. Thus, LXA_4 displays three modes of action that are cell type specific. They include: (1) LXA_4 binding to specific LXA_4 receptors (LXA_4-r) that transduce its signal and are likely to be a member of the superfamily of serpentine 7 spanner transmembrane receptors; (2) LXA_4 interacts with LTD_4 binding sites, for example on endothelial cells, that represent a receptor subtype or subset (i.e., LTD_4/LXA_4 receptors); and (3) LXA_4 can be taken up by specific LX carriers and act at intracellular sites[18,32] (reviewed in Ref. 3). The structural elucidation of these sites is in progress and will reveal the tissue display and potential additional sites of action. Thus, the present evidence indicates that LX are generated *in vivo* and can act at specific receptors to impact processes of interest in human disease.

REFERENCES

1. Serhan, C. N., M. Hamberg & B. Samuelsson. 1984. Lipoxins: Novel series of biologically active compounds formed from arachidonic acid in human leukocytes. Proc. Natl. Acad. Sci. USA **81:** 5335-5339.
2. Samuelsson, B., S.-E. Dahlén, J. Å. Lindgren, C. A. Rouzer & C. N. Serhan. 1987. Leukotrienes and lipoxins: Structures, biosynthesis, and biological effects. Science **237:** 1171-1176.
3. Serhan, C. N. 1991. Lipoxins: Eicosanoids carrying intra- and intercellular messages. J. Bioenerg. Biomembr. **23:** 105-122.
4. Pettitt, T. R., A. F. Rowley, S. E. Barrow, A. I. Mallet & C. J. Secombes. 1991. Synthesis of lipoxins and other lipoxygenase products by macrophages from the rainbow trout, Oncorhynchus mykiss. J. Biol. Chem. **266:** 8720-8726.
5. Serhan, C. N. & K.-A. Sheppard. 1990. Lipoxin formation during human neutrophil-platelet interactions. J. Clin. Invest. **85:** 772-780.
6. Edenius, C., M. Kumlin, T. Björk, A. Ånggård & J. Å. Lindgren. 1990. Lipoxin formation in human nasal polyps and bronchial tissue. FEBS Lett. **272:** 25-28.
7. Lee, T. H., A. E. G. Crea, V. Gant, B. W. Spur, B. E. Marron, K. C. Nicolaou, E. Reardon, M. Brezinski & C. N. Serhan. 1990. Identification of lipoxin A_4 and its relationship to the sulfidopeptide leukotrienes C_4, D_4, and E_4 in the bronchoalveolar

lavage fluids obtained from patients with selected pulmonary diseases. Am. Rev. Respir. Dis. **141:** 1453-1458.
8. BREZINSKI, D. A., R. W. NESTO & C. N. SERHAN. 1992. Angioplasty triggers intracoronary leukotrienes and lipoxin A_4: Impact of aspirin therapy. Circulation **86:** 56-63.
9. STENKE, L., M. MANSOUR, C. EDENIUS, P. REIZENSTEIN & J. Å. LINDGREN. 1991. Formation and proliferative effects of lipoxins in human bone marrow. Biochem. Biophys. Res. Commun. **180:** 255-261.
10. LEE, T. H., C. E. HORTON, U. KYAN-AUNG, D. HASKARD, A. E. G. CREA & B. W. SPUR. 1989. Lipoxin A_4 and lipoxin B_4 inhibit chemotactic responses of human neutrophils stimulated by leukotriene B_4 and N-formyl-L-methionyl-L-leucyl-L-phenylalanine. Clin. Sci. **77:** 195-203.
11. GRANDORDY, B. M., H. LACROIX, E. MAVOUNGOU, S. KRILIS, A. E. G. CREA, B. W. SPUR & T. H. LEE. 1990. Lipoxin A_4 inhibits phosphoinositide hydrolysis in human neutrophils. Biochem. Biophys. Res. Commun. **167:** 1022-1029.
12. BADR, K. F., D. K. DEBOER, M. SCHWARTZBERG & C. N. SERHAN. 1989. Lipoxin A_4 antagonizes cellular and in vivo actions of leukotriene D_4 in rat glomerular mesangial cells. Evidence for competition at a common receptor. Proc. Natl. Acad. Sci. USA **86:** 3438-3442.
13. HEDQVIST, P., J. RAUD, U. PALMERTZ, J. HAEGGSTRÖM, K. C. NICOLAOU & S.-E. DAHLÉN. 1989. Lipoxin A_4 inhibits leukotriene B_4-induced inflammation in the hamster cheek pouch. Acta Physiol. Scand. **137:** 571-572.
14. CHRISTIE, P. E., B. W. SPUR & T. H. LEE. 1992. The effects of lipoxin A_4 on airway responses in asthmatic subjects. Am. Rev. Respir. Dis. **145:** 1281-1284.
15. LEVY, B. D., S. BERTRAM, H. H. TAI, E. ISRAEL, A. FISCHER, J. M. DRAZEN & C. N. SERHAN. 1993. Agonist-induced lipoxin A_4 generation: Detection by a novel lipoxin A_4-ELISA. Lipids **28:** 1047-1053.
16. NIGAM, S., S. FIORE, F. W. LUSCINSKAS & C. N. SERHAN. 1990. Lipoxin A_4 and lipoxin B_4 stimulate the release but not the oxygenation of arachidonic acid in human neutrophils: Dissociation between lipid remodeling and adhesion. J. Cell. Physiol. **143:** 512-523.
17. FIORE, S., S. W. RYEOM, P. F. WELLER & C. N. SERHAN. 1992. Lipoxin recognition sites: Specific binding of labeled lipoxin A_4 with human neutrophils. J. Biol. Chem. **267:** 16168-16176.
18. FIORE, S., M. ROMANO, E. M. REARDON & C. N. SERHAN. 1993. Induction of functional lipoxin A_4 receptors in HL-60 cells. Blood **81:** 3395-3403.
19. IMAIZUMI, M. & T. R. BREITMAN. 1986. A combination of a T cell-derived lymphokine differentiation-inducing activity and a physiologic concentration of retinoic acid induces HL-60 to differentiate to cells with functional chemotactic peptide receptors. Blood **67:** 1273-1280.
20. COLLINS, S. J. 1987. The HL-60 promyelocytic leukemia cell line: Proliferation, differentiation, and cellular oncogene expression. Blood **70:** 1233-1244.
21. FRÖHLICH, J. C., M. SAWADA, G. BOCHMANN & O. OELZ. 1986. Advances in the analysis of eicosanoids by ELISA and GC/MS/MS. Adv. Prostaglandin Thromboxane Leukotriene Res. **16:** 363-372.
22. MILLER, D. K., S. SADOWSKI, D. DESOUSA, A. L. MAYCOCK, D. L. LOMBARDO, R. N. YOUNG & E. C. HAYES. 1985. Development of enzyme-linked immunosorbent assays for measurement of leukotrienes and prostaglandins. J. Immunol. Methods **8:** 169-186.
23. TONAI, T., K. YOKOTA, T. YANO, Y. HAYASHI, S. YAMAMOTO, K. YAMASHITA & H. MIYAZAKI. 1985. Enzyme immunoassay of 6-ketoprostaglandin $F_{1\alpha}$ in a solid phase. Biochim. Biophys. Acta **836:** 335-343.

24. Del Vecchio, R. P., K. M. Maxey & G. S. Lewis. 1992. A quantitative solid-phase enzyme immunoassay for 13,14-dihydro-15-keto-prostaglandin $F_{2\alpha}$ in plasma. Prostaglandins **43:** 321-330.

25. Shono, F., K. Yokota & S. Yamamoto. 1985. A solid-phase enzyme immunoassay of thromboxane B_2. J. Biochem. **98:** 1069-1077.

26. Fiore, S. & C. N. Serhan. 1990. Formation of lipoxins and leukotrienes during receptor-mediated interactions of human platelets and recombinant human granulocyte/macrophage colony-stimulating factor-primed neutrophils. J. Exp. Med. **172:** 1451-1457.

27. Edenius, C., J. Haeggström & J. Å. Lindgren. 1988. Transcellular conversion of endogenous arachidonic acid to lipoxins in mixed human platelet-granulocyte suspensions. Biochem. Biophys. Res. Commun. **157:** 801-807.

28. Shure, D., R. M. Senior, G. L. Griffin & T. F. Deuel. 1992. PDGF AA homodimers are potent chemoattractants for fibroblasts and neutrophils, and for monocytes activated by lymphocytes or cytokines. Biochem. Biophys. Res. Commun. **186:** 1510-1514.

29. Tzeng, D. Y., T. F. Deuel, J. S. Huang, R. M. Senior, L. A. Boxer & R. L. Baehner. 1984. Platelet-derived growth factor promotes polymorphonuclear leukocyte activation. Blood **64:** 1123-1128.

30. Harada, Y. 1990. Effect of leukotriene B_4 on enhancement of superoxide production evoked by formyl-methionyl-leucyl-phenylalanine in myeloid differentiated HL-60 cells: Possible involvement of intracellular calcium influx and high affinity receptor for leukotriene B_4. Hiroshima J. Med. Sci. **39:** 89-94.

31. Serhan, C. N., S. Fiore, D. A. Brezinski & S. Lynch. 1993. Lipoxin A_4 metabolism by human promyelocytic leukemia cells and monocytes: Conversion to novel 15-oxo and dihydro products. Biochemistry **32:** 6313-6319.

32. Simchowitz, L., S. Fiore & C. N. Serhan. Carrier-mediated transport of lipoxin A_4 in human neutrophils. Am. J. Physiol. In press.

Cytokine Regulation of Eicosanoid Generation

PETER ISAKSON,[a] SCOTT HAUSER, YAN ZHANG, KATHLEEN LEAHY, JAIME MASFERRER, AND KAREN SEIBERT

Inflammatory Diseases Research
G. D. Searle and Monsanto Corporate Research
700 Chesterfield Village Parkway
St. Louis, Missouri 63198

INTRODUCTION

For many years it was believed that prostaglandins (PGs) were formed via the activity of a single enzyme, cyclooxygenase (COX), which is present constitutively in most cells (for review see Ref. 1). Inhibition of COX would lead to decreased production of proinflammatory PGs at the site of inflammation, while at the same time inhibiting the formation of PGs that play a role in sustaining normal cellular or tissue function. In a number of cell types and tissues COX activity was induced following incubation with cytokines, growth factors, or endotoxin, a process that required new protein synthesis.[2,3] The induction of *de novo* COX synthesis was inhibited by antiinflammatory glucocorticoids such as dexamethasone, but these drugs did not alter basal COX activity observed *in vitro* and *in vivo*.[4–6] Based on these data it was hypothesized that there is an inducible form of the COX enzyme that is sensitive to cytokine stimulation and selective inhibition by glucocorticoids. This hypothesis was supported by the recent isolation of a second COX gene whose expression is induced by a number of inflammatory cytokines, and whose mRNA expression is selectively blocked by dexamethasone.[7–14]

The identification of two forms of COX, a constitutively expressed COX-1 and an inducible COX-2, led us to examine the relative distribution of these two forms of the enzyme in normal and inflamed tissues. We report that COX-1 is expressed constitutively throughout normal tissues, while COX-2 mRNA and proinflammatory PGs are induced in parallel following stimulation of an acute inflammatory reaction. Based on these data we propose that selective COX-2 inhibitors will block inflammatory PG production without affecting constitutive PG in peripheral tissues and thus provide superior nonsteroidal antiinflammatory drugs.

RESULTS AND DISCUSSION

The distribution of COX-1 and COX-2 in normal murine tissues was examined by quantitative analysis of mRNA using the nuclease protection method. In all tissues

[a] To whom correspondence should be addressed.

examined there was detectable mRNA for COX-1, including colon, kidney, spleen, stomach, liver, lung, heart, and brain. When the same samples were assayed for COX-2 expression, virtually no mRNA was detectable, with the exception of lung and liver, where a small amount of message was measurable. From these results we conclude that normal tissue is dominated by constitutive COX-1 expression.

Given the low abundance of COX-2 mRNA detectable in normal tissue, inflammatory tissue was evaluated for relative expression of COX-1 and COX-2 mRNA. The model chosen for study was edema induced by injection of carrageenan into the footpad of the mouse. In this model induction of edema over time corresponded to the production of proinflammatory thromboxane B (TxB_2) extractable from the paw tissue; TxB_2 production and swelling were both maximal at three hours. Quantitative analysis of mRNA by RNase protection from these paws showed that maximal induction of COX-2 mRNA was at three hours, coinciding with the maximal effect on edema and thromboxane production.

Recently Futaki *et al.* reported data obtained on a compound that was antiinflammatory *in vivo* but did not cause lesions in rats,[15,16] a pharmacological profile expected of a selective COX-2 inhibitor. The effect of this compound was examined on murine recombinant COX-1 and COX-2 *in vitro.* NS-398 inhibited recombinant COX-2 with an IC_{50} of approximately 30 nM, but did not inhibit COX-1 activity at concentrations exceeding 10 μM. This compares with indomethacin, a potent antiinflammatory that inhibited both COX-1 (IC_{50} = 10nM) and COX-2 (IC_{50} = 200nM) in the *in vitro* enzyme assay.

Since NS-398 is a selective COX-2 inhibitor *in vitro,* we examined the ability of the compound to inhibit edema and thromboxane production *in vivo* in the rat carrageenan paw model. Both indomethacin and NS-398 inhibited edema and thromboxane production in the paw, with the maximal effect of the drug coinciding with maximal induction of COX-2 mRNA.

In conclusion, we report that normal tissue expression of COX enzyme is due to the expression of constitutive COX-1 mRNA, while COX-2 mRNA is abundant at the site of inflammation. Selective inhibitors of COX-2 are antiinflammatory *in vivo* without causing gastric lesions,[15,16] while traditional NSAIDs like indomethacin are both antiinflammatory and ulcerogenic, consistent with their ability to inhibit COX-1 as well as COX-2. Development of selective inhibitors of COX-2 that spare gastric PG production may represent a significant advance for the treatment of acute and chronic inflammatory disorders.

REFERENCES

1. DeWitt, D. 1991. Biochim. Biophys. Acta **1083:** 121–134.
2. Raz, A., A. Wyche, N. Siegel & P. Needleman. 1988. J. Biol. Chem. **263:** 3022–3028.
3. Fu, J., J. L. Masferrer, K. Seibert, A. Raz & P. Needleman. 1990. J. Biol. Chem. **265:** 16737–16740.
4. Raz, A., A. Wyche & P. Needleman. 1989. Proc. Natl. Acad. Sci. USA **86:** 1657–1661.
5. Masferrer, J. L., K. Seibert, B. S. Zweifel & P. Needleman. 1992. Proc. Natl. Acad. Sci. USA **89:** 3917–3921.
6. Masferrer, J. L., B. S. Zweifel, K. Seibert & P. Needleman. 1990. J. Clin. Invest. **86:** 1375–1379.

7. Xie, W., J. G. Chipman, D. L. Robertson, R. L. Erikson & D. L. Simmons. 1991. Proc. Natl. Acad. Sci. USA **88:** 2692–2696.
8. Kujubu, D. A., B. S. Fletcher, B. C. Varnum, R. W. Lim & H. R. Herschman. 1991. J. Biol. Chem. **266:** 12866–12872.
9. O'Banion, M. K., H. B. Sadowski, V. Winn & D. A. Young. 1991. J. Biol. Chem. **266:** 23261–23267.
10. Fletcher, B. S., D. A. Kujubu, D. M. Perrin & H. R. Herschman. 1992. J. Biol. Chem. **267:** 4338–4344.
11. Sirois, J. & J. S. Richards. 1992. J. Biol. Chem. **267:** 6382–6388.
12. Kujubu, D. A. & H. R. Herschman. 1992. J. Biol. Chem. **267:** 7991–7994.
13. O'Banion, M. K., V. Winn & D. A. Young. 1992. Proc. Natl. Acad. Sci. USA **89:** 4888–4892.
14. Kujubu, D. A., S. T. Reddy, B. S. Fletcher & H. Herschman. 1993. J. Biol. Chem. **268:** 5425–5430.
15. Futaki, N., I. Arai, S. Hamasaki, S. Takahashi, S. Higuchi & S. Otomo. 1992. J. Pharm. Pharmacol. **45:** 753–755.
16. Futaki, N., K. Yoshikawa, H. Yumiko, I. Arai, S. Higuchi, H. Iizuka & S. Otomo. 1993. Gen. Pharmacol. **24:** 105–110.

Priming of Monocytes with Interferon-Gamma Is Associated with Up-Regulation of Expression of the Receptor for Platelet-Activating Factor[a]

SYLVIE OUELLET, DENIS GINGRAS, SYLVIE TURCOTTE, AND MAREK ROLA-PLESZCZYNSKI[b]

Immunology Division
Department of Pediatrics
Faculty of Medicine
University of Sherbrooke
Sherbrooke, Québec, Canada

Platelet-activating factor (PAF) is a potent phospholipid mediator of inflammation and immunoregulation.[1] It is produced by many cell populations, including monocytes, macrophages, neutrophils, eosinophils, basophils, platelets, and vascular endothelial cells in response to a variety of cell-specific stimuli. These same cell populations, as well as other cells such as B lymphocytes, NK cells, and vascular smooth muscle cells, can become targets of PAF bioactivity. In recent years, we and others have shown PAF to modulate immune cell functions and, in particular, to up-regulate their production of cytokines, such as interleukin (IL)-1, IL-6, IL-8, and tumor necrosis factor-alpha (TNFα).[2-8] Most if not all biological actions of PAF are thought to be mediated via specific receptors in the plasma membranes of responsive cells.[9] A PAF receptor (PAFR) was first cloned from the guinea pig lung by Honda *et al.*,[10] and this was followed by the cloning of a human PAFR by several groups soon afterwards.[11-14]

Interferon-gamma (IFNγ) is a product of activated T lymphocytes and NK cells. It possesses numerous and pleiotropic activities, among which are the activation of monocytes and macrophages.[15,16] Moreover, IFNγ was shown to synergize with PAF in stimulating IL-1 production by human monocytes.[17] We have also shown that priming with IFNγ enhanced the effect of PAF on muramyl dipeptide (MDP)-driven TNFα production in the human promyelocytic HL-60 cell line differentiated toward the monocyte lineage with vitamin D_3.[7] In this paper, we show that human monocytes can also be primed with IFNγ, that such priming yields enhanced responsiveness to

[a]This work was supported by grants to M.R.-P. and a studentship to S.O. from the Medical Research Council of Canada.

[b]To whom correspondence should be addressed at: Immunology Division, Faculty of Medicine, University of Sherbrooke, 3001 N. 12th Avenue, Sherbrooke, Québec, J1H 5N4 Canada.

PAF in terms of TNFα production, and that such priming is associated with a transcriptional up-regulation of PAFR gene expression.

MATERIALS AND METHODS

Preparation of Monocytes

Human venous blood from healthy, medication-free volunteers was collected on citrate/dextrose/adenine. Cell preparation was performed at room temperature. The platelet-rich plasma was removed after centrifugation of the blood at 250 *g* for 15 minutes and replaced by an equal volume of phosphate-buffered saline (PBS). After dextran sedimentation (Dextran T-500, Pharmacia Fine Chemicals, Uppsala, Sweden) of red blood cells, the leukocyte-rich supernatant was layered over a Ficoll-Hypaque (1.077 g/mL; Pharmacia, Uppsala, Sweden) cushion and centrifuged at 400 *g* for 20 minutes. The peripheral blood mononuclear leukocytes (PBML) were collected at the interface and were washed twice in Ca^{2+}/Mg^{2+}-free Dulbecco's PBS and then resuspended in RPMI 1640 medium with 10% heat-inactivated fetal bovine serum (FBS). Monocytes were then purified by adherence to the surface of plastic petri dishes previously coated with baby hamster kidney cell microexudate as described.[18] This was effective in enriching the adherent population to greater than 90% monocytes as assessed by nonspecific esterase staining. This population was removed with EDTA (0.01 M) in RPMI 1640 medium and resuspended in RPMI 1640 medium containing gentamicin (40 mg/L; Schering, Pointe Claire, Québec, Canada) and 5% FBS, at a final concentration of 5×10^6 cells/mL. This concentration proved to be optimal for our assays. Cell viability was >95%, as determined by trypan blue exclusion. Cells were allowed to rest overnight in polypropylene tubes to allow them to return to baseline activity following initial activation by adherence. The medium was then removed by pipetting and gently replaced by fresh RPMI 1640, 5% FBS. The cells were allowed a 30-minute rest at 37 °C before stimulation.

TNFα Production and Measure

L 929 cells were used as target cells in the TNF assay. This bioassay was performed essentially as described.[19] Briefly, L 929 cells were seeded into flat-bottomed 96-well microtiter plates (Costar) at a density of 5×10^4 cells per well. After overnight incubation, spent medium was removed, and 100 μl of twofold dilutions of human monocyte supernatants were added to each well. Medium (100 μl) containing actinomycin D was then added at a final concentration of 5 μg/ml. Plates were incubated for 20 h, medium removed, and cells stained with crystal violet (0.5%) in methanol/water (1 : 4 v : v) for 10 min. The amount of cell lysis was determined by measuring absorbance at 540 nm using a Bio-Rad microplate reader (Bio-Rad Laboratories, Richmond, CA). A standard of human recombinant (r)TNFα was used (Cistron), and results are expressed as units of TNF, based on probit analysis of the rTNF dilution curve.

mRNA Studies

After appropriate treatment, cells were pelleted in 15-mL polypropylene tubes (Falcon, Becton-Dickinson Labware, Lincoln Park, NJ), and total RNA was isolated by acid guanidinium thiocyanate-phenol chloroform extraction according to Chomczynski and Sacchi.[20] RNA (10 μg) was separated by electrophoresis on 1% agarose and transferred onto a Hybond-N (Amersham, Arlington Heights, IL) membrane for Northern analysis. The cDNA probes used were: a 0.7-kb *Eco*RI insert of human PAFR[21] and a 1-kb *Pst*I insert of the control "housekeeping" gene glyceraldehyde-3-phosphate dehydrogenase (GAPDH; ATCC). The probes were labeled with a multiprime DNA labeling system (Amersham) using [α-^{32}P]dCTP (specific activity > 3,000 Ci/mmol; Amersham).

Membranes were prehybridized for 4 h in a mixture containing Tris 120 mM, NaCl 600 mM, EDTA 8 mM, sodium pyrophosphate (NaPP) 0.1%, sodium dodecyl sulfate (SDS) 0.2%, and heparin 100 μg/mL; hybridization was performed overnight at 68 °C in the same mixture in which the concentration of heparin was increased to 625 μg/mL, and dextran sulfate at 10% was added. The membranes were then washed once at room temperature for 20 minutes in 2× SSC (1× SSC: 0.15 M NaCl, 0.015 M sodium citrate, pH 7); once with 0.1× SSC, 0.5% SDS at 68 °C for 60 minutes; and then rinsed at room temperature with 0.1× SSC. The membranes were exposed to Kodak XAR-5 film (Eastman Kodak, Rochester, NY) with intensifying screens for 24 h at −80 °C. Signal intensity was quantitated by densitometry with a Pharmacia LKB Ultrascan XL (Pharmacia, Canada). The membranes were then stripped of the probe by washing for 60 minutes with 0.1% SDS at 90 °C and rehybridized with the GAPDH probe. Densitometric values are expressed as ratios of hPAFR/GAPDH densitometry quantification.

RESULTS AND DISCUSSION

Overnight (18-h) treatment of human monocytes with IFNγ induced a modest increase in their production of TNFα in response to bacterial lipopolysaccharide (LPS) (FIG. 1). However, when graded concentrations of PAF were added to monocytes that had been treated overnight with IFNγ, their production of TNFα was markedly enhanced, especially at lower concentrations of PAF.

Priming of monocytes by IFNγ can be associated with a number of intracellular events, including activation of one or several elements of PAF signal transduction pathways. The first and foremost of these elements would be the PAFR itself. We therefore planned a series of experiments to test the hypothesis that IFNγ could up-regulate PAFR in monocytes.

Human monocytes isolated from peripheral blood constitutively express PAFR mRNA as well as PAFR protein on the cell surface. Treatment of monocytes for 3 h with graded concentrations of IFNγ induced a significant up-regulation of hPAFR mRNA accumulation (FIG. 2). Maximal effect was observed at 400-1000 U/ml of IFNγ, but as little as 10 U/ml were effective. In contrast to IFNγ, IFNα was totally inactive at concentrations up to 2000 U/ml. IFNγ-induced up-regulation of hPAFR mRNA accumulation was associated with a significant increase in PAFR protein

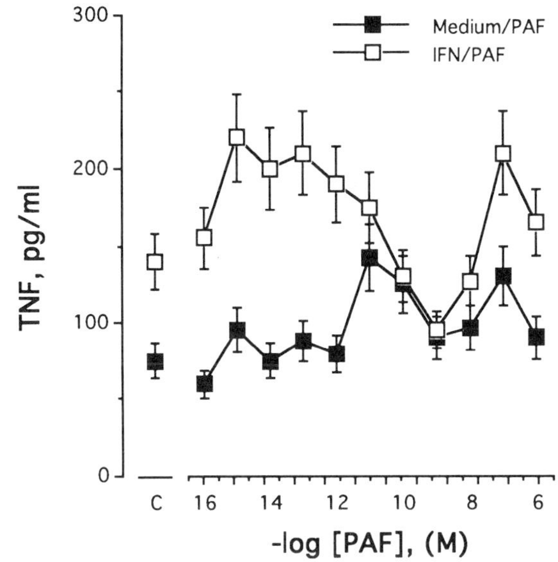

FIGURE 1. Priming of human monocytes with IFNγ enhances their responsiveness to PAF in terms of TNFα production. Monocytes were pretreated for 18 h with 10 U/ml IFNγ, washed, and incubated in the absence or presence of graded concentrations of PAF for 24 h. Cell-free supernatants were analyzed for TNFα content using the L 929 cell bioassay, as described in MATERIALS AND METHODS. Results indicate means ±SE, $n = 4$.

expression at 24 h, as assessed by flow cytometry, using a polyclonal anti-hPAFR antibody.[22] It was also associated with markedly enhanced responsiveness to PAF in terms of $[Ca^{2+}]_i$ flux at 24 and 48 h. This latter effect could involve additional signal transduction elements, in analogy to the findings of Klein *et al.*[23] These investigators reported that IFNγ not only increased the number of binding sites for formyl-methionyl-leucyl-phenylalanine (fMLP), but also up-regulated G protein expression, thus effectively enhancing responsiveness of the HL-60 granulocytes to the chemotactic peptide.

Kinetic studies of IFNγ-stimulated PAFR mRNA up-regulation revealed a rapid induction, evident as early as 1 h after stimulation. Peak activity was seen at 2 h and was maintained for at least 24 h.[22] Interestingly, the baseline level of PAFR mRNA expression decreased with time, while IFNγ-stimulated levels remained elevated. At this time, we cannot rule out additional events following stimulation with IFNγ that could participate in or be responsible for the prolonged effect on PAFR mRNA, such as induction of secondary cytokines, for instance.

PAFR mRNA stability was not affected by IFNγ, remaining at approximately 95 min for both control and treated cells. On the other hand, pretreatment of monocytes with actinomycin D for 20 min prior to addition of IFNγ completely abolished the effect of the cytokine, suggesting that IFNγ-stimulated increase in PAFR gene expression was under transcriptional control.[22] Recently, Mutoh *et al.*[24] reported the sequence of two PAFR promoters, with tissue-specific differential splicing. The

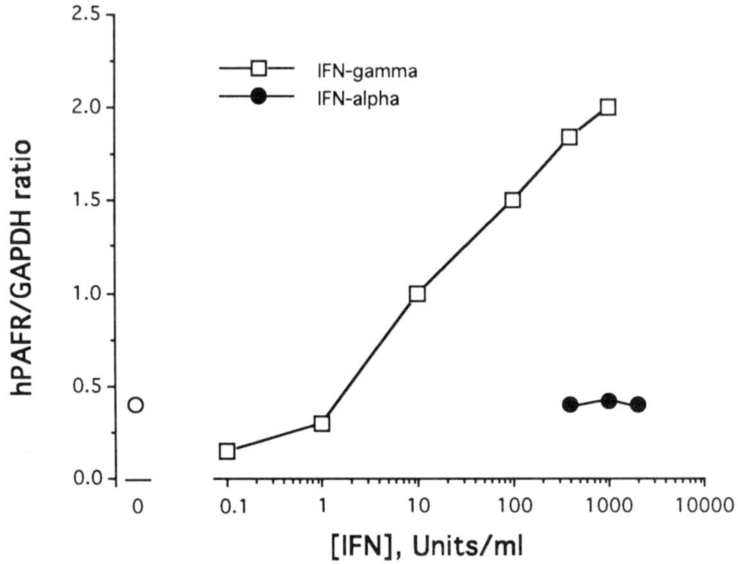

FIGURE 2. IFNγ but not IFNα up-regulates hPAFR mRNA accumulation in monocytes in a dose-dependent fashion. Human monocytes were incubated in the absence or presence of graded concentrations of IFNγ or IFNα for 2 h. Total cellular RNA was then extracted, and hPAFR mRNA was quantified as indicated in MATERIALS AND METHODS.

leukocyte hPAFR promoter region contains several potential IFNγ-responsive elements, and it will be interesting to assess their eventual involvement in the transcriptional regulation of PAFR gene expression by IFNγ.

The importance of this up-regulation of hPAF-R expression by IFNγ in monocytes may be examined in the context of the inflammatory reaction. Monocytes/macrophages are generally considered to be the central contributors to the complex reaction of inflammation, and lymphocyte-derived IFNγ is considered the principal activator of these cells.[25] PAF has also many important proinflammatory activities. In monocytes/macrophages, PAF causes chemotaxis, aggregation, superoxide anion production, IL-1, TNFα, and other cytokine production, as well as enhanced cytotoxicity.[1–8] Our present studies suggest that IFNγ can prime human monocytes for enhanced responses to PAF by up-regulating their hPAFR expression. This IFNγ/PAF interaction would result in synergistic responses of the activated monocytes/macrophages. Such an interaction may play a major role in the inflammation process as well as in pathological situations in which these two mediators have been implicated, such as endotoxic shock and graft rejection.

In summary, the studies reported here suggest a mechanism by which IFNγ can act to enhance the responses to PAF in human monocytes, namely by up-regulating the expression of the PAF receptor. This cannot, however, exclude other, additional mechanisms mediated by IFNγ, such as modulation of the PAF signal transduction cascade. Our studies were conducted in human monocytes, and it will be of interest to determine if other cell types, such as human macrophages, neutrophils, or endothelial cells, also respond to IFNγ by up-regulating their hPAF-R expression. This is currently under investigation in our laboratory.

REFERENCES

1. Braquet, P. & M. Rola-Pleszczynski. 1987. Platelet-activating factor and cellular immune responses. Immunol. Today **8:** 345-352.
2. Dubois, C., E. Bissonnette & M. Rola-Pleszczynski. 1989. Platelet-activating factor (PAF) stimulates tumor necrosis factor production by alveolar macrophages: Prevention by PAF receptor antagonists and lipoxygenase inhibitors. J. Immunol. **143:** 964-971.
3. Rola-Pleszczynski, M. 1990. Priming of human monocytes with PAF augments their production of tumor necrosis factor. J. Lipid Mediators **2:** S77-S82.
4. Pignol, B., S. Hénane, B. Sorlin, M. Rola-Pleszczynski, J. M. Mencia-Huerta & P. Braquet. 1990. Effect of long-term treatment with platelet-activating factor on IL 1 and IL 2 production by rat spleen cells. J. Immunol. **145:** 980-984.
5. Poubelle, P., D. Gingras, C. Demers, C. Dubois, D. Harbour & M. Rola-Pleszczynski. 1991. Platelet activating factor (PAF-acether) enhances the concomitant production of tumor necrosis factor alpha and interleukin 1 by subsets of human monocytes. Immunology **72:** 181-187.
6. Thivierge, M. & M. Rola-Pleszczynski. 1992. Platelet-activating factor (PAF) enhances interleukin-6 production by alveolar macrophages. J. Allergy Clin. Immunol. **90:** 796-802.
7. Rola-Pleszczynski, M. & J. Stankova. 1992. Differentiation-dependent modulation of TNF production by PAF in human HL-60 myeloid leukemia cells. J. Leukocyte Biol. **51:** 609-616.
8. Denault, S., M.-J. April, S. Kocsis & J. Stankova. 1993. Platelet-activating factor (PAF) up-regulates IL-8 gene expression in human monocytes: Transcriptional and post-transcriptional mechanisms. J. Immunol. **150** (II): 125a. (Abstr.)
9. Hwang, S.-B. 1990. Specific receptors of platelet-activating factor, receptor heterogeneity, and signal transduction mechanisms. J. Lipid Mediators **2:** 123.
10. Honda, Z.-I., M. Nakamura, I. Miki, M. Minami, T. Watanabe, Y. Seyana, H. Okado, H. Toh, K. Ito, T. Miyamota & T. Shimizu. 1991. Cloning by functional expression of platelet-activating factor receptor from guinea-pig lung. Nature **349:** 342.
11. Nakamura, M., Z.-I. Honda, T. Izumi, C. Sakanaka, H. Mutoh, M. Minami, H. Bito, Y. Seyama, T. Matsumoto, M. Noma & T. Shimizu. 1991. Molecular cloning and expression of platelet-activating factor receptor from human leukocytes. J. Biol. Chem. **266:** 20400.
12. Ye, R. D., E. R. Prossnitz, A. Zou & C. G. Cochrane. 1991. Characterization of a human cDNA that encodes a functional receptor for platelet-activating factor. Biochem. Biophys. Res. Commun. **180:** 105.
13. Kunz, D., N. P. Gerard & C. Gerard. 1992. The human leukocyte platelet-activating factor receptor. J. Biol. Chem. **267:** 9101.
14. Trinchieri, G. & B. Perussia. 1985. Immune interferon: A pleiotropic lymphokine with multiple effects. Immunol. Today **6:** 131.
15. Nathan, C. F. 1990. Coordinate actions of growth factors in monocytes/macrophages. *In* Peptide Growth Factors and Their Receptors II. M. B. Sporn & A. B. Roberts, Eds.: 427-453. Springer-Verlag. Berlin.
16. Barthelson, R. & F. Valone. 1990. Interaction of platelet-activating factor with interferon-γ in the stimulation of interleukin-1 production by human monocytes. J. Allergy Clin. Immunol. **86:** 193.

17. Rola-Pleszczynski, M. & J. Stankova. 1992. Leukotriene B4 enhances interleukin-6 (IL-6) production and IL-6 messenger RNA accumulation in human monocytes in vitro: Transcriptional and posttranscriptional mechanisms. Blood **80:** 1004.
18. Flick, D. A. & G. E. Gifford. 1984. Comparison of *in vitro* cell cytotoxic assays for tumor necrosis factor. J. Immunol. Methods **68:** 167.
19. Chomczynski, P. & N. Sacchi. 1987. Single-step method of RNA isolation by acid guanidinium thiocyanate-phenol-chloroform extraction. Anal. Biochem. **162:** 156.
20. Müller, E., P. Dagenais, N. Alami & M. Rola-Pleszczynski. 1993. Identification and functional characterization of platelet-activating factor receptors in human leukocyte populations using polyclonal anti-peptide antibody. Proc. Natl. Acad. Sci. USA **90:** 5818-5822.
21. Ouellet, S., E. Müller & M. Rola-Pleszczynski. 1994. IFN-α up-regulates platelet-activating factor receptor gene expression in human monocytes. J. Immunol. **152:** 5092-5099.
22. Klein, J. B., J. A. Scherzer & K. R. McLeish. 1992. IFNγ enhances expression of formyl peptide receptors and guanine nucleotide-binding proteins by HL-60 granulocytes. J. Immunol. **148:** 2483.
23. Mutoh, H., H. Bito, M. Minami, M. Nakamura, Z. Honda, T. Izumi, R. Nakata, Y. Kurachi, A. Terano & T. Shimizu. 1993. Two different promoters direct expression of two distinct forms of mRNAs of human platelet-activating factor receptor. FEBS Lett. **322:** 129.
24. Billiau, A. & R. Dijkmans. 1990. Interferon-γ: Mechanism of action and therapeutic potential. Biochem. Pharmacol. **40:** 1433.

Interactions between Histamine and Leukotrienes in the Microcirculation

Aspects of Relevance to Acute Allergic Inflammation[a]

J. RAUD, J. THORLACIUS, X. XIE, L. LINDBOM, AND P. HEDQVIST

Department of Physiology and Pharmacology
Karolinska Institutet
S-171 77, Stockholm, Sweden

INTRODUCTION

Acute allergic inflammation is characterized by increased blood flow, extravasation of plasma, and recruitment of circulating leukocytes. These events are triggered by locally released inflammatory mediators, including histamine and leukotrienes. In the microcirculation, the cysteinyl-leukotrienes (LTC_4, LTD_4, LTE_4) evoke dose-dependent extravasation of plasma with a potency exceeding that of histamine by approximately 1000 times.[1] The permeability increase is localized to postcapillary venules,[1] and ultrastructural studies of tissues exposed to histamine or cysteinyl-leukotrienes have shown that plasma proteins leak out via gaps between contracted endothelial cells.[2–4] On the other hand, local administration of leukotriene B_4 (LTB_4) causes leukocytes to adhere to the venular endothelium, followed by emigration to the perivascular interstitium.[5–7] This process is accompanied by plasma leakage that is unrelated to endothelial gap formation and is entirely dependent on the presence and/or intact adhesive function of leukocytes.[5,8]

At sites of acute inflammation, the chemical and cellular complexity provides a basis for multiple interactions, some of which involve synergism between inflammatory mediators that have different profiles of action in the microcirculation. The classical example in this respect is potentiation by specific vasodilators such as prostaglandin E_2 (PGE_2) of plasma extravasation induced, for example, by bradykinin or histamine.[9] In this case, the enhancement of plasma leakage is believed to result from the increased blood flow and hydrostatic pressure in venules where the integrity of the endothelial barrier has been disrupted.[10] Leukocyte recruitment into tissues in response to chemotactic factors has also been shown to be potentiated by specific vasodilators;[11] however, the detailed microvascular mechanisms behind this type of synergism involving inflammatory cells have not been clarified.

[a]This work was supported by the Swedish Medical Research Council (14X-4342, 04P-10738), the Swedish Society for Medical Research, the Swedish Association against Asthma and Allergy, and Karolinska Instituet.

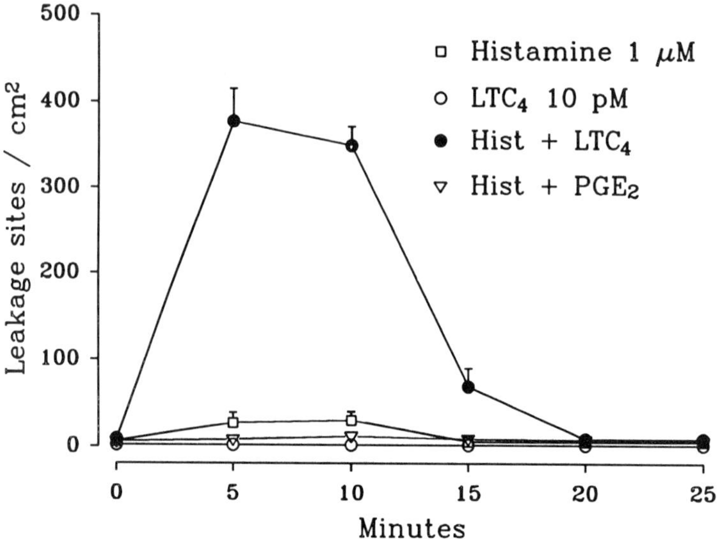

FIGURE 1. Plasma extravasation in hamster cheek pouch evoked by topical challenge with histamine, LTC_4, or PGE_2 for 10 min, either alone or in combination as defined by legends. Mean values ± SEM (n = 3–7). (For details, see Ref. 21.)

In this report, based on intravital microscopic observation of the terminal vascular beds of the hamster cheek pouch and rat mesentery, we summarize findings related to interactions between histamine and LTC_4 (vascular permeability), and histamine and LTB_4 (leukocyte recruitment). The findings indicate that: (1) independent of changes in blood flow, threshold concentrations of histamine and LTC_4 can act in concert to activate/contract endothelial cells; (2) histamine has the potential to greatly enhance LTB_4-induced leukocyte recruitment, partly by increasing leukocyte rolling in a blood flow–dependent manner.

RESULTS AND DISCUSSION

Synergism between Histamine and Leukotriene C_4

Vasodilating agents such as PGE_2 effectively potentiate LTC_4-induced extravasation of plasma in the hamster cheek pouch if the concentration of LTC_4 is high enough per se to increase vascular permeability.[12] This potentiation occurs in spite of PGE_2 having no effect on vascular permeability by itself, and, as already mentioned, the synergism is believed to be a direct consequence of the vasodilating properties of PGE_2. Here we report that histamine and LTC_4 can act synergistically to greatly increase vascular permeability by a mechanism that apparently is unrelated to changes in vessel tone and blood flow.

Topical challenge with histamine or LTC_4 at concentrations higher than 3 μM and 1 nM, respectively, causes marked postcapillary plasma extravasation in the

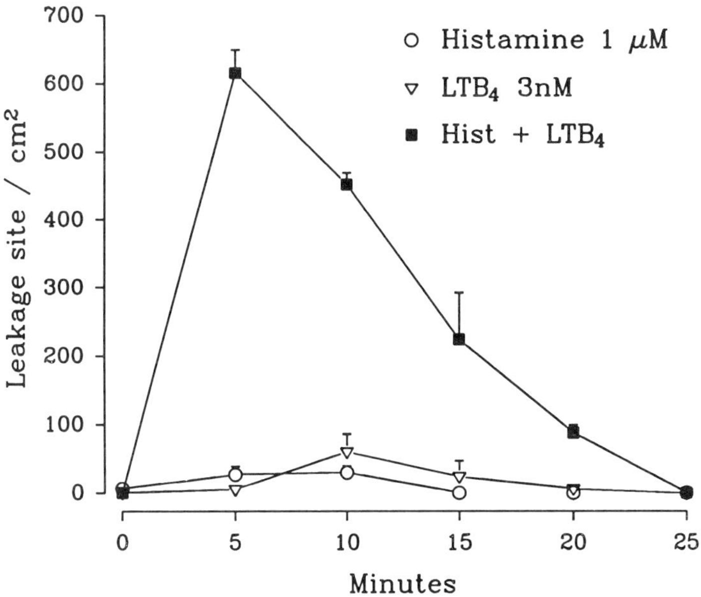

FIGURE 2. Plasma extravasation in hamster cheek pouch evoked by topical challenge with histamine or LTB_4 for 10 min, either alone or in combination as defined by legends. Mean values ± SEM (n = 4–5). (For methods, see Ref. 10.)

hamster cheek pouch.[1] In the experiments described here, the concentrations of histamine and LTC_4 were chosen to cause no or only minimal increases in venular permeability when applied individually to the cheek pouch (10 min topical application). Thus, 1 μM histamine caused a weak plasma leakage response, and 10 pM LTC_4 was inactive in this respect (FIG. 1). However, when LTC_4 was administered together with histamine in these low concentrations, the plasma leakage was strikingly enhanced (FIG. 1). Histamine (1 μM) alone caused moderate arteriolar dilatation (26 ± 6% peak increase in inner diameter), whereas LTC_4 (10 pM), which causes vasoconstriction in higher concentrations,[1] evoked no change in arteriolar diameter. The arteriolar dilatation in response to coadministration of histamine and LTC_4 did not differ from histamine challenge alone (not shown), indicating that changes in local blood flow and hydrostatic pressure did not contribute to the enhanced plasma extravasation. This notion was further supported by the finding that PGE_2 (3 nM), which caused a 36 ± 13% peak increase in arteriolar diameters with a time course very similar to the vasodilation caused by histamine, failed to enhance the vascular permeability response to LTC_4 (FIG. 1). Interestingly, the plasma extravasation evoked by combined challenge with histamine and LTC_4 was abolished by the H_1-receptor antagonist chlorpheniramine (3 μM topically) (not shown). Apparently, synergism between histamine and LTC_4 with respect to vascular permeability involves a previously unrecognized mode of action in addition to the well-known blood flow–dependent potentiation. The mechanism behind the blood flow–independent mediator

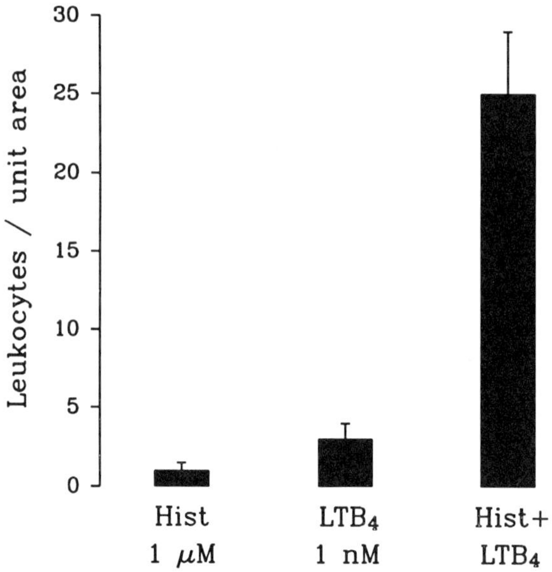

FIGURE 3. Leukocyte emigration in hamster cheek pouch evoked by topical challenge with histamine or LTB_4 for 10 min, either alone or in combination as defined below bars. Mean values ± SEM (n = 4–5). (For methods, see Ref. 10.)

synergism is presently not known; however, our working hypothesis is that threshold concentrations of histamine and LTC_4 act in concert to activate/contract endothelial cells.

Synergism between Histamine and Leukotriene B_4

Pretreatment with antihistamines has been demonstrated to significantly reduce allergen-induced leukocyte accumulation in both man and animals.[13,14] However, it is important to emphasize that histamine alone does not recruit leukocytes into human or animal tissues,[8,15] suggesting that endogenous histamine rather contributes to inflammatory leukocyte recruitment by acting in synergy with chemotactic mediators. Yet, the mechanisms by which histamine interacts with chemotactic factors in intact microvessels are largely unknown.

In the hamster cheek pouch, topically applied LTB_4 in concentrations above 1 nM causes dose-dependent leukocyte accumulation and leukocyte-dependent plasma leakage.[1] At 3 nM locally, LTB_4 evokes a threshold reaction with regard to plasma and leukocyte extravasation (FIGS. 2 and 3). However, in the presence of a low dose of histamine (1 μM), these microvascular responses to 1 nM LTB_4 are strikingly enhanced (FIGS. 2 and 3).

The process of leukocyte recruitment during inflammation is a sequential, multistep process involving slow rolling along the venular endothelium, followed by

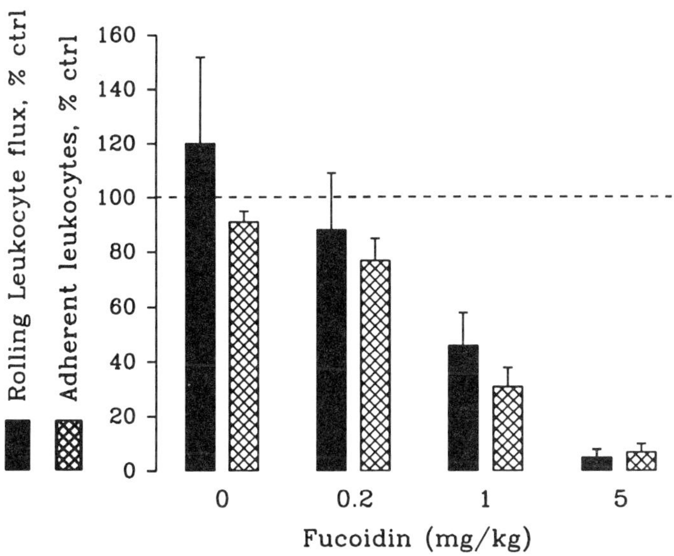

FIGURE 4. Firm leukocyte adhesion induced by fMLP (*cross-hatched bars*) in rat mesenteric venules is critically dependent on the degree of prestimulatory rolling leukocyte flux (*solid columns*). Values on rolling and adhesion illustrate effects of iv treatment with saline (*bars* on **far left** labeled "0") or increasing doses of the sulfated polysaccharide fucoidin, expressed as percent of the corresponding paired value before treatment (*dashed line*). Challenge with fMLP (1 μM) was performed before and 10 min after fucoidin or saline treatment. Rolling leukocyte flux is defined as the number of leukocytes per min rolling slowly past a reference point in the venule. Mean values ± SEM (*n* = 5 in each group). (For details, see Ref. 17.)

firm adhesion and extravasation. In this cascade of events, leukocyte rolling has been shown to be a precondition for the subsequent steps in the process of leukocyte emigration.[16–18] In fact, we have demonstrated a direct relationship between the degree of rolling and the extent of firm adhesion induced by chemotactic stimuli in the rat mesenteric microcirculation.[17] In other words, graded and selective reduction of leukocyte rolling with the sulfated polysaccharide fucoidin prior to chemotactic stimulation was paralleled over a wide range by a proportional decrease in chemoattractant-induced leukocyte adhesion (FIG. 4). Thus, knowing that the degree of rolling may determine the magnitude of firm adhesion in response to chemotactic factors, we next examined if histamine can increase rolling in rat mesenteric microvessels. It was found that histamine (0.01 to 1 mM topically) indeed caused significant, dose-dependent increases in leukocyte rolling by up to 50% (not shown). Moreover, this increase in rolling correlated with the histamine-induced vasodilatation (not shown) and thereby with the increase in total flux (rolling + free-flowing cells) of leukocytes (FIG. 5). In this series of experiments we also confirmed that histamine significantly potentiates LTB_4-induced firm leukocyte adhesion in the mesentery as well. Taken together, these observations provide a physiological basis for explaining the well-established enhancement of leukocyte recruitment caused by vasodilating inflammatory mediators such as histamine.

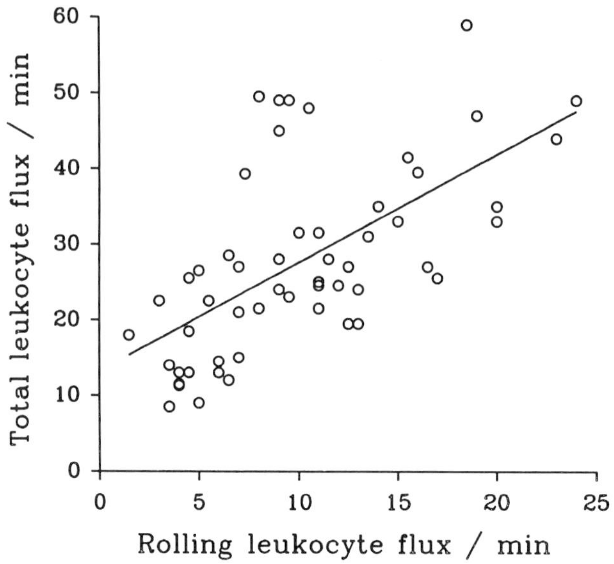

FIGURE 5. Relationship between total and rolling leukocyte flux in small venules of the rat mesentery. Figure includes data from unstimulated preparations and values after topical application of 0.1 mM histamine for 5 and 10 min. $r = 0.668$, $p < 0.001$ (Spearman rank-order correlation) for all values collectively ($n = 56$).

CONCLUDING REMARKS

We have described that histamine in low concentration has the potential to greatly enhance the inflammatory actions of threshold doses of leukotrienes in the hamster cheek pouch and the rat mesentery, apparently by making endothelial cells more sensitive to LTC_4 and by promoting LTB_4-induced firm leukocyte adhesion via a blood flow–dependent increase in leukocyte rolling. Interestingly, the doses of histamine, LTB_4 and LTC_4 used here correspond rather closely to the small amounts of endogenous leukotrienes and histamine released from cheek pouches of immunized hamsters challenged with antigen.[12,19,20] These findings suggest that apparently insignificant or even undetectable formation of leukotrienes may have a great impact on the expression of inflammation in the presence of histamine. Moreover, because antihistamine treatment abolished the synergism between histamine and LTC_4, it is worth considering that complete inhibition of an inflammatory reaction by antihistamines does not necessarily exclude an important mediator role for leukotrienes.

REFERENCES

1. Dahlén, S.-E., J. Björk, P. Hedqvist, K.-E. Arfors, S. Hammarström, J. Å. Lindgren & B. Samuelsson. 1981. Leukotrienes promote plasma leakage and leukocyte adhesion in postcapillary venules: In vivo effects with relevance to the acute inflammatory response. Proc. Natl. Acad. Sci. USA **78:** 3887–3891.

2. Hedqvist, P., Å. Thureson-Klein, A. Öhlén, J. Raud, L. Lindbom & S.-E. Dahlén. 1987. Neuropeptides and arachidonic acid derivatives as messengers in microvascular function. *In* Neuronal Messengers in Vascular Function. A. Nobin, C. Owman, & B. Arneklo-Nobin, Eds.: 435–446. Elsevier, Amsterdam.
3. Joris, I., G. Majno, E. J. Corey & R. A. Lewis. 1987. The mechanism of vascular leakage induced by leukotriene E_4. Endothelial contraction. Am. J. Pathol. **126:** 19–24.
4. Wu, N. Z. & A. L. Baldwin. 1992. Transient venular permeability increase and endothelial gap formation induced by histamine. Am. J. Physiol. **262:** H1238–H1247.
5. Björk, J., P. Hedqvist & K.-E. Arfors. 1982. Increase in vascular permeability induced by leukotriene B_4 and role of polymorphonuclear leukocytes. Inflamation **6:** 189–200.
6. Lindbom, L., P. Hedqvist, S.-E. Dahlén, J.-Å. Lindgren & K.-E. Arfors. 1982. Leukotriene B_4 induces extravasation and migration of polymorphonuclear leukocytes in vivo. Acta Physiol. Scand. **116:** 105–108.
7. Thureson-Klein, Å., P. Hedqvist & L. Lindbom. 1986. Leukocyte diapedesis and plasma extravasation after leukotriene B_4: Lack of structural injury to the endothelium. Tissue Cell **18:** 1–12.
8. Arfors, K.-E., C. Lundberg, L. Lindbom, K. Lundberg, P. G. Beatty & J. M. Harlan. 1987. A monoclonal antibody to the membrane glycoprotein complex CD 18 inhibits polymorphonuclear leukocyte accumulation and plasma leakage in vivo. Blood **69:** 338–340.
9. Williams, T. J. & J. Morley. 1973. Prostaglandins as potentiators of increased vascular permeability in inflammation. Nature **246:** 215–217.
10. Raud, J. & L. Lindbom. 1994. Studies by intravital microscopy of basic inflammatory mechanisms and acute allergic inflammation. *In* The Handbook of Immunopharmacology: Immunopharmacology of the Microcirculation. S. Brain, Ed: 127-170. Academic Press. London.
11. Bray, M. A., F. M. Cunningham, A. W. Ford-Hutchinson & M. J. Smith. 1981. Leukotriene B_4: A mediator of vascular permeability. Br. J. Pharmacol. **72:** 483–486.
12. Raud, J., S. E. Dahlén, A. Sydbom, L. Lindbom & P. Hedqvist. 1988. Enhancement of acute allergic inflammation of indomethacin is reversed by prostaglandin E_2: Apparent correlation with in vivo modulation of mediator release. Proc. Natl. Acad. Sci. USA **85:** 2315–2319.
13. Woodward, D. F., C. S. Spada, S. B. Hawley & A. L. Nieves. 1985. Histamine H_1- and H_2-receptor involvement in eosinophil infiltration and the microvascular changes associated with cutaneous anaphylaxis. Agents Actions **17:** 121–125.
14. Ciprandi, G., S. Buscaglia, A. Iudice & G. W. Canonica. 1992. Protective effect of different doses of terfenadine on the conjunctival provocation test. Allergy **47:** 309–312.
15. Atkins, P., G. R. Green & B. Zweiman. 1973. Histologic studies of human skin test responses to ragweed, compound 48/80, and histamine. J. Allergy Clin. Immunol. **51:** 263–273.
16. Lawrence, M. B. & T. A. Springer. 1991. Leukocytes roll on a selectin at physiologic flow rates: Distinction from and prerequisite for adhesion through integrins. Cell **65:** 859–873.
17. Lindbom, L., X. Xie, J. Raud & P. Hedqvist. 1992. Chemoattractant-induced firm adhesion of leukocytes to vascular endothelium in vivo is critically dependent on initial leukocyte rolling. Acta Physiol. Scand. **146:** 415–421.
18. Von Andrian, U. H., P. Hansell, J. D. Chambers, E. M. Berger, I. Torres Filho, E. C. Butcher & K.-E. Arfors. 1992. L-selectin function is required for beta 2-integrin–mediated neutrophil adhesion at physiological shear rates in vivo. Am. J. Physiol. **263:** H1034–H1044.

19. Hedqvist, P., J. Raud, U. Palmertz, M. Kumlin & S.-E. Dalén. 1990. Eicosanoids as mediators of and modulators of inflammation. Adv. Prostaglandin Thromboxane Leukotriene Res. **21:** 537–543.
20. Hedqvist, P., L. Lindbom & J. Raud. 1994. Microvascular mechanisms in inflammation. Adv. Prostaglandin Thromboxane Leukotriene Res. In press.
21. Raud, J., L. Lindbom & P. Hedqvist. 1992. Histamine and leukotriene C_4 act synergistically via a blood flow–independent mechanism to enhance microvascular plasma leakage. Acta Physiol. Scand. **146:** 545–546.

12-Lipoxygenase, 12(S)-HETE, and Cancer Metastasis[a]

D. G. TANG[b] AND K. V. HONN[b-f]

Departments of [b]Radiation Oncology, [c]Chemistry, and [d]Pathology
Wayne State University
Detroit, Michigan 48202

[e]Gershenson Radiation Oncology Center
Harper Hospital
Detroit, Michigan 48201

INTRODUCTION

During metastasis tumor cells undergo multiple cell-host interactions, such as tumor cell-platelet, tumor cell-endothelial cell, and tumor cell-matrix protein interactions. These interactions are influenced by positive and negative regulatory factors. Eicosanoids and other bioactive lipids have been shown to be involved in various aspects of neoplasia including cell transformation, tumor promotion, tumor cell growth, and metastasis. Platelets and endothelial cells with which tumor cells interact during hematogenous metastasis are capable of producing a vast array of lipid mediators by either direct or transcellular metabolism of precursors. Platelets are capable of converting arachidonic acid (AA) through the 12-lipoxygenase (12-LOX) pathway into 12(S)-hydroxyeicosatetraenoic acid [12(S)-HETE]. Endothelial cells metabolize AA and linoleic acid (LA) through the 15-LOX pathway, producing 15(S)-HETE and 13(S)-hydroxoctadecadienoic acid [13(S)-HODE], respectively (see FIG. 1 for an overview of AA and LA metabolism). The generation of these fatty acids occurs in response to cell activation by various cytokines and growth factors. The production of these fatty acids [i.e., 12(S)-HETE, 13(S)-HODE] in response to cell activation may suggest their role in as yet unidentified cell signaling pathway(s).

Mammalian lipoxygenases have been described that display varying degrees of substrate specificity for insertion of molecular oxygen into arachidonic acid at carbon positions 5, 12, and 15. Based on the abundance of the majority product, the enzymes have been termed 5, 12, and 15 lipoxygenases, respectively. The major metabolic products catalyzed by these lipoxygenases are presented in FIGURE 1. The 12-LOX catalyzes the transformation of AA into 12(S)-hydroperoxyeicosatetraenoic acid and its 12(S)-hydroxy derivatives [i.e. 12(S)-HETE, FIG. 1A]. Three types of mammalian 12-LOX enzymes have so far been reported. The first is human platelet-type 12-

[a]This work is supported by National Institutes of Health Grants CA47115 and CA29997, Harper Hospital Development grants, NATO Linkage Grant LG92311, and the Fogarty International Research grant awarded to K.V.H.

[f]Address for correspondence: K. V. Honn, Department of Radiation Oncology, Wayne State University, 431 Chemistry, Detroit, Michigan 48202.

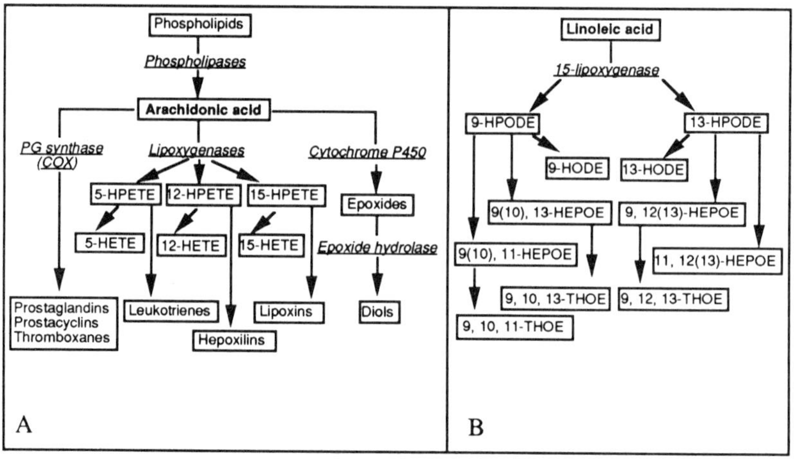

FIGURE 1. Arachidonic and linoleic acid metabolism. **A.** Arachidonic acid is released from cell membrane phospholipids mainly by the action of phospholipase A_2 and is metabolized by various lipoxygenases (LOX), prostaglandin endoperoxide synthase (COX), and cytochrome P-450 monooxygenase (epoxygenase). The activation of these enzymes following ligand-receptor coupling gives rise to a variety of metabolites, including prostaglandins, leukotrienes, lipoxins, hepoxilins, thromboxanes, epoxides, hydroperoxyeicosatetraenoic acids (HPETE), and hydroxyeicosatetraenoic acids (HETEs). **B.** Linoleic acid is metabolized to 13-hydroperoxyoctadecadienoic acid (HPODE) and 9-HPODE by 15-lipoxygenase. Subsequently, HPODEs are converted to hydroxyoctadecadienoic acids (HODEs), hydroepoxyoctadecaenoic acids (HEPOEs), and trihydroxyoctadecaenoic acids (THOEs).

lipoxygenase expressed normally in platelets, HEL (human erythroleukemia) cells, and umbilical vein endothelial cells.[1,2] Platelet-type 12-LOX metabolizes only AA (but not C-18 fatty acids such as linoleic acid) to form exclusively 12(S)-HETE.[3,4] The second is porcine leukocyte-type 12-lipoxygenase, which metabolizes both arachidonic acid and linoleic acid, thus generating 12(S)-HETE as well as small amounts of 15(S)-HETE.[5] The third type of 12-lipoxygenase (sometimes termed epithelial 12-lipoxygenase), which shares more homology with 15-lipoxygenase and leukocyte-type 12-lipoxygenase than with platelet-type 12-lipoxygenase has been isolated from bovine tracheal epithelial cells[6] and rat brain.[7] This type of 12-lipoxygenase, like reticulocyte 15-lipoxygenase and leukocyte-type 12-lipoxygenase, catalyzes the formation of both 12(S)-HETE and 15(S)-HETE. Recently, mouse homologues of both platelet-type and leukocyte-type 12-lipoxygenases have been reported.[8] In the following discussion, we will present evidence demonstrating the expression of 12-LOX by tumor cells and linking 12(S)-HETE to tumor cell metastasis.

TUMOR CELLS POSSESS 12-LIPOXYGENASE AND METABOLIZE ARACHIDONIC ACID TO 12(S)-HETE

The metabolic profiles of AA metabolites were analyzed in several tumor cell lines by a combination of high performance liquid chromotography (HPLC), gas chromatography mass spectral analysis (GC/MS) and chiral phase HPLC. Low metastatic (LM180) and high metastatic (HM340) subpopulations of murine B16a melanoma isolated by centrifugal elutriation[9] were labeled with [^{3}H]arachidonic acid, the products extracted and analyzed by reverse-phase (RP) HPLC for LOX products. Different metabolite profiles for LOX products of AA were obtained with extracts from LM180 and HM340 cells. Specifically, LM180 cells converted AA to equal amounts of two major products that cochromatographed with authentic standards of 12(S)-HETE and 5(S)-HETE in addition to a minor product that comigrated with the 15(S)-HETE standard.[10] In contrast, in HM340 cells, products with retention times corresponding to authentic standards of 15(S)-, 11(S)-, 12(S)-, and 5(S)-HETE were detected. The predominant product coeluted with the 12(S)-HETE standard.[10]

Evaluation of AA metabolites also was performed in several other tumor cell lines. Products of AA metabolism by Lewis Lung carcinoma cells (3LL) were analyzed by both straight-phase and RP-HPLC. Results demonstrated that 12-HETE was the predominant lipoxygenase metabolite in 3LL cells.[3] In addition, 12-HPETE (the metabolic precursor of 12-HETE) also was detected in extracts of 3LL cells,[3] suggesting that 3LL cells possess a 12-LOX pathway. Radioimmunoassays demonstrated that cultured B16a[11] as well as 3LL[12] cells biosynthesize 12(S)-HETE. Rat Walker 256 carcinosarcoma (W256) cells also were analyzed by RP-HPLC. The metabolic profile indicated the production of both 12-HETE and 15-HETE.[13]

The structural identity and stereochemical configuration of the 12-HETE peak was further characterized in B16a cells and 3LL cells by chiral phase HPLC and GC/MS analysis.[10,11] The peak with a retention time of authentic 12(S)-HETE standard was collected, converted to its methyl ester/trimethylsilyl ether derivative, and analyzed by GC/electron impact MS analysis. Both authentic 12-HETE and the hydrogenated form of 12-HETE were analyzed by mass spectral analysis. GC/MS analysis of authentic 12-HETE revealed typical "saddle effect" peaks in the reconstructed gas chromatograms. The mass spectral data in conjunction with HPLC profiles confirmed the structure of the major AA metabolite in tumor cells as 12-HETE.[10,11] 12-HETE can exist in two (i.e., S and R) enantiomeric forms. Since only the S-enantiomer of 12-HETE is biologically active (see below), it was thus important to determine the stereochemistry of the biosynthetic 12-HETE by chiral phase HPLC analysis. A single peak at a retention time of 20.5 minutes was observed when an Me-12-HETE sample derived from biosynthetic 12-HETE by B16a cells was injected onto the chiral-OD phase HPLC column.[10] The peak that corresponded to the S-enantiomer was determined by retention time and coinjection with racemic Me-12-HETE. Similar analysis was performed on 12-HETE derived from 3LL cells.[11] Such analyses revealed that the 12-HETE biosynthesized by tumor cells was the S-enantiomer.[10,11]

B16a and W256 cells also appeared to metabolize linoleic acid. In the cellular lipid extracts from both LM180 and HM340 subpopulations of B16a melanoma, a major product was observed that coeluted with authentic 13(S)-HODE.[10] No product was detected coeluting with authentic 9(S)-HODE in cells from either subpopulation. However, both 13-HODE and 9-HODE were detected in extracts of W256 cells (data not shown). Chiral phase analysis revealed that 13-HODE produced by tumor cells was in the S configuration (data not shown).

The above experimental observations suggest that solid tumor cells express the 12-LOX enzyme, which is capable of metabolizing AA to 12(S)-HETE. Subsequently, molecular biology experiments were performed to examine the message for 12-LOX. Messenger RNAs from human, rat, and mouse platelets as well as human colon carcinoma (Clone A), rat W256, and murine B16a and 3LL cells were reverse transcribed and amplified by PCR with platelet-type 12-LOX specific primers. RT-PCR detected expected bands in these tumor cell lines, and the identity of the PCR products as platelet-type 12-LOX was confirmed by sequencing.[14] Also, the presence of 12-LOX proteins in these tumor cells was revealed by Western blotting.[14,15] Finally, Northern blotting was performed on cultured W256 and elutriated B16a cells using platelet-type 12-LOX specific cDNA[1] as the probe. As shown in FIGURE 2, a ~3.2-kb band was detected in W256 cells, and a smaller species (~3.0 kb) of mRNA was detected in B16a cells, thus proving that solid tumor cells express the message for platelet-type 12-LOX.

EXPRESSION OF 12-LOX AND GENERATION OF 12(S)-HETE BY TUMOR CELLS ARE CORRELATED WITH THEIR METASTATIC POTENTIAL

Elutriated B16a cells from the LM180 and HM340 subpopulations[9] were incubated with [^{3}H]arachidonic acid or [^{14}C]linoleic acids (LA), lipids extracted and analyzed by reverse-phase (for AA) and normal-phase (for LA) HPLC, respectively. The amounts of biosynthesized HETE and HODEs in these tumor cell subpopulations were quantified and compared to their metastatic ability. In LM180 cells the production of the minor product 15-HETE varied slightly with increasing substrate concentration. The production of two major lipoxygenase metabolites, i.e., 12-HETE and 5-HETE, increased linearly and concomitantly to 40 and 34 pmol/10^6 cells, respectively, at 10 μM arachidonic acid (FIG. 3A).[10] In contrast, production of both 15- and 5-HETE by B16a cells from the HM340 subpopulation reached a plateau at 5 μM of arachidonic acid (FIG. 3A). However, the production of the major lipoxygenase metabolite, 12(S)-HETE, by HM340 cells increased exponentially to 123 pmol/10^6 cells at 10 μM arachidonic acid (FIG. 3A). The finding that HM340 cells produced three- to fourfold more 12-HETE than the LM180 cells from exogenously added arachidonic acid (FIG. 3A)[10] is consistent with our previous observation that the level of endogenous 12(S)-HETE (as determined by RIA) in HM340 cells is twice as much as that in LM180 cells.[17] Contrary to AA metabolism the substrate concentration-dependent production of 13-HODE by LM180 cells was not significantly different from that in HM340 cells (FIG. 3B). *In vitro,* adhesion of B16a cells to vascular endothelium is accompanied or immediately followed by a surge of 12(S)-HETE biosynthesis by tumor cells, which was correlated with tumor cell-induced endothelial cell retraction.[11,12] LM180 B16a cells generated little 12(S)-HETE upon adhesion and did not induce endothelial cell retraction. In contrast, HM340 B16a cells adhering to endothelium biosynthesized large amounts of 12(S)-HETE and induced prominent retraction of endothelial cell monolayers.[11]

Collectively, the above data demonstrated that a greater amount of 12-HETE was produced in high metastatic B16a cells (i.e., HM340) than by low metastatic (i.e., LM180) cells, whereas similar amounts of 13-HODE were synthesized. To determine whether higher biosynthesis of 12(S)-HETE in HM340 cells results from a higher

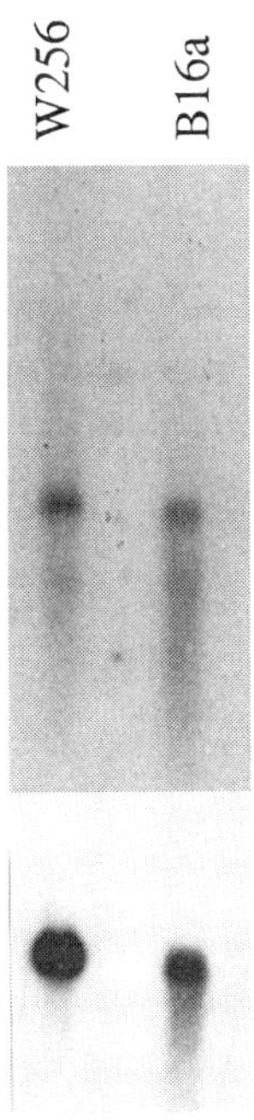

FIGURE 2. Detection of 12-LOX mRNA in solid tumor cells by Northern blotting. Total cellular RNA was obtained from cultured HEL, clone A (not shown), B16a, and W256 cells using the guanidinium-isothiocyanate method. Poly (A^+)-RNA was isolated using PolyATtract mRNA Isolation Systems (Promega),[16] and 4 μg of denatured (glyoxal/DMSO, 50 °C, 1 h) mRNA was loaded onto a 1.0% agarose gel. A 0.24-9.5 kb RNA Ladder (Gibco BRL) was used as the MW standard. Separated RNA was transferred to GeneScreen plus membrane (NEN, DuPont) using a PosiBlot Pressure Blotter (Stratagene, La Jolla, CA). RNA was then UV-linked to the membrane. Radioactive cDNA (i.e., platelet-type 12-LOX cDNA)[1] probe was prepared with ^{32}P-dCTP using Prime-it random primer labeling kit (Stratagene). The membrane was prehybridized in the solution containing 50% formamide, 5× Denhardt's, 10% dextran sulfate, 5× SSPE, 0.1% SDS, and 100 μg/ml salmon sperm DNA (42 °C, 4 h). Hybridization was performed with the addition of labeled platelet-type 12-LOX probe (42 °C, overnight). The membrane was washed under high stringency conditions (0.1 × SSC and 0.1% SDS for 15 min at room temperature and several washes at 65 °C) and exposed to Kodak XOMAT x-ray film at −80 °C using an enhancing screen.[16] After hybridization, the membrane was deprobed and then reprobed with radiolabeled β-actin cDNA probe.

amount of 12-LOX mRNA, Northern blotting was performed using elutriated B16a melanoma cell fractions 100, 180, 260, and 340 (with increasing metastatic potential). As shown in Fig. 4, hybridization of mRNA of these subpopulations with labeled platelet-type 12-LOX cDNA[1] detected increasing amounts of 12-LOX message from LM100 fraction to HM340 fraction. Densitometric scanning revealed a greater than fourfold difference between LM180 and HM340 cells in the amount of 12-LOX mRNA (Fig. 4). Whether different amounts of 12-LOX proteins and/or different

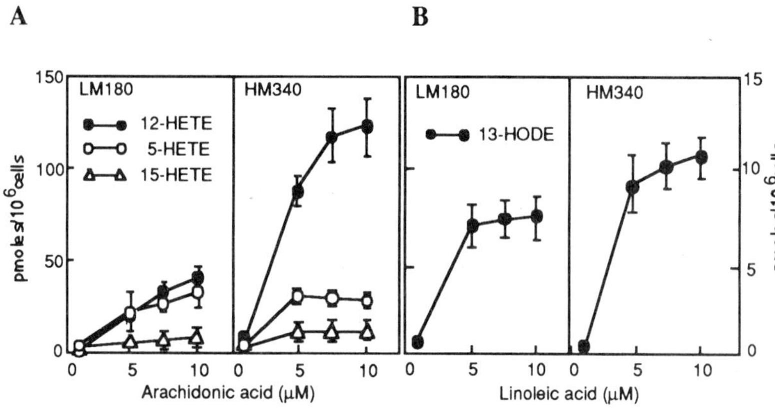

FIGURE 3. AA and LA metabolism in subpopulations of elutriated B16a melanoma cells. B16a cells were incubated (15 min) with indicated concentrations of AA (**A**) or LA (**B**), and the metabolites were analyzed either by reverse phase (**A**) or normal phase (**B**) HPLC. For experimental details, refer to Ref. 10.

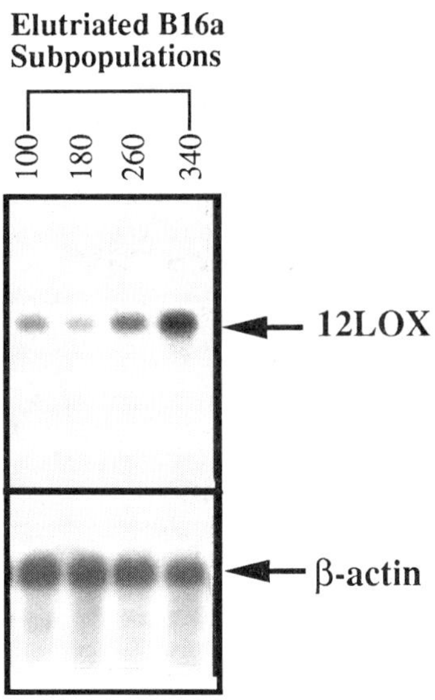

FIGURE 4. Differential production of 12(S)-HETE by low and high metastatic subpopulations of B16a melanoma cells may result from differential expression of 12-LOX mRNA. B16a tumors maintained *in vivo* were fractionated by centrifugal elutriation[9] into four distinct subpopulations, i.e., 100, 180, 260, and 340, with increasing metastatic potential. Total and messenger RNA were isolated and Northern blotting performed as described in the legend to FIGURE 2. The 12-LOX mRNA (~3.0 kb) was indicated by the *upper arrow*. The membrane was stripped and reprobed with β-actin cDNA probe (*lower arrow*).

enzymatic activities of the protein contribute to differential production of 12(S)-HETE in low and high metastatic subpopulations of B16a cells is not yet known.

The correlation of 12-HETE production and metastatic potential also was evaluated in low (AT 2.1 and GP 9F3) and high (MAT Lu and MLL) metastatic Dunning rat prostatic carcinoma cells. As shown in FIG. 5, the high metastatic cells produced ~20-fold more 12-HETE than the low metastatic AT 2.1. The LM line GP 9F3 produced a detectable amount of 12-HETE. Interestingly, GP 9F3 cells proliferate rapidly both *in vitro* and *in vivo.* Treatment of GP 9F3 cells with a select 12-LOX inhibitor (BHPP; 1 μM, see below) inhibited proliferation of GP 9F3 cells by 50% (data not shown), suggesting a possible role for 12-HETE in proliferation.

12-HETE production was correlated with metastatic potential in the B16F1 (low metastatic) and F10 (high metastatic) cell lines. B16F10 cells produced a significant amount of 12-HETE (135 pmoles/10^6 cells), while 12-HETE production from exogenous arachidonic acid was reduced (13 pmoles/10^6 cells) in B16F1 cells. A minor amount of 15-HETE metabolism was also observed (data not shown).

Finally, correlation between 12-HETE production and metastatic capability was demonstrated in another system, the K-1735 melanoma. This line was developed in a C3H mouse treated with a short course of exposure to UV radiation followed by chronic painting of the skin with croton oil.[18] Two clones were developed from the parental line. The C1-11 clone was obtained from the fifth *in vitro* passage of the parental cells, and the M1 cell line was derived from pulmonary metastasis after iv inoculation of the parent tumor cells. The M1 cell line demonstrates increased motility and lung colonization compared to the C1-11 line.[19] These two clones were analyzed for 12-HETE production by incubation with 10 μM AA followed by extraction of the metabolites and purification by RP-HPLC. The M1 cell line produced 13.6 pmoles of 12(S)-HETE/10^6 cells, while the C1-11 clone produced only 0.67 pmoles/10^6 cells during a 15-min incubation period. Therefore, in four separate models of high and low metastatic cells (elutriated B16a cells, Dunning prostate carcinoma, B16 F1 vs. F10, and K1735 M1 vs. K1735 C1-11), we consistently observed higher 12-HETE production in the high metastatic cell lines. In addition, this is demonstrated in two species (mouse and rat) and two separate tumor systems (melanoma and carcinoma).

The close association of 12-LOX/12(S)-HETE and metastatic potential of tumor cells also is revealed by inhibitor studies. BHPP(*N*-benzyl-*N*-hydroxy-5-phenyl-pentanamide) belongs to a group of hydroxyamic acids that was first reported to inhibit 5-LOX activity derived from rat leukocytes.[20] Subsequently it was demonstrated that one compound from this class (i.e., BHPP) effectively inhibited 12-LOX activity in cytosol prepared from 3LL and porcine leukocyte type 12-LOX.[3] Elutriated B16a cells from the HM340 subpopulation were incubated with BHPP (0.01-1 μM; 10 min) and labeled with exogenous AA or LA. Metabolites were extracted and analyzed by reverse-phase or normal-phase HPLC. A dose-dependent inhibition of 12-HETE biosynthesis from AA by BHPP was observed, and 92% inhibition of 12-HETE biosynthesis was observed at 1 μM of BHPP, while 15- and 5-HETE metabolism from AA and 13-HODE metabolism from LA were only slightly inhibited (Ref. 10 and data not shown). These data demonstrate that BHPP is a selective inhibitor of 12-HETE biosynthesis. BHPP was thus used to treat HM340 cells, which were then tested for adhesion to endothelium *in vitro* or injected intravenously to test for lung colonization. A dose-dependent inhibition of both HM340 cell adhesion to microvessel

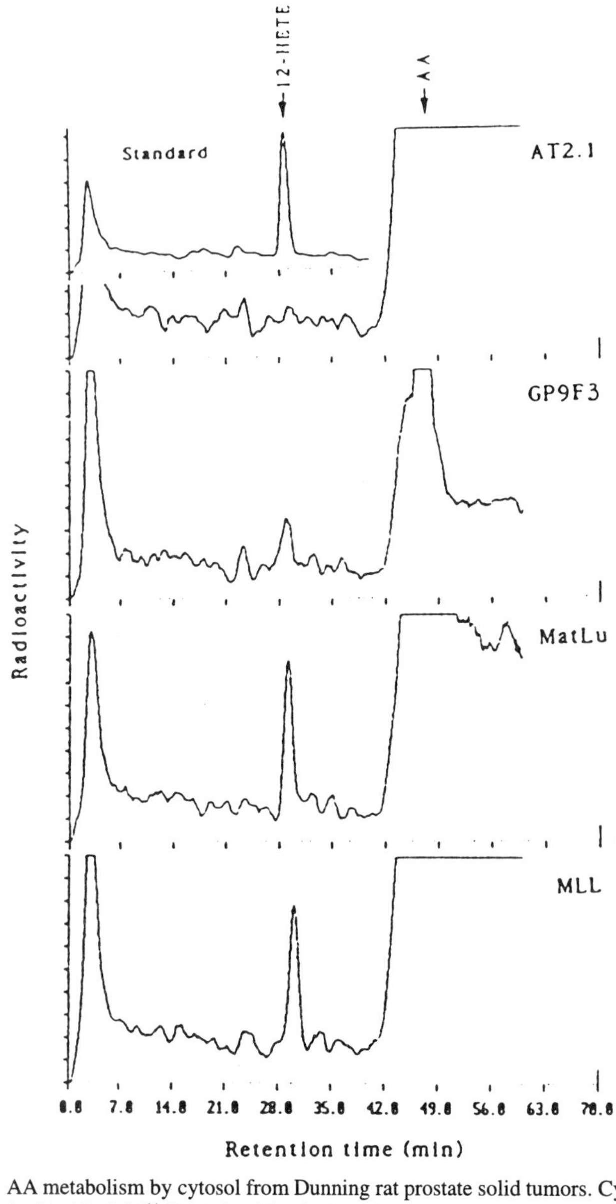

FIGURE 5. AA metabolism by cytosol from Dunning rat prostate solid tumors. Cytosol (2mg) was incubated with 10 μM [^{14}C]AA for 15 min. Metabolites were extracted and analyzed by reverse phase HPLC. Although low metastatic variants (i.e., AT2.1 and GP9F3) produced little 12(S)-HETE, high metastatic tumor sublines (i.e., MatLu and MLL) metabolized AA to high amounts of 12(S)-HETE.

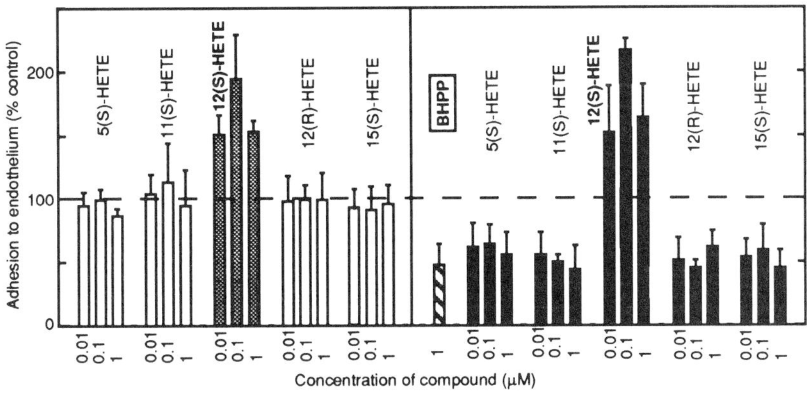

FIGURE 6. Effect of HETEs on the adhesion to endothelium of B16a cells pretreated with BHPP. B16a cells were pretreated with diluent ethanol (**left panel**) or BHPP (1 μM; 10 min) (**right panel**) followed by indicated HETE (15 min). Treated cells were assayed for their adhesion to microvessel endothelial cell monolayers.

endothelium and lung colonization *in vivo* was observed.[10,14] BHPP at the doses tested was not toxic to the cells. Oral administration of BHPP (25 mg/kg) to mice 30 min prior to iv injection of HM340 B16a cells decreased lung colonization by 60%.[14]

To confirm that the greater capacity of adhesion to endothelium by HM340 cells was due to their higher level of biosynthesis of 12-HETE, HM340 cells were pretreated with BHPP and then incubated with 5(S)-, 11(S)-, 12(S)-, 12(R)- or 15(S)-HETE (0.01 - 1 μM) and assayed for adhesion to microvessel endothelium. As demonstrated in FIG. 6, cells pretreated with BHPP and then incubated with 12(S)-HETE demonstrated an enhanced ability to adhere to endothelium comparable to 12(S)-HETE alone. No other monohydroxy fatty acid tested produced this effect.

WIDE-SPECTRUM ACTIVITY OF 12(S)-HETE: AN IMPORTANT MODULATOR OF METASTASIS

Physiologically 12(S)-HETE has been shown to induce platelet aggregation, stimulate insulin secretion, suppress renin production, chemoattract leukocytes, facilitate macrophage adhesion, and inhibit prostacyclin biosynthesis by vascular endothelial cells.[21] In addition, more and more signaling functions have been ascribed to 12(S)-HETE.[22,23] Experiments performed in our laboratory and others' have demonstrated that 12(S)-HETE possesses versatile biological activities in modulating the phenotypic properties of tumor cells and the metastatic processes such as tumor cell-platelet-endothelial cell interactions. The widespread effect of 12(S)-HETE in tumor metastasis has been summarized in several of our recent reviews;[24–26] therefore, we select to present only some new aspects of this eicosanoid in modulating the metastatic phenotypes of tumor cells in the present writing.

12(S)-HETE Modulates Tumor Cell Interactions with Matrix

Our early studies observed that 12(S)-HETE stimulated metastatic murine tumor cell adhesion to fibronectin and subendothelial matrix in a dose- and time-dependent fashion, with a maximal effect observed at 0.1 μM of 12(S)-HETE 15 min after stimulation.[27,28] The 12(S)-HETE enhanced tumor cell adhesion appeared to result from an increased surface expression of integrin receptors αIIbβ3, which was a consequence of receptor translocation to the cell surface from the intracellular pool without involving up-regulated gene transcription.[27] Recently, it was demonstrated that 12(S)-HETE also promotes tumor cell spreading on this matrix protein, by inducing reestablishment of the filamentous cytoskeleton system and enhancing the formation of focal adhesion plaques, a process dependent on the function of protein kinase C (PKC).[29]

12(S)-HETE Promotes Tumor Cell Motility

Several hydroxy fatty acid metabolites of AA including 12(S)-HETE have been shown to modulate the chemotaxis and motility of leukocytes.[21] Therefore, we examined the possibility that 12(S)-HETE may modulate tumor cell movement. Studies on murine melanoma cell lines with different metastatic potential (B16a, K1735-Cl.11, and K1735 M1) indicated that 12(S)-HETE induces tumor cell motility comparable to that induced by autocrine motility factor.[30] The 12(S)-HETE effect appears to be mediated by increasing the expression at the cell surface of gp78, a receptor for autocrine motility factor.[30] More recently, 12(S)-HETE is observed to increase the invasive potential of AT 2.1 rat prostate tumor cells.[31] The tumor cell motility (as assessed by phagokinetic track assay) and invasion (as assessed by Matrigel invasion assay) were significantly augmented by 12(S)-HETE.[31] The 12(S)-HETE effect was dramatically inhibited by a selective PKC inhibitor calphostin C as well as by a cell membrane-permeable Ca^{2+}-chelator BAPTA,[31] suggesting the involvement of PKC (see below).

12(S)-HETE Promotes Tumor Cell Release of Proteolytic Enzyme Cathepsin B

Cathepsin B is a cysteine proteinase normally stored in lysosomes. However, in tumor cells this proteolytic enzyme is found to be associated constitutively with the cell surface.[32] 12(S)-HETE treatment resulted in cathepsin B release from highly malignant cells, i.e., B16a murine melanoma cells and MCF10AneoT human mammary carcinoma cells.[33,34] 12(S)-HETE appears to trigger the release first of mature cathepsin B and then the immature form, suggesting that the mature form of the enzyme is at or near the cell surface, while the immature form is stored intracellularly. Morphologic studies indicated that, following 12(S)-HETE stimulation, the intracellular cathepsin B-containing vesicles are directed to specialized cell surface areas, i.e., lamellipodia and filopodia.[33]

12(S)-HETE Reorganizes Tumor Cell Cytoskeleton

The above presentation suggests that 12(S)-HETE has a regulatory effect on diverse cellular processes (adhesion, spreading, motility, lysosomal enzyme release, receptor translocation, etc.) associated with cytoskeletal functions. Therefore, we examined the effect of exogenous 12(S)-HETE on cytoskeleton in tumor cells.[35,36] These studies indicated that 12(S)-HETE induces translocation of organelles from the perinuclear space to the cell periphery paralleled by a reversible rearrangement of cytoskeletal filaments.[35] The earliest event post 12(S)-HETE treatment is a decreased labeling of cytoplasmic stress fibers with a concomitantly enhanced cortical actin labeling. These alterations in microfilaments are accompanied or immediately followed by microtubule polymerization and vimentin intermediate filament bundling, and also by rearrangements of the focal adhesion protein vinculin.[35,36] The 12(S)-HETE effect on tumor cell cytoskeleton is PKC dependent since pretreatment of tumor cells with a selective PKC inhibitor, i.e., calphostin C, abolished the 12(S)-HETE effects.[35] Interestingly, 12(S)-HETE-induced cytoskeletal alterations generally abate by ~30 min. However, simultaneous treatment of tumor cells with 12(S)-HETE and okadaic acid, a general serine/threonine protein phosphatase inhibitor, prevented disappearance of the 12(S)-HETE effects by 30 min.[36] These observations thus raise the possibility that a phosphatase(s), activated due to either the direct effect of 12(S)-HETE or homeostatic mechanisms, is responsible for dampening the 12(S)-HETE-triggered, PKC-dependent, and phosphorylation-mediated cytoskeleton rearrangement.

12(S)-HETE Modulates Endothelial Cell Phenotypes and Tumor Cell-Endothelial Cell Interactions

Tumor cell endothelial cell interactions, which are spatiotemporally regulated by a large array of bioactive mediators (growth factors, circulating hormones, cytokines, chemokines, lipid mediators, etc.) and can be analyzed at the molecular level with the "docking and locking" hypothesis,[25] constitute one of the most important factors in the organ preference of metastasis. 12(S)-HETE plays an essential modulatory role ranging from regulating tumor cell "locking" onto endothelial cells by modulating integrin expression to promoting tumor cell extravasation by inducing endothelial cell retraction.

Most tumor cells exit from the vessel by inducing endothelial cell retraction. The biological mediators responsible for tumor cell-induced endothelial cell retraction remain largely unknown. Five years ago, our laboratory presented the first experimental evidence that 12(S)-HETE is capable of inducing a nondestructive, reversible retraction of cultured large vessel endothelial cell monolayers.[37] Subsequent studies established that 12(S)-HETE also induces retraction of microvessel endothelial cells.[38–40] More importantly, we recently demonstrated that tumor cell adhesion to microvessel endothelium induced a nondenuding retraction of endothelial cell monolayers[11] that in many aspects mimicked the 12(S)-HETE-induced endothelial cell retraction. Tumor cell-induced endothelial cell retraction can be potentiated by homologous platelets,[12] thus providing an experimental rationale for the platelet contribution to tumor cell metastasis. Further investigations revealed that both platelet-augmented and tumor cell-induced endothelial cell retraction were mediated through 12(S)-HETE.[11,12]

We subsequently investigated biochemical mechanisms for the 12(S)-HETE-induced endothelial cell retraction. The series of studies indicated that 12(S)-HETE activates PKC in endothelial cells, which subsequently phosphorylates several major cytoskeletal elements, including myosin light chain, actin, and vimentin, leading to a reorganization of the cytoskeleton network (microfilament, intermediate filament, and focal adhesions) and apparent endothelial cell retraction.[38,39] 12(S)-HETE-induced endothelial cell retraction may also involve alterations in adhesion molecules that normally are concentrated at cell-cell junctions. Integrin $\alpha v\beta 3$ and PECAM-1 (platelet endothelial cell adhesion molecule-1), an immunoglobulin supergene family member that is expressed on platelets, vascular endothelial cells, certain subtypes of neutrophils, as well as in some solid tumor cells, are localized to the cell junctional areas in confluent endothelial cell cultures.[16,40,41] 12(S)-HETE treatment results in a series of well-defined changes in $\alpha v\beta 3$-containing focal adhesions, leading to an eventual decrease in $\alpha v\beta 3$ localization to focal adhesion plaques at both cell body and cell-cell borders.[40] In contrast to its effect on integrin $\alpha v\beta 3$, 12(S)-HETE does not appear to alter the normal cellular distribution patterns of another integrin receptor, i.e., $\alpha 5\beta 1$, an authentic fibronectin receptor. The 12(S)-HETE-induced reorganization of $\alpha v\beta 3$-enriched focal adhesions resembles its effect on vinculin and again appears to be dependent on PKC activation.[40] PECAM-1 has been proposed to be involved in maintaining the integrity of endothelial cell monolayers.[41] This molecule is concentrated at the cell-cell junctions of both large-vessel (i.e., rat aortic endothelial cells, or RAEC) and microvessel (i.e., CD3 pulmonary capillary endothelial cells) endothelial cells (FIG. 7). Treatment of postconfluent RAEC and CD3 cultures dispersed cell border-localized PECAM-1 molecules to the whole cell body (FIG. 7). Whether this alteration of PECAM-1 is a trigger or a consequence of 12(S)-HETE-induced endothelial cell retraction remains to be established.

12(S)-HETE also modulates the surface expression of integrin receptors on endothelial cells. 12(S)-HETE stimulation up-regulates the surface expression of integrin $\alpha v\beta 3$ (but not $\alpha 5\beta 1$) in both large-vessel and microvessel endothelial cells.[40,42] The increased $\alpha v\beta 3$ surface expression appears to be concomitant with a decrease in $\alpha v\beta 3$-enriched focal adhesions and is a posttranscriptional, PKC- and cytoskeleton-dependent process.[40,43] The end result of 12(S)-HETE enhanced surface expression of $\alpha v\beta 3$ integrin receptors is increased tumor cell adhesion to the activated endothelium.[40,42,43]

Studies by others have shown that 12(S)-HETE may be involved in regulating endothelial cell growth. DNA synthesis as well as proliferation of cultured fetal bovine aortic endothelial cells were inhibited by nordihydroguaiaretic acid (NDGA), a general lipoxygenase inhibitor, but were not affected by the cyclooxygenase inhibitor indomethacin,[44] suggesting lipoxygenase metabolites as endogenous regulators of endothelial cell growth. These endothelial cells generate as their major lipoxygenase metabolite 15(S)-HETE, with a minor amount of 12(S)-HETE.[45] Application of either of these two lipoxygenase metabolites promoted significant DNA synthesis as well as proliferation of endothelial cells.[45] Very recently, the vital role of lipoxygenase products, 12(S)-HETE in particular, in controlling endothelial cell growth has been

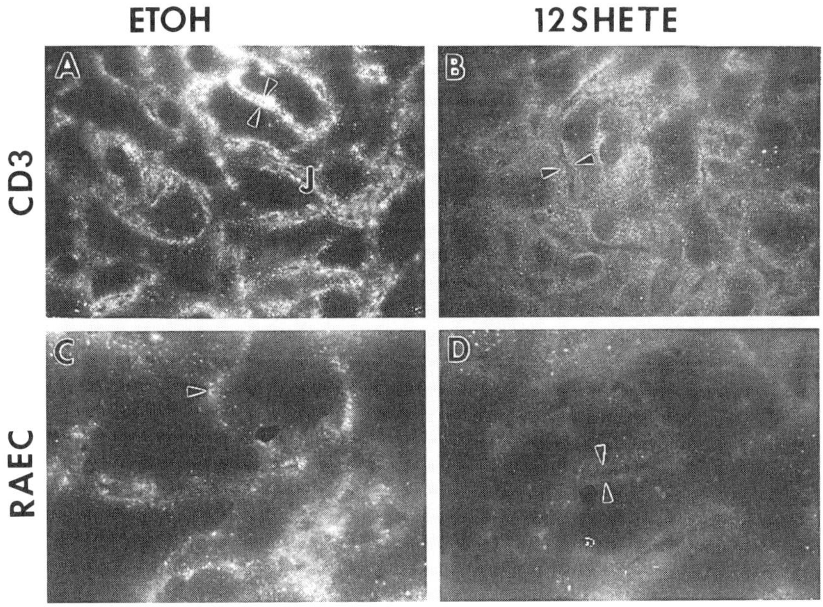

FIGURE 7. 12(S)-HETE disrupts the localization of PECAM-1 molecules in the cell-cell borders. Postconfluent large vessel RAEC (rat aortic endothelial cells) (**C and D**) or CD3 microvessel (murine capillary endothelial cells) (**A and B**) endothelial cells grown in Vitrogen were treated with either diluent ethanol (**A and C**) or 0.1 μM 12(S)-HETE (**B and D**) for 15 min. Thereafter, cells were permeabilized and then stained for PECAM-1 using SEW-3 polyclonal anti-PECAM-1 antibody (kindly provided by Dr. P. J. Newman). PECAM-1 in control cultures (**A and C**) is typically localized to cell-cell junctional (J) areas (indicated by *arrowheads*), while in 12(S)-HETE-treated cells it is dispersed to the cell bodies leading to a diffused staining and wider cell-cell borders (indicated by *arrowheads*). (Magnification: 600 ×.)

established in bovine capillary endothelial cells.[46] Beta FGF-, PDGF-, and serum-stimulated vascular cell (including both endothelial cells and smooth muscle cells) growth is dose-dependently inhibited by NDGA and a 12-lipoxygenase inhibitor baicalein, but not by indomethacin and a 5-lipoxygenase inhibitor 5,6-dehydroarachidonic acid.[46] The growth of unstimulated endothelial cells was also inhibited by NDGA and baicalein,[46] lending further support to the concept that 12(S)-HETE and perhaps some other lipoxygenase metabolites such as 15(S)-HETE are necessary growth signals for vascular cells. PDGF and bFGF are well-known angiogenic factors. The observations that their cell growth-promoting effects are to a certain extent mediated through lipoxygenase metabolites such as 12(S)-HETE[46] strongly suggest that the latter by themselves may be important physiological angiogenic factors.

12(S)-HETE Mediates its Biological Activity by Activating PKC

Our early experiments observed that 12(S)-HETE mimicked the tumor promoter PMA in enhancing tumor cell integrin expression and adhesion,[28] suggesting that

12(S)-HETE may work via activating PKC. This hypothesis was supported by subsequent experiments, demonstrating that in W256 cells 12(S)-HETE induced a 100% increase in membrane-associated PKC activity and that down-regulation of PKC by prolonged treatment with PMA abolished the 12(S)-HETE-enhanced W256 cell adhesion to endothelium.[13] To further support PKC activation by 12(S)-HETE, our later work revealed that essentially all of the 12(S)-HETE-evoked biological responses (in both tumor cells and endothelial cells), encompassing rearrangement of cytoskeleton, promotion of cell adhesion and spreading, induction of endothelial cell retraction, release of proteolytic enzymes, and enhancement of cell motility, can be abrogated by selective PKC inhibitors (see Refs. 24-26 for reviews). Experiments by others also have shown that 12(S)-HETE and some other HETEs (such as 15-HETE) activate PKC.[47]

How 12(S)-HETE activates PKC has not been clearly characterized. The physiological activators of PKC are diacylglycerol (DAG) and Ca^{2+}. It appears that 12(S)-HETE in different cell types activates different isoform(s) of PKC. In murine B16a melanoma cells, PKCα is the only isoform that can be detected by immunoblotting using isoform-specific anti-peptide antibodies (Ref. 26 and data not shown). 12(S)-HETE stimulation leads to an increased translocation of PKC from the diffuse cytoplasmic pool to the plasma membrane.[26] When B16a cells plated on fibronectin were stimulated with 12(S)-HETE, increased membrane-associated PKC was observed to colocalize with vinculin to the focal adhesion plaques.[26] In rat AT 2.1 prostate carcinoma cells, two isoforms of PKC, i.e., α and δ, were identified.[31] Interestingly, 12(S)-HETE preferentially activates PKCα by increasing the membrane association of this isoform, resulting in an up-regulated tumor cell motility and invasion.[31] Treating tumor cells with thymelea toxin, a selective activator of PKCα over Ca^{2+}-independent PKCδ, also promoted tumor cell motility and invasion,[26,31] thus confirming the preferential activation of PKCα by 12(S)-HETE in this tumor cell line. Further, BAPTA, a Ca^{2+}-chelator, dose-dependently inhibited 12(S)-HETE-promoted tumor cell motility and invasion,[26,31] suggesting that the PKC isoform activated by 12(S)-HETE is Ca^{2+} dependent. In murine microvessel endothelial cells (i.e., CD3 cells), at least two isoforms of PKC are expressed (Tang, Liu, and Honn, unpublished observations). Which isoform of PKC is activated or whether both isoforms of PKC are activated by 12(S)-HETE is not yet known. In rat adrenal glomerulosa cells, the α and ϵ isoforms of PKC are expressed.[48] In sharp contrast to the situation in tumor cells, 12(S)-HETE and 15(S)-HETE alike predominantly activate PKCϵ, although angiotensin II increases the membrane-bound levels of both PKCα and PKCϵ.[48]

Interestingly, 12(S)-HETE does not appear to directly activate PKC. Stimulation of tumor cells with 12(S)-HETE is observed to be followed by a rapid accumulation of DAG and IP3,[49] therefore suggesting PIP_2 metabolism through the action of PLC. To support this, blocking the functions of G proteins by pertussis toxin in B16a melanoma cells as well as inhibiting the PLC activity by specific antagonists also suppress the effects of exogenous 12(S)-HETE,[49] indicating that 12(S)-HETE may function via activating an upstream G protein. Another apparent alternative for the accumulation of DAG following 12(S)-HETE treatment is through down-regulating the DAG kinase activity, as proposed by Setty *et al.*[45] In the phosphoinositide cycle, DAG derived from various sources undergoes rapid phosphorylation by DAG kinase to phosphatidic acid.[45] In bovine aortic endothelial cells, 12(S)-HETE as well as

TABLE 1. Biological Activities of 12(S)-HETE

Effect on tumor cells
- Activates protein kinase C
- Reorganizes tumor cell cytoskeleton
- Increases tumor cell adhesiveness by up-regulating integrin αIIbβ3 expression
- Facilitates tumor cell spreading on extracellular matrix
- Enhances tumor cell motility
- Promotes release of cathepsin B from tumor cells
- Stimulates tumor cell invasion

Effect on endothelial cells
- Activates protein kinase C
- Reorganizes cytoskeletal network
- Induces reversible endothelial cell retraction
- Modulates integrin αvβ3 expression
- Increases adhesiveness of endothelial cells
- Regulates endothelial cell growth and angiogenic properties

Effect on tumor cell-endothelial cell interactions
- Promotes tumor cell adhesion to endothelial cells
- Mediates tumor cell-induced endothelial cell retraction

15(S)-HETE were found to inhibit the DAG kinase activity, leading to an accumulation of DAG mass in these cells.[45]

Based on the above discussions, a general scheme can be summarized for the effects of 12(S)-HETE (see TABLE 1). 12(S)-HETE, possibly by binding to a G protein-linked cell surface binding site(s) or receptor, activates PKC, which subsequently phosphorylates and reorganizes the cytoskeletal elements that mobilize αIIbβ3 integrin-containing vesicles to the plasma membrane. Cytoskeletal rearrangements induced by 12(S)-HETE also are involved in modulation of tumor cell spreading on matrix and tumor cell motility. Tumor cell-derived 12(S)-HETE likewise acts on endothelial cells, resulting in enhanced αvβ3 integrin surface expression, tumor cell adhesion, and endothelial cell retraction. In addition, 12(S)-HETE may promote tumor cell release of cathepsin B to degrade the subendothelial matrix. *In vivo,* tumor cell interactions with platelets will lead to platelet aggregation, activation, and release reactions. Platelet-derived 12(S)-HETE similarly acts on endothelial cells to initiate the diverse biological responses described above, thus contributing to tumor cell metastasis.

ACKNOWLEDGMENTS

We wish to express our heartfelt thanks to many of our colleagues for allowing the use and citation of their work, published and unpublished, in this writing. Special thanks go to Dr. B. Liu for his experiments on tumor cell metabolism of AA and LA, Dr. S. Duniec for performing HPLC analysis on prostatic carcinoma cells, and

Ms. C. Renaud for performing adhesion assays. The generous gifts of the platelet-type 12-LOX cDNA clone from Dr. C. Funk (Vanderbilt University) and of antibodies against PECAM-1 from Dr. P. J. Newman (The Blood Center of Southeastern Wisconsin) are highly appreciated.

REFERENCES

1. Funk, C. D., L. Furchi & G. A. Fitzgerald. 1990. Proc. Natl. Acad. Sci. USA **87:** 5638-5642.
2. Funk, C. D., L. B. Funk, G. A. Fitzgerald & B. Samuelsson. 1992. Proc. Natl. Acad. Sci. USA **89:** 3962-3966.
3. Marnett, L. J., M. T. Leithauser, K. M. Richards, I. Blair, K. V. Honn, S. Yamamoto & T. Yoshimoto. 1990. Adv. Prostaglandin Thromboxane Leukotriene Res. **21:** 895-900.
4. Funk, C. D., L. Furchi & G. A. Fitzgerald. 1990. Adv. Prostaglandin Thromboxane Leukotriene Res. **21:** 33-38.
5. Hada, T., N. Ueda, Y. Takahashi & S. Yamamoto. 1991. Biochim. Biophys. Acta **1083:** 89-96.
6. De Marzo, N., D. L. Sloane, S. Dicharry, E. Highland & E. Sigal. 1992. Am. J. Physiol. **262:** L198-L207.
7. Watanabe, S., J. F. Medina, J. Z. Haeggstrom, O. Radmark & B. Samuelsson. 1993. Eur. J. Biochem. **212:** 605-612.
8. Chen, X. S., U. Kurre, N. A. Jenkins, N. G. Copeland & C. D. Funk. 1994. J. Biol. Chem. **269:** 13979-13987.
9. Honn, K. V., K. K. Nelson, C. Renaud, R. Bazaz, C. A. Diglio & J. Timar. 1992. Prostaglandins **44:** 413-429.
10. Liu, B., L. J. Marnett, A. Chaudhary, C. Ji, I. A. Blair, C. R. Johnson, C. A. Diglio & K. V. Honn. 1994. Lab. Invest. **70:** 314-323.
11. Honn, K. V., D. G. Tang, I. Grossi, Z. M. Duniec, J. Timar, C. Renaud, M. Leithauser, I. Blair, C. R. Johnson, C. A. Diglio, V. A. Kimler, J. D. Taylor & L. J. Marnett. 1994. Cancer Res. **54:** 565-574.
12. Honn, K. V., D. G. Tang, I. M. Grossi, C. Renaud, Z. M. Duniec, C. R. Johnson & C. A. Diglio. 1994. Exp. Cell Res. **210:** 1-9.
13. Liu, B., J. Timar, J. Howlett, C. A. Diglio & K. V. Honn. 1991. Cell Regul. **2:** 1045-1055.
14. Chen, Y. Q., Z. M. Duniec, B. Liu, W. Hagmann, X. Gao, K. Shimoji, L. J. Marnett, C. R. Johnson & K. V. Honn. 1994. Cancer Res. **54:** 1574-1579.
15. Hagmann, W., X. Gao, J. Timar, Y. Q. Chen, C. Fahrenkopf, D. Kagawa, M. Lee, A. Zacgarek & K. V. Honn. Submitted for publication.
16. Tang, D. G., Y. Q. Chen, P. J. Newman, L. Shi, X. Gao, C. A. Diglio & K. V. Honn. 1993. J. Biol. Chem. **268:** 22883-22894.
17. Liu, B., C. Renaud, K. K. Nelson, Y. Q. Chen, R. Bazaz, J. Kowynia, J. Timar, C. A. Diglio & K. V. Honn. 1992. Int. J. Cancer **52:** 147-152.
18. Kripke, M. L. 1979. J. Natl. Cancer Inst. **63:** 541-548.
19. Watanabe, H., I. R. Nabi & A. Raz. 1991. Cancer Res. **51:** 2699-2705.
20. Huang, F. C., S. Shopue, C. J. Lin, T. D. Y. Lee, W. K. Chan, M. Schnapper, J. T. Suh, R. J. Gordon, P. A. Sonnino, C. A. Sutherland, R. G. Vaninwegen & S. M. Coutts. 1989. J. Med. Chem. **32:** 1836-1842.
21. Spector, A. A., J. A. Gordon & S. A. Moore. 1988. Prog. Lipid Res. **27:** 271-323.
22. Piomelli, D., A. Volterra, N. Dale, S. A. Siegelbaum, E. R. Kandel, J. H. Schwartz & F. Belardetti. 1987. Nature **328:** 38-43.

23. NADLER, J. L., R. NATARAJIAN & N. STERN. 1987. J. Clin. Invest. **80:** 1763-1769.
24. CHEN, Y. Q., B. LIU, D. G. TANG & K. V. HONN. 1992. Cancer Metastasis Rev. **11:** 389-410.
25. HONN, K. V. & D. G. TANG. 1992. Cancer Metastasis Rev. **11:** 353-375.
26. HONN, K. V., D. G. TANG, X. GAO, I. A. BUTOVICH, B. LIU, J. TIMAR & W. HAGMANN. 1994. Cancer Metastasis Rev. In press.
27. CHOPRA, H., J. TIMAR, Y. Q. CHIEN, X. RONG, I. M. GROSSI, L. A. FITZGERALD, J. D. TAYLOR & K. V. HONN. 1991. Int. J. Cancer **49:** 774-786.
28. GROSSI, I. M., L. A. FITZGERALD, L. A. UMBARGER, K. K. NELSON, C. A. DIGLIO, J. D. TAYLOR & K. V. HONN. 1989. Cancer Res. **49:** 1029-1037.
29. TIMAR, J., Y. Q. CHEN, B. LIU, R. BAZAZ, J. D. TAYLOR & K. V. HONN. 1992. Int. J. Cancer **52:** 594-603.
30. TIMAR, J., S. SILLETTI, R. BAZAZ, A. RAZ & K. V. HONN. 1993. Int. J. Cancer **55:** 1003-1010.
31. LIU, B., R. J. MAHER, Y. A. HANNUN, A. T. PORTER & K. V. HONN. 1994. J. Natl. Cancer Inst. **86:** 1145-1151.
32. SLOANE, B. F., J. ROZHIN, K. JOHNSON, H. TAYLOR, J. D. CRISSMAN & K. V. HONN. 1986. Proc. Natl. Acad. Sci. USA **83:** 2483-2487.
33. HONN, K. V., J. TIMAR, J. ROZHIN, R. BAZAZ, M. SAMENI, G. ZIEGLER & B. F. SLOANE. 1994. Exp. Cell Res. In press.
34. SLOANE, B. F., J. ROZHIN, A. P. GOMEZ, I. M. GROSSI & K. V. HONN. 1991. *In* Eicosanoids and Other Bioactive Lipids in Cancer and Radiation Injury. K. V. Honn, L. J. Marnett, S. Nigam & T. Walden, Eds.: 373-377. Kluwer Academic Publishers. Boston, MA.
35. TIMAR, J., D. TANG, R. BAZAZ, M. M. HADDARD, V. A. KIMLER, J. D. TAYLOR & K. V. HONN. 1993. Cell Motil. Cytoskeleton **26:** 49-65.
36. TANG, D. G. & K. V. HONN. 1994. *In* Eicosanoids and Other Bioactive Lipids in Cancer, Inflammation, and Radiation Injury. K. V. Honn, L. J. Marnett & S. Nigam, Eds. Kluwer Academic Publishers. Boston, MA. In press.
37. HONN, K. V., I. M. GROSSI, C. A. DIGLIO, M. WOITUKIEWICZ & J. D. TAYLOR. 1989. FASEB J. **3:** 2285-2293.
38. TANG, D. G., J. TIMAR, I. M. GROSSI, C. RENAUD, V. A. KIMLER, C. A. DIGLIO, J. D. TAYLOR & K. V. HONN. 1993. Exp. Cell Res. **207:** 361-375.
39. TANG, D. G., C. A. DIGLIO & K. V. HONN. 1993. Prostaglandins **45:** 249-268.
40. TANG, D. G., Y. Q. CHEN, C. A. DIGLIO & K. V. HONN. 1993. J. Cell Biol. **121:** 689-704.
41. ALBELDA, S. M., W. A. MUELLER, C. A. BUCK & P. J. NEWMAN. 1991. J. Cell Biol. **114:** 1059-1068.
42. TANG, D. G., I. M. GROSSI, Y. Q. CHEN, C. A. DIGLIO & K. V. HONN. 1993. Int. J. Cancer **54:** 102-111.
43. TANG, D. G, C. A. DIGLIO & K. V. HONN. 1994. Cancer Res. **54:** 102-111.
44. SETTY, B. N. Y., R. L. DUBOWY & M. J. STUART. 1987. Biochem. Biophys. Res. Commun. **144:** 345-351.
45. SETTY, B. N. Y., J. E. GRAEBER & M. J. STUART. 1987. J. Biol. Chem. **262:** 17613-17622.
46. DETHLEFSEN, S. M., D. SHEPRO & P. D'AMORE. 1994. Exp. Cell Res. **212:** 262-273.
47. HAGERMAN, R. A., T. J. SMITH & M. F. LOCNISKAR. 1993. FASEB J. **7:** A601.
48. NATARAJIAN, R., L. LANTING & X. J. NADLER. 1994. Mol. Cell. Endocrinol. **101:** 59-66.
49. LIU, B., W. A. KHAN, Y. A. HANNUN, R. BAZAZ, C. RENAUD, S. STOJAKOVIC, J. TIMAR, J. D. TAYLOR & K. V. HONN. Submitted for publication.

Lipoxygenase Products in Normal and Diseased Glomeruli[a]

KAMAL F. BADR[b] AND FADI G. LAKKIS

Division of Nephrology
Department of Medicine
Emory University
and
Atlanta Veterans Administration Medical Center
Atlanta, Georgia 30033

In our studies regarding the role of eicosanoids in immune-mediated inflammatory glomerular injury, we have focused our efforts on those events that immediately follow the deposition and/or *in situ* formation of immune complexes in the glomerulus. As depicted in FIGURE 1, a generalized paradigm of immune injury, presence of immune complexes in the glomerulus, regardless of their sources, routes of formation, or intraglomerular localization, inevitably and necessarily provokes a complement-mediated influx and activation of polymorphonuclear leukocytes (PMN). PMNs are detected frequently when biopsies are performed during ongoing acute injury, such as in patients with postinfectious glomerulonephritides.[1] Characteristically, this initial wave of neutrophil infiltration/activation is replaced by monocyte infiltration and macrophage proliferation and activation. While strategies aimed at arresting glomerular injury by targeting the mediators of matrix expansion and scar formation show promise,[2,3] the complexity of the "mediator soup" during this phase of injury and the various cell populations involved (including tubulointerstitial elements[4]) present serious theoretical and practical obstacles to the development of effective therapeutic interventions.

Targeting the mechanisms that govern the severity of early immune-mediated injury rests on the premise that those diseases that most commonly lead to renal failure due to immune deposition are, for the most part, progressive over months to years, suggesting incremental phases of nephron loss. Evidence from pathologic examinations in several forms of glomerulonephritis indicates that injury is heterogeneous: the number of affected versus healthy glomeruli varies among patients, as well as over time in individual patients. Moreover, within individual glomeruli lesions are often segmental, with inflammatory reactions present in certain lobules while others are totally normal.[1] These data, as well as a clinical course characterized by steadily diminishing renal reserve over highly varying periods of time, suggest strongly that in an individual patient "early" injury is occurring continuously in some fixed proportion of nephrons. It is therefore reasonable to predict that institution of therapy

[a] This work was supported by National Institutes of Health Grant DK-43883.

[b] Address for correspondence: Kamal F. Badr, Division of Nephrology, Atlanta VA Medical Center, 1670 Clairmont Road, Atlanta, Georgia 30033.

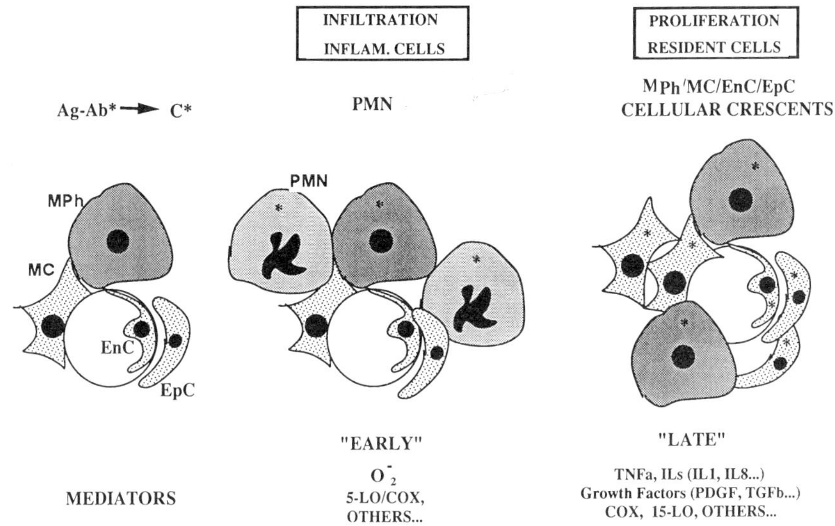

FIGURE 1. General scheme of immune injury in the glomerulus. Following antigen-antibody (Ag-Ab*) complex formation/deposition, complement activation (C*) provokes an early PMN infiltrate and activation (*) of resident macrophages. Mediators of injury at this stage include reactive oxygen molecules (O_2^-) and 5-LO cyclooxygenase (COX) arachidonate metabolites. This initial reaction is replaced by a proliferative stage of injury in which indigenous glomerular mesangial (MC), epithelial (EpC), and endothelial (EnC) cells are stimulated and may undergo proliferative reactions, possibly including epithelial cell crescent formation. At this stage, which precedes matrix deposition and fibrosis, the inflammatory mediator profile expands to include cytokines, such as tumor necrosis factor (TNF), various interleukins (IL), growth factors (such as platelet-derived growth factor [PDGF], transforming growth factor-beta [TGFβ], and others), as well as COX and 15-LO products of arachidonic acid.

that specifically targets these early events will arrest initial injury in those nephrons, albeit small in number, in which it is underway and, more importantly, prevent or abort its development in intact nephrons, despite the potential continued deposition or formation of immune complexes in these normal glomeruli. This latter assumption is based on the dramatic evidence from experimental studies indicating that mere deposition of antigen-antibody complexes in the glomerular capillary wall or mesangium, in the absence of cellular infiltration (as in leukocyte- or complement-depleted animals) or the capacity to generate arachidonate metabolites (as in fatty acid-deficient animals), is without any detrimental acute or chronic consequences to glomerular structure and functions.[5]

LEUKOTRIENES AS MEDIATORS OF INJURY DURING GLOMERULAR INFLAMMATION

The sulfidopeptide leukotrienes (LTs:LTC_4, LTD_4, and LTE_4) and the nonpeptidyl LTB_4 elicit potent biological responses: LTC_4 and LTD_4 contract vascular, pulmonary, and gastrointestinal smooth muscle and increase vascular permeability to macromolecules.[6,7] LTB_4 has minimal spasmogenic properties. Its primary target appears to be PMNs, which express specific high- and low-affinity receptors for LTB_4.[8] Through the former, LTB_4 is the most potent chemotactic substance yet described for this cell and also increases PMN aggregation and adhesion to endothelium.[7] Through the latter, it acts as a calcium ionophore,[9] leading to PMN activation,[10–12] stimulation of phosphoinositide turnover,[13,14] release of lysosomal enzymes, and an increase in oxidative metabolism.[9–14] In turn, activated PMNs are the best-studied source of LTB_4,[7] where its synthesis is coupled to activation of protein kinase C[15].

Biologic Activities in the Normal Glomerulus: LTC_4 and LTD_4

Systemic administration of LTC_4 in the rat leads to reductions in renal blood flow (RBF) and glomerular filtration rate (GFR).[16] Since the responses to LTC_4 in vascular smooth muscle results from its bioconversion to LTD_4,[17] we evaluated the glomerular microcirculatory actions of LTD_4 in the intact kidney *in vivo*, using glomerular micropuncture techniques.[18] Intrarenal administration of LTD_4 decreased RBF, due to augmentation of postglomerular arteriolar resistance (Re). Notably, LTD_4 administration reduced markedly the glomerular capillary ultrafiltration coefficient (K_f), thereby decreasing GFR.[18] *In vitro,* LTC_4 and LTD_4 contract glomerular mesangial cells,[19,20] a presumed mechanism for the fall in K_f *in vivo*.

In addition to its vasoactive properties in arteriolar and mesangial smooth muscle, LTD_4 exerts additional proinflammatory actions via its promotion of PMN adhesion to mesangial cells, an effect likely mediated through the augmented expression of leukocyte adhesion proteins on the mesangial cell surface.[21] Binding of $[^3H]LTD_4$ to mesangial cells in culture is stereoselective, specific, saturable, and rapidly reversible.[22] Two binding sites were recognized with dissociation constants and binding site densities at equilibrium of 2.2 and 16.8 nM and 1.1×10^4 and 3×10^4 binding sites/cell. LTD_4 also induced time- and concentration-dependent formation of inositol triphosphate (IP_3), increased intracellular pH, and stimulated $[^3H]$thymidine incorporation and mitogenesis in glomerular mesangial cells. These findings were confirmed in *human* mesangial cells, where the K_d for LTD_4 binding (12 nM) was remarkably similar to that of the rat.[23]

Biologic Activities in the Normal Glomerulus: LTB_4

In the context of studies examining the pathophysiologic role of LTB_4 in experimental rat glomerulonephritis, Yared and colleagues tested the potential biologic actions of LTB_4 in normal kidneys.[24] Intrarenal administration of LTB_4 resulted in mild vasorelaxant and natriuretic responses. No increase in PMN infiltration was noted. Hence, LTB_4 as such is without acute detrimental effects on glomerular functions or structure.

Biologic Activities in the Inflamed Glomerulus: LTC_4 and LTD_4

Increased generation rates of LTC_4 and LTD_4 have been documented in glomeruli from rats with nephrotoxic serum nephritis and mice with spontaneously developing lupus nephritis.[25,26] Moreover, results from numerous physiologic studies utilizing specific LTD_4 receptor antagonists have provided strong evidence for the release of these eicosanoids during glomerular inflammation: In four animal models of glomerular immune injury (anti-GBM nephritis,[27] anti-Thy 1 Ag mesangiolysis,[28] passive Heymann nephritis [PHN],[29] and murine lupus nephritis[26]) acute antagonism of LTD_4 by receptor binding competition or inhibition of LTD_4 synthesis led to highly significant increases in GFR in nephritic animals. The principal mechanism underlying the improvement in GFR was reversal of the depressed values of the glomerular ultrafiltration coefficient, which is characteristically compromised in immune injured glomeruli.[27,28,29–32] In more recent studies in PHN,[29] Katoh and colleagues provide evidence that endogenous LTD_4 not only mediates reductions in K_f and GFR, but that LTD_4-evoked increases in intraglomerular pressure underlie, to a large extent, the accompanying proteinuria.

Biologic Activities in the Inflamed Glomerulus: LTB_4

LTB_4 synthesis, measured in the supernates of isolated glomeruli, is enhanced markedly early in the course of several forms of glomerular immune injury.[33–38] Cellular sources of LTB_4 in injured glomeruli include PMNs and macrophages.[29,36–38] All studies concur as to the *transient* nature of LTB_4 release:[33–38] LTB_4 levels are undetectable beyond the first 24 hours of injury in both proliferative as well as "nonproliferative" glomerulopathies. The suppression of LTB_4 synthesis beyond the first 24 h of injury is rather surprising, since both PMN and macrophages are capable of effecting the total synthesis of LTB_4 (they contain the two necessary enzymes that convert arachidonic acid to LTB_4, namely, 5-lipoxygenase [5-LO] and LTA_4-hydrolase).[6,7] It has therefore been suggested, based on *in vitro* evidence, that the major route for LTB_4 synthesis in inflamed glomeruli is through transcellular metabolism of leukocyte-generated LTA_4 to LTB_4 by LTA_4-hydrolase present in glomerular mesangial, endothelial, and epithelial cells.[39] Since the transformation of LTA_4 to LTB_4 is rate limiting, regulation of the LTB_4 synthesis rate might relate to regulation of LTA_4-hydrolase gene expression or catalytic activity in these parenchymal cells, rather than to the number of infiltrating leukocytes. The recent cloning of rat leukotriene A_4 hydrolase should allow for elucidation of these mechanisms in animal models of injury.[40] In any case, leukocytes represent an indispensable source for LTA_4, the initial 5-LO product and the precursor for LTB_4, since indigenous glomerular cells do not express the 5-LO gene.[41] Thus, it was demonstrated that the PMN cell-specific activator *N*-formyl-Met-Leu-Phe stimulated LTB_4 production from isolated perfused kidneys harvested from NTS-treated rats to a significantly greater degree than from control animals treated with nonimmune rabbit serum.[24] The renal production of LTB_4 correlated directly and strongly ($R = 0.79$, $p < 0.01$) with renal myeloperoxidase activity, suggesting interdependence of LTB_4 generation and PMN infiltration. Similarly, macrophage depletion in PHN is associated with abrogation of stimulated LTB_4 synthesis.[29,37]

The acute and long-term significance of LTB_4 generation in conditioning the extent of glomerular structural and functional deterioration has been highlighted in studies in which LTB_4 was exogenously administered[24] or in which its endogenous synthesis was inhibited.[24,42] Intrarenal administration of LTB_4 to rats with mild nephrotoxic serum nephritis (NTS)-induced injury was associated with an increase in PMN infiltration, reduction in renal plasma flow rate, and marked exacerbation of the fall in glomerular filtration rate, the latter correlating strongly with the number of infiltrating PMNs/glomerulus; while inhibition of 5-lipoxygenase led to preservation of GFR and abrogation of proteinuria.[24] Of interest, 5-LO inhibition limited to the initial two days following induction of NTS nephritis resulted in total absence of proteinuria two weeks later.[42] Thus, while devoid of vasoconstrictor actions in the normal kidney, increased intrarenal generation of LTB_4 during early glomerular injury amplifies leukocyte-dependent reductions in glomerular perfusion and filtration rates, likely due to enhancement of PMN recruitment/activation.

ARACHIDONATE 15-LIPOXYGENASE PRODUCTS: CHEMISTRY AND CELLULAR SOURCES

Hydroperoxidation of arachidonic acid by 15-LO leads to the formation of 15-(S)-hydroxyeicosatetraenoic acid [15-(S)-HETE]. Dual lipoxygenation at both the 5 and 15 positions in activated neutrophils and macrophages yields a class of "lipoxygenase interaction products" [lipoxins (LXs)].[7] Like 5-LO, 15-LO gene expression is restricted largely to leukocyte cell lines, but has also been detected in reticulocytes and airway epithelial cells.[43] Macrophages are a particularly rich source of 15-LO and hence of 15-S-HETE and LXs.[7] Three biologically active lipoxins have been identified–LXA_4, LXB_4, and 7-*cis*-11-*trans*-LXA_4;[7,44] the pharmacological profile of their renal actions has been characterized.[45] Lipoxin synthesis, like that of leukotrienes, can also occur via transformation of leukocyte-generated LTA_4 by either 15-LO or 12-LO in adjoining cells, such as mesangial cells[46] or platelets.[47]

ARACHIDONATE 15-LIPOXYGENASE PRODUCTS: "ANTIINFLAMMATORY" ACTIONS

Evidence for a generalized antiinflammatory role for 15-LO products has been derived from clinical observations and experimental studies *in vivo* and *in vitro*.[48–55] Administration of 15-S-HETE causes regression of psoriatic lesions in humans[49] and significantly reduces the clinical severity of a canine arthritis model.[50]

15-LO Products Antagonize LTB_4 Generation and Actions in Human and Rat Neutrophils

Upon exposure of human neutrophils to 15-S-HETE, it is rapidly incorporated and stored in the phosphatidyl inositol fraction of membrane lipids and subsequently released upon cellular activation to generate alternate profiles of intracellular messengers and eicosanoids.[51] These intracellular second messengers, such as 15-S-HETE-substituted diacylglycerol and/or LXA_4, appear to markedly dampen the capacity of PMN to undergo a respiratory burst and other classical activation responses associated with PMN stimulation during immune reactions.[51] Since LTB_4 generation is a consequence of leukocyte activation, LTB_4 synthesis is markedly depressed in PMNs preincubated with 15-S-HETE and subsequently activated *in vitro*.[51,52] Furthermore, preincubation of rat PMN with 15-S-HETE abolishes their chemotactic responses to LTB_4 but not to other chemoattractants (see below), suggesting specific inhibitory interactions between 5- and 15-LO products in the regulation of PMN recruitment and activation.[52] In addition to these intracellular actions, exogenously administered LXA_4 and LXB_4 exert antagonistic actions on LTB_4-induced chemotaxis of human neutrophils,[53] suggesting that following extracellular export LXs exert biological actions via interactions with cell surface receptors on PMN and other cell types.

CONTRASTING ACTIONS OF LEUKOTRIENES AND 15-LO PRODUCTS IN THE GLOMERULUS

Single Nephron Function in Vivo *and Mesangial Cell Actions* in Vitro*: LTD_4/LXA_4*

In the normal kidney, administration of LXA_4 leads to vasorelaxation and augmentation of GFR.[54] These effects are opposite to those elicited by LTD_4.[18] Glomerular micropuncture studies revealed that in dramatic contrast to the vasoconstrictor action of LTD_4 on postglomerular arterioles (Re), LXA_4 elicited selective vasorelaxant responses in preglomerular arteriolar resistance vessels (Ra), thereby increasing RBF, transcapillary hydraulic pressure difference (ΔP), and GFR. The vasodilator actions of LXA_4 are mediated by vasodilator prostaglandins.[45] To our surprise and despite its relaxant actions on Ra, administration of LXA_4 was consistently associated with a mild, but significant, reduction in K_f. This observation led us to examine the interactions of LXA_4 with mesangial cells in culture.[55] LXA_4 competed for $[^3H]LTD_4$ binding to mesangial cells, its presence prevented LTD_4-induced IP_3 generation, and its own stimulation of mesangial cell IP_3 was blocked by an LTD_4 receptor antagonist. *In vivo*, LXA_4 antagonized LTD_4-induced falls in GFR but not RPF, implying selective prevention of LTD_4-mediated reductions in K_f.

Thus, LTD_4 and LXA_4 either activate a common receptor on rat mesangial cells or share common intracellular signalling pathways such that LXA_4 provokes partial agonist responses and antagonizes both the cellular and physiological actions of LTD_4. The basis for this interaction is in part related to the shared spatial orientation [(S), (R)] of the polar substituents at C5 and C6 in both eicosanoids, since reversal of this configuration leads to loss of inhibitory interactions as well as biological activities for both eicosanoids. Subsequent to our observations in mesangial cells, strikingly similar interactions for receptor recognition and stereospecificity between LTD_4 and LXA_4 were reported in pulmonary smooth muscle.[56]

Brady and colleagues have unraveled another facet of LTD_4/LXA_4 interactions in mesangial cells with direct relevance to immune injury, namely the capacity of LTD_4 to specifically promote mesangial cell-PMN adhesion and the inhibitory effect of LXA_4 on this phenomenon.[21] Interestingly, LTD_4-induced augmentation of PMN adhesion to mesangial cells is mediated via increased expression of leukocyte adhesion proteins on mesangial cell surface membrane.[21]

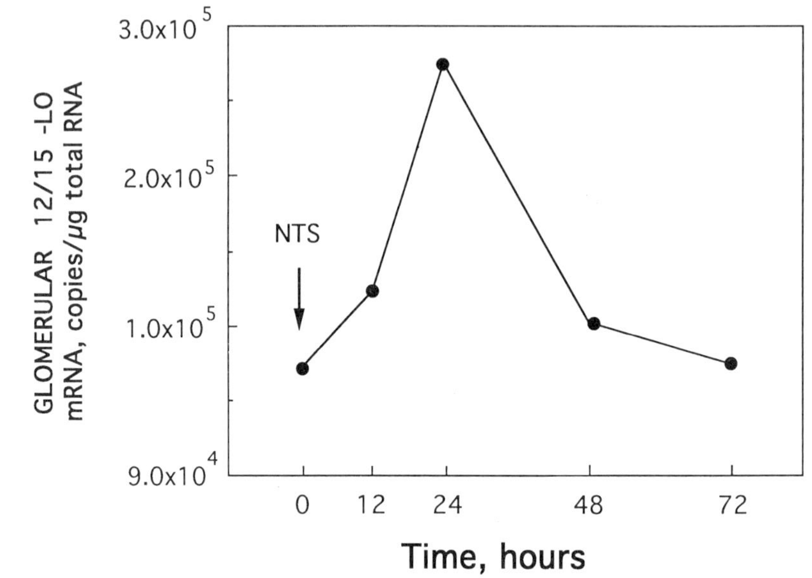

FIGURE 2. Sequential changes in 12/15-LO gene expression in glomeruli isolated from rats before (0 h) and after (12, 24, 48, and 72 h) the injection of nephrotoxic serum, as determined by competitive PCR method.

LTB_4-15-(S)-HETE Interactions in Experimental Glomerulonephritis

When rat neutrophils were pretreated with 15-S-HETE, chemotaxis to LTB_4 was inhibited.[52] The inhibitory effect of 15-S-HETE was concentration-dependent and not due to a generalized toxic effect on PMN function, since the response of these cells to the chemotactic action of *n*-formyl-*l*-methionyl-*l*-phenylalanine (FMP) was unaffected.[52] Further, the inhibitory action of 15-S-HETE appears to be specific to this eicosanoid, since a structurally similar compound, 12-S-HETE, was without effect on LTB_4-induced chemotaxis.[52] Thus, both LTB_4 and 15-S-HETE display reciprocal specificity in their biological interactions.

To explore the physiologic relevance of 15-S-HETE during experimental glomerulonephritis, we determined the production of 15-S-HETE by glomeruli isolated from rats 3 h, 72 h, and 2 weeks after administration of nephrotoxic serum and stimulated with 2 μM calcium ionophore A23187. The production of LTB_4 by isolated glomeruli follows the pattern described previously:[33,34,36] an early peak in the first three hours following injury and a subsequent return to baseline levels. The production of 15-S-HETE, however, is fivefold greater than LTB_4 production and increases steadily over the two-week period following initial injury. The relatively high glomerular generation of 15-S-HETE correlates well with the higher levels required *in vitro* to achieve the observed antagonistic effects of 15-S-HETE on the actions of LTB_4. The cellular source of 15-S-HETE produced by isolated glomeruli could not be determined from these experiments and may very well be the macrophage infiltrate as opposed

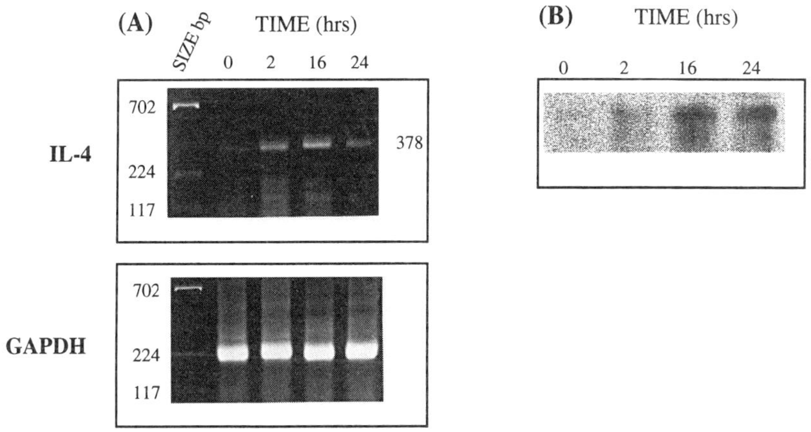

FIGURE 3. (**A**) Agarose gel electrophoresis of IL-4 RT-PCR products. Glomerular mRNA from NTS rats was amplified using primers specific for rat IL-4 and GAPDH as described above. (**B**) Southern analysis (autoradiogram) of rat IL-4 RT-PCR product. The amplified DNA shown in (**A**) was transferred to nylon membrane and probed with labeled rat IL-4 cDNA.

to the endogenous glomerular cells. Indeed, the enhanced production of 15-S-HETE is greatest beyond 72 h after the lesion is induced and as such correlates well with the predominance of macrophages in the lesion at that time.

Since 15-S-HETE had been shown to decrease LTB_4 generation from activated neutrophils,[51] we examined whether the antiinflammatory action of 15-S-HETE was restricted to the antagonism of LTB_4-evoked PMN attraction or extended to the regulation of LTB_4 biosynthesis. Nephrotoxic serum nephritis resulted in increased production of LTB_4 at three hours. Preincubation of glomeruli with 15-S-HETE for 20 minutes resulted in marked abrogation of LTB_4 production. The mechanism underlying the inhibition of LTB_4 synthesis by 15-S-HETE may well be similar to those described above for its abrogation of LTB_4-mediated chemotactic responses.

In nephrotoxic serum nephritis, suppression of LTB_4 synthesis accompanies the shift in the glomerular infiltrate from one composed mainly of neutrophils to a macrophage/monocyte infiltrate, localized in the mesangium.[2,33,34,36] Macrophages represent a potentially major source of 15-LO products.[57,58]

We therefore sought to examine 15-LO gene expression in glomeruli from nephrotoxic serum nephritis rats. A 0.76-kbp fragment of rat LO cDNA was cloned from leukocytes stimulated by interleukin-4 (IL-4) by PCR amplification using primers based on the cloned human 15-LO gene.[43] The deduced amino acid sequence showed 71.0% and 60.1%, homology to human 15-LO and 12-LO, respectively but, surprisingly, was 100% identical to a portion of a recently cloned rat LO that possesses 12- as well as 15-lipoxygenase activities in a ratio of 5 : 1 (henceforth referred to as rat 12/15-LO).[59] A deletion mutant was utilized to generate internal standard cRNA in quantitative PCR assays. Glomerular 12/15-LO mRNA increased significantly over controls 24 and 48 h after NTS injection, then decreased at 72 h (FIG. 2). RNA from NTS glomeruli contained higher levels of rat 12/15-LO mRNA than that from

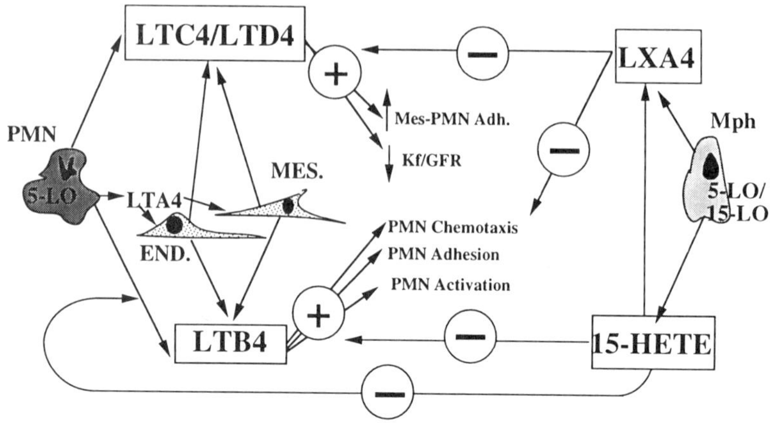

FIGURE 4. Proposed scheme for proinflammatory actions of leukotrienes in glomerular immune injury and their specific antagonism by LXA_4 and 15-S-HETE. See text for abbreviations. LTB_4, C_4, and D_4 can be generated directly from leukocytes or through transcellular transfer of PMN-derived LTA_4 to adjacent mesangial (MES) or endothelial (END) cells and subsequent synthesis of LTs.

unstimulated peripheral leukocytes, suggesting that its transcription is up-regulated locally in native and/or infiltrating glomerular cells. IL-4 specifically stimulates human 15-LO gene expression. Glomerular IL-4 mRNA increased markedly 16 h post-NTS and was then reduced (FIG. 3), a time course consistent with a potential role for T-cell-derived IL-4 in directing the expression of 12/15-LO during glomerulonephritis.

SUMMARY AND CONCLUSION

FIGURE 4 is a schematic summary of current evidence implicating products of the 15-lipoxygenase pathway of arachidonic acid metabolism, principally 15-S-HETE and LXA_4, as endogenous antagonists for the proinflammatory actions of leukotrienes. Here, we have presented evidence for the pathophysiologic relevance of leukotrienes in glomerular immune injury and the emerging data on the multifaceted counterinflammatory actions of 15-lipoxygenase products as they relate specifically to the renal glomerulus. Clearly, these concepts are of a broader nature and would be expected to pertain to inflammatory reactions in general, be they in the glomerulus, the renal interstitium, or in extrarenal sites. The extent to which these early observations can be exploited to design strategies for the control of self-destructive inflammatory reactions in the kidney and elsewhere will be determined by future studies. Imaginative design of molecular tools for the manipulation of these enzyme systems *in vivo*, however, represents a potentially fruitful area of research towards the attainment of a highly worthwhile goal: the cure of glomerulonephritis.

REFERENCES

1. WILSON, C. B. & F. J. DIXON. 1986. Renal response to immunological injury. *In* The Kidney. B. M. Brenner & F. C. Rector, Jr., Eds.: 800-890. Saunders. Philadelphia, PA.
2. BORDER, A. W., B. NOBLE, T. YAMAMOTO & E. RUOSLAHTI. 1992. Natural inhibitors of transforming growth factor b1 protect against scarring in experimental kidney disease. Nature **360:** 361-364.
3. BORDER, W. A., S. OKUDA, L. R. LANGUINO, M. B. SPORN & E. RUOSLAHTI. 1990. Suppression of experimental glomerulonephritis by antiserum against transforming growth factor β1. Nature **346:** 371-374.
4. NEILSON, E. G. 1991. Cell biology of the tubulointerstitium. Kidney Int. **39:** 369-550.
5. TAKAHASHI, K., T. KATOH, G. F. SCHREINER & K. F. BADR. 1992. Essential fatty acid deficiency normalizes function and histology in rat nephrotoxic serum nephritis. Kidney Int. **41:** 1245-1253.
6. LEWIS, R. A. & K. F. AUSTEN. 1984. The biologically active leukotrienes. J. Clin. Invest. **73:** 889-897.
7. SAMUELSSON, B., S-E. DAHLÉN, J-Å. LINDGREN, C. A. ROUZER & C. N. SERHAN. 1987. Leukotrienes and lipoxins: Structures, biosynthesis and biological effects. Science **237:** 1171-1176.
8. GOLDMAN, D. W., L. A. GIFFORD, T. MAROTTI, C. H. KOO & E. J. GOETZL. 1987. Molecular and cellular properties of human polymorphonuclear leukocyte receptors for leukotriene B_4. Fed. Proc. **46:** 200-203.
9. SERHAN, C. N., J. E. FRIDOVICH, E. J. GOETZL, P. B. DUNHAM & G. WEISMANN. 1982. Leukotriene B_4 and phosphatidic acid are calcium ionophores. J. Biol. Chem. **257:** 4746-4752.
10. FEINMARK, S. J., J-Å. LINDGREN, H-E. CLAESSON, C. MALMSTEN & B. SAMUELSSON. 1981. Stimulation of human leukocyte degranulation by leukotriene B_4 and its ω-oxidized metabolites. FEBS Lett. **136:** 141-144.
11. PALMBLAD, J., C. L. MALMSTEN, A-M. UDÉN, O. RÄDMARK, L. ENGSTEDT & B. SAMUELSSON. 1981. Leukotriene B_4 is a potent and stereospecific stimulator of neutrophil chemotaxis and adherence. Blood **58:** 658-661
12. NACCACHE, P. H., N. FAUCHER, S. THERRIEN & P. BORGEAT. 1988. Calcium mobilization, actin polymerization and right-angle light scatter responses to leukotriene B_4, 12(R)- and 12(S)-hydroxyeicosatetraenoic acid in human neutrophils. Life Sci. **42:** 727-733.
13. ANDERSSON, T., W. SCHLEGEL, A. MONOD, K-H. KRAUSE, O. STENDAHL & D. P. LEW. 1986. Leukotriene B_4 stimulation of phagocytes results in the formation of inositol 1,4,5-trisphosphate. A second messenger for Ca2+ mobilization. Biochem. J. **240:** 333-340.
14. MONG, S., G. CHI-ROSSO, J. MILLER, K. HOFFMAN, K. A. RAZGAITIS, P. BENDER & S. T. CROOKE. 1986. Leukotriene B_4 induces formation of inositol phosphates in rat peritoneal polymorphonuclear leukocytes. Mol. Pharmacol. **30:** 235-242.
15. MCINTYRE, T. M., S. L. REINHOLD, S. M. PRESCOTT & G. A. ZIMMERMAN. 1987. Protein kinase C activity appears to be required for the synthesis of platelet-activating factor and leukotriene B_4 by human neutrophils. J. Biol. Chem. **262:** 15370-15376.
16. BADR, K. F., C. BAYLIS, J. M. PFEFFER, M. A. PFEFFER, R. J. SOBERMAN, R. A. LEWIS, K. F. AUSTEN, E. J. COREY & B. M. BRENNER. 1984. Renal and systemic hemodynamic responses to intravenous infusion of leukotriene C4 in the rat. Circ. Res. **54:** 492-499.
17. SUN, F. F., L. Y. CHAU & K. F. AUSTEN. 1987. Binding of leukotriene C4 by glutathione transferase: A reassessment of biochemical and functional criteria for leukotriene receptors. Fed. Proc. **46:** 204-207.

18. BADR, K. F., B. M. BRENNER & I. ICHIKAWA. 1987. Effects of leukotriene D4 on glomerular dynamics in the rat. Am. J. Physiol. **22:** F239-F243.
19. BARNETT, R., P. GOLDWASSER, L. A. SCHARSCHMIDT & D. SCHLONDORFF. 1986. Effects of leukotrienes on isolated rat glomeruli and cultured mesangial cells. Am. J. Physiol. **19:** F838-F844.
20. SIMONSON, M. S. & M. J. DUNN. 1986. Leukotriene C4 and D4 contract rat glomerular mesangial cells. Kidney Int. **30:** 524-531.
21. BRADY, H. R., U. PERSSON, B. J. BALLERMAN, B. M. BRENNER & C. N. SERHAN. 1990. Leukotrienes stimulate neutrophil adhesion to mesangial cells: Modulation with lipoxins. Am. J. Physiol. **259:** F809-F815.
22. BADR, K. F., R. L. HOOVER, S. MONG, J. EBERT, M. SCHWARTZBERG, H. R. JACOBSON & R. C. HARRIS. 1989. Leukotriene D4 binding and signal transduction in rat glomerular mesangial cells. Am. J. Physiol. (Renal Fluid Electrolyte Physiol. **26**): F280-F287.
23. SIMONSON, M. S., P. MENE, G. R. DUBYAK & M. J. DUNN. 1988. Identification and transmembrane signaling of leukotriene D4 receptors in human mesangial cells. Am. J. Physiol. **255:** C771-C780.
24. YARED, A., C. ALBRIGHTSON-WINSLOW, D. GRISWOLD, K. TAKAHASHI, A. FOGO & K. F. BADR. 1991. Functional significance of leukotriene B4 in normal and glomerulonephritic kidneys. J. Am. Soc. Nephrol. **2:** 45-56.
25. MUNE, M., T. GOTO, K. MATOBA, S. YUKAWA & H. NOMOTO. 1987. Glomerular leukotriene synthesis in rat nephrotoxic serum nephritis (Abstr.). Xth International Congress of Nephrology, London.
26. SPURNEY, R. F., P. RUIZ, D. S. PISETSKY & T. M. COFFMAN. 1991. Enhanced renal leukotriene production in murine lupus: Role of lipoxygenase metabolites. Kidney Int. **39:** 95-102.
27. BADR, K. F., G. F. SCHREINER, M. WASSERMAN & I. ICHIKAWA. 1988. Preservation of the glomerular capillary ultrafiltration coefficient during rat nephrotoxic serum nephritis by a specific leukotriene D4 receptor antagonist. J. Clin. Invest. **81:** 1702-1709.
28. BRESNAHAN, B. A., W. SHENGUA & E. A. LIANOS. 1991. Role of platelet and leukocyte-derived eicosanoids in mesangial cell immune injury (Abstr.). J. Am. Soc. Nephrol. **2:** 426.
29. KATOH, T., E. A. LIANOS, M. FUKUNAGA, K. TAKAHASHI & K. F. BADR. 1993. Leukotriene D4 is a mediator of proteinuria and glomerular hemodynamic deterioration in passive heymann nephritis. J. Clin. Invest. **91:** 1507-1515.
30. BLANTZ, R. C. & C. B. WILSON. 1976. Acute effects of antiglomerular basement membrane antibody on the process of glomerular filtration in the rat. J. Clin. Invest. **58:** 899-911.
31. BLANTZ, R. C., B. J. TUCKER & C. B. WILSON. 1977. The acute effect of antiglomerular basement membrane antibody upon glomerular filtration in the rat. The influence of dose and complement depletion. J. Clin. Invest. **59:** 910-921.
32. MADDOX, D. A., C. M. BENNET, W. M. DEEN, R. J. GLASSOCK, D. KNUTSON, T. M. DAUGHARTY & B. M. BRENNER. 1975. Determinants of glomerular filtration in experimental glomerulonephritis in the rat. J. Clin. Invest. **55:** 305-318.
33. LIANOS, E. A. 1988. Synthesis of hydroxyeicosatetraenoic acids and leukotrienes in rat nephrotoxic serum glomerulonephritis: Role of antiglomerular basement membrane antibody dose, complement, and neutrophils. J. Clin. Invest. **82:** 427-435.
34. FAULER, J., A. WIEMEYER, K. H. MARX, K. KÜHN, K. M. KOCH & J. C. FRÖHLICH. 1989. LTB_4 in nephrotoxic serum nephritis in rats. Kidney Int. **36:** 46-50.
35. RAHMAN, M. A., M. NAKAZAWA, S. N. EMANCIPATOR & M. J. DUNN. 1988. Increased leukotriene B_4 synthesis in immune injured rat glomeruli. J. Clin. Invest. **81:** 1945-1952.

36. Schreiner, G. F., B. Rovin & J. B. Lefkowith. 1989. The antiinflammatory effects of essential fatty acid deficiency in experimental glomerulonephritis: The modulation of macrophage migration and eicosanoid metabolism. J. Immunol. **143:** 3192-3199.
37. Lianos, E. A. & B. A. Bresnahan. 1990. Origin of leukotrienes and HETE in glomerular epithelial and mesangial cell immune injury (Abstr.). Kidney Int. **37:** 421.
38. Lianos, E. A., B. Noble & B. Hucke. 1989. Glomerular leukotriene synthesis in Heymann nephritis. Kidney Int. **36:** 998-1002.
39. Badr, K. F. 1992. Five-lipoxygenase products in glomerular immune injury. J. Am. Soc. Nephrol. **3:** 907-915.
40. Makita, N., C. D. Funk, E. Imai, R. L. Hoover, M. Frazer & K. F. Badr. 1992. Molecular cloning and functional expression of rat leukotriene A4 hydrolase from glomerular mesangial cells using polymerase chain reaction (PCR). FEBS Lett. **299:** 273-277.
41. Imai, E., R. L. Hoover, N. Makita, C. D. Funk & K. F. Badr. 1990. Localization and relative abundance of 5-lipoxygenase, 15-lipoxygenase, 12-lipoxygenase and leukotriene A4-hydrolase gene expression in cultured glomerular cells. J. Am. Soc. Nephrol. **1:** 751.
42. Fischer, D., K. Takahashi, J. Ebert & K. F. Badr. 1990. Limited early therapy with a novel 5-lipoxygenase (5-LO) activating protein (FLAP) antagonist, MK 886, during heterologous rat nephrotoxic serum (NTS) nephritis totally prevents proteinuria in the autologous phase. J. Am. Soc. Nephrol. **1:** 628.
43. Sigal, E., D. Grunberger, E. Highland, C. Goprsst, R. A. F. Dixon & C. S. Craik. 1990. Expression of cloned human reticulocyte 15-lipoxygenase and immunological evidence that 15-lipoxygenases of different cell types are related. J. Biol. Chem. **265:** 5113-5120.
44. Nicolau, K. C., B. E. Marron, C. A. Veale, S. H. Webber, S. E. Dahlen, B. Samuelsson & C. N. Serhan. 1989. Identification of a novel 7-cis-11-trans-lipoxin A4 generated by human neutrophils: Total synthesis, spasmogenic activities, and comparison with other geometric isomers of lipoxin A4 and B4. Biochim. Biophys. Acta **1003:** 44-53.
45. Katoh, T., K. Takahashi, D. K. DeBoer, C. N. Serhan & K. F. Badr. 1992. Renal hemodynamic actions of lipoxins in rats: A comparative physiological study. Am. J. Physiol. (Renal Fluid Electrolyte Physiol. **32**). F436-F442.
46. Garrick, R., S. Y. Shen, S. Ogunc & P. Y. K. Wong. 1989. Transformation of leukotriene A4 to lipoxins by rat kidney mesangial cell. Biochem. Biophys. Res. Commun. **162:** 626-633.
47. Serhan, C. N. & K. A. Sheppard. 1990. Lipoxin formation during human neutrophil-platelet interactions. Evidence for the transformation of leukotriene A4 by platelet 12-lipoxygenase in vitro. J. Clin. Invest. **85:** 772-780.
48. Hedqvist, P., J. Raud, U. Palmertz, J. Haeggstrom, K. C. Nicolau & S. E. Dahlén. 1989. Lipoxin A4 inhibits leukotriene B4-induced inflammation in the hamster cheek pouch. Acta Physiol. Scand. **137:** 571-572.
49. Fogh, K., H. Sogaard, T. Herlin & K. Kragballe. 1988. Improvement of psoriasis vulgaris after intralesional injections of 15-hydroxyeicosatetraenoic acid (15-S-HETE). J. Am. Acad. Dermatol. **18:** 279-285.
50. Fogh, K., S. E. Hansen, V. Herlin, T. Knudsen, T. B. Henriksen, H. Ewald, C. Bungerand & K. Kragballe. 1989. 15-S-HETE inhibits caragheen-induced experimental arthritis and reduces synovial fluid leukotriene B4 (LTB4). Prostaglandins **37:** 213-228.
51. Brezinski, M. E. & C. N. Serhan. 1990. Selective incorporation of 15-S-hydroxyeicosatetraenoic acid in phosphotidylinositol of human neutrophils: Agonist induced deacylation

and transformation of stored hydroxyeicosanoids. Proc. Natl. Acad. Sci. U.S.A. **87:** 6248-6252.

52. FISCHER, D. B., J. W. CHRISTMAN & K. F. BADR. 1992. Fifteen-S-hydroxyeicosatetraenoic acid (15-S-HETE) specifically antagonizes chemotactic responses to LTB4 in rat neutrophils and suppresses leukotriene B4 synthesis following glomerular inflammatory injury. Kidney Int. **41:** 1155-1160.
53. LEE, T. H., C. E. HORTON, U. KYAN-AUNG, D. HASKARD, A. E. G. CREA & B. W. SPUR. 1989. Lipoxin A4 and lipoxin B4 inhibit chemotactic responses of human neutrophils stimulated by leukotriene B4 and N-formyl-L-methionyl-L-leucyl-L-phenylalanine. Clin. Sci. **77:** 195-203.
54. BADR, K. F., C. N. SERHAN, K. C. NICOLAU & B. SAMUELSSON. 1987. The action of lipoxin A on glomerular microcirculatory dynamics in the rat. Biochem. Biophys. Res. Commun. **145:** 408-414.
55. BADR, K. F., D. DEBOER, M. SCHWARTZBERG & C. N. SERHAN. 1989. Lipoxin A4 antagonizes cellular and in vivo actions of leukotriene D4 in rat glomerular mesangial cells: Evidence for competition at a common receptor. Proc. Natl. Acad. Sci. U.S.A. **86:** 3438-3442.
56. LEFER, A. M., G. L. STAHL, D. J. LEFER, M. E. BREZINSKI, K. C. NICOLAU, C. A. VEALE, Y. ABE & B. J. SMITH. 1988. Lipoxins A4 and B4: Comparison of eicosanoids having bronchoconstrictor and vasodilator actions but lacking platelet aggregatory activity. Proc. Natl. Acad. Sci. U.S.A. **85:** 8340-8344.
57. KOUZAN, S., A. R. BRADY, P. NETTESHEIM & T. ELING. 1985. Production of arachidonic acid metabolites by macrophages exposed in vitro to asbestos, carbonyl iron particles or calcium ionophore. Am. Rev. Respir. Dis. **131:** 624-632.
58. TAYLOR, S. M., D. H. LIGGIT, W. W. LAEGREID, R. SIFLOW & R. W. LEID. 1986. Arachidonic acid metabolism in bovine alveolar macrophages: Effect of calcium ionophore or lipoxygenase products. Inflammation **10:** 157-165.
59. WATANABE, T., J. F. MEDINA, J. Z. HAEGGSTROM, O. RADMARK & B. SAMUELSSON. 1993. Molecular cloning of a 12-lipoxygenase cDNA from rat brain. Eur. J. Biochem. **212:** 605-612.

Eicosanoid Biosynthesis and Metabolism in Myeloproliferative Disorders[a]

CARLO PATRONO,[b,c] GIOVANNI CIABATTONI,[d] PAOLA PATRIGNANI,[c] BIANCA ROCCA,[e] AND RAFFAELE LANDOLFI[e]

[c]*Department of Pharmacology*
University of Chieti "G. D'Annunzio" School of Medicine
Chieti, Italy
Departments of [d]Pharmacology and [e]Medicine
Catholic University School of Medicine
Rome, Italy

INTRODUCTION

An unusual combination of thrombotic and hemorrhagic complications punctuates the clinical course of myeloproliferative disorders and represents the main cause of morbidity and mortality in most benign diseases such as polycythemia vera and essential thrombocythemia.[1] The mechanism(s) underlying the hemostatic overshooting or failure is currently thought to be related to opposite changes in platelet function. However, extensive *in vitro* studies of platelet biochemistry and function in myeloproliferative disorders have yielded conflicting results[2-9] that do not provide a sound basis for preventive strategies.

More recently, *in vivo* studies of eicosanoid biosynthesis and metabolism in polycythemia vera[10] and essential thrombocythemia[11] have provided clues that may help in designing a rational antithrombotic approach while minimizing the inherent hemorrhagic risk. In this paper, we discuss the methodological aspects of eicosanoid measurements and review recently completed studies of eicosanoid biosynthesis and metabolism in these disorders.

THROMBOXANE AND LEUKOTRIENE BIOSYNTHESIS AND METABOLISM IN MAN

There is a wide discrepancy between the capacity of human platelets to synthesize and release thromboxane (Tx) A_2 when maximally stimulated *ex vivo* (for example,

[a]This work was supported by grants from Consiglio Nazionale delle Ricerche (CNR), Progetto Finalizzato "Prevenzione e Controllo dei Fattori di Malattia" (SP8: 92.00152.PF41 and 93.00712.PF41).

[b]Address for correspondence: Carlo Patrono, M.D., Cattedra di Farmacologia I, Universitá degli Studi "G. D'Annunzio" Palazzina del Centro Servizi Biomedici, Via dei Vestini 31, 66013 Chieti, Italy.

by endogenous thrombin during whole blood clotting at 37 °C) and the actual rate of TxA_2 secretion into the human circulation.[12] Thus, the platelets present in 1 mL of human whole blood have the capacity to generate the same amount (300-400 ng) of TxB_2 during 1 hour clotting *in vitro* as the total body synthesis *in vivo* over the same period of time. It appears likely that the low biosynthetic rate measured in healthy subjects (0.11 ng/kg · min^{-1}) reflects the low frequency and/or intensity of stimuli to TxA_2 production under physiological circumstances.[12] As a consequence, measurements of *ex vivo* TxB_2 production in whole blood or platelet-rich plasma (PRP) are unlikely to be relevant to changes occurring *in vivo*, except to reflect drug-induced changes in cyclooxygenase or Tx-synthase activity.[13]

TxB_2 has two major metabolic pathways in humans,[14] one involving β-oxidation resulting in the formation of 2,3-dinor-TxB_2, the other involving dehydrogenation of the hemiacetal alcohol group at C-11, resulting in the formation of a series of metabolites with a delta-lactone ring structure. Although the β-oxidation pathway was originally described as the prevailing form of enzymatic degradation of infused TxB_2 in humans,[14] the dehydrogenation pathway has received considerable attention following the demonstration that 11-dehydro-TxB_2 has a substantially longer plasma half-life[15] and is excreted at a higher rate[16] than 2,3-dinor-TxB_2.

Because the relative contribution of these two pathways of TxB_2 metabolism might vary under pathophysiologic circumstances associated with platelet activation, we have characterized the fractional conversion of TxB_2 to urinary 11-dehydro-TxB_2 and 2,3-dinor-TxB_2 over a 50-fold range of infusion rates in healthy volunteers.[17] In this study, the fractional elimination of 11-dehydro-TxB_2 was independent of the rate of TxB_2 infusion and averaged 6.8 ± 0.7%. This figure was not significantly different from the corresponding value of 6.4 ± 1.2% for the conversion of TxB_2 to 2,3-dinor-TxB_2 simultaneously measured in the same healthy volunteers.[17] Moreover, our estimate of the fractional conversion of systemically administered cold TxB_2 into urinary 11-dehydro-TxB_2 is in agreement with the 7.6% figure obtained following the infusion of radiolabeled TxB_2 at a much higher rate (65 ng/kg · min^{-1}).[14]

This finding of a relatively constant fractional conversion of TxB_2 to 11-dehydro-TxB_2 over a wide range of entry rates of the parent compound into the circulation (0.1-65 ng/kg · min^{-1}) should be contrasted with the apparent discrepancy between the fractional elimination of 2,3-dinor-TxB_2 at 0.1-5.0 ng/kg · min^{-1}—approximately 6 percent[12]—and that measured at 65 ng/kg · min^{-1}, which is 23 percent.[14] This might imply that one enzyme system may saturate faster than the other. Thus, measurement of 11-dehydro-TxB_2 might more accurately than 2,3-dinor-TxB_2 reflect changes in TxB_2 production associated with intense and/or long-lasting platelet activation. Measurement of both would obviate the theoretical possibility of drug- and/or disease-dependent alteration in the metabolic disposition of TxB_2.

Both leukotriene (LT) E_4 and its ω- and β-oxidation products have been identified in urine, following the infusion of LTC_4[18] or LTE_4[19] in healthy subjects. We have recently measured the excretion of LTE_4 during the infusion of exogenous LTC_4 over a 25-fold dose range to enable estimation of the rate of entry of endogenous LTC_4 into the bloodstream of normal humans.[20] The fractional elimination of LTE_4 was independent of the rate of LTC_4 infusion and averaged 4.3 ± 0.9%. Interpolation of LTE_4 values obtained under basal conditions onto the linear relationship between the quantities of infused LTC_4 and the amount of LTE_4 excreted in excess of control values permitted calculation of the mean rate of entry of LTC_4 into the circulation as 0.06 pmol/kg · min^{-1}. Statistically significant changes in the urinary excretion of LTE_4 were measured in association with simulated short-term (10–30 min) increases of LTC_4 entry into the bloodstream. In addition, we characterized further metabolism of LTE_4 by infusing [^{14}C]LTC_4 in two of the same subjects. The two major urinary metabolites were the ω- and β-oxidation products identified as the conjugated tetraene 16-COOH-LTE_4 and 14-COOH-LTE_3, which accounted for 6 to 8% of the total infused amount of [^{14}C]LTC_4. Although the excretion of LTE_4 fell to almost undetectable levels 2 h after cessation of the infusion, the presence of the ω/β-oxidized metabolites persisted for up to 15 h.[20]

EICOSANOID BIOSYNTHESIS AND METABOLISM IN POLYCYTHEMIA VERA

Polycythemia vera is a clonal, chronic, progressive myeloproliferative disorder characterized by an absolute increase in red-cell mass and usually leukocytosis, thrombocytosis, and splenomegaly. In this disease, thrombotic complications are more prevalent than in other myeloproliferative disorders and represent the major cause of morbidity and mortality. Increased TxA_2 production[6,21,22] and modified aspirin sensitivity[23] have been detected *in vitro* in platelets isolated from patients with polycythemia vera. To verify the relevance of these capacity-related measurements to the actual rate of TxA_2 biosynthesis *in vivo* and to its suppression by oral aspirin, we have investigated the urinary excretion of 11-dehydro-TxB_2 and 2,3-dinor-TxB_2 in 17 patients with polycythemia vera and 23 sex- and age-matched healthy subjects.[10] The rate of excretion of both metabolites was 2- to 40-fold higher in patients than controls, with all 17 patients having metabolite excretion rates in excess of 2 SD above the normal means (FIG. 1). Thromboxane metabolite excretion was not related to the current treatment or to a clinical history positive for thrombotic complications. The latter finding is not surprising, inasmuch as these thrombotic events had occurred 6 months to 5 years before the study.

The finding of a highly significant linear correlation between the excretion rates of the two metabolites, originating through the 11-hydroxy-dehydrogenase and the β-oxidation pathways, respectively, suggests that the enhanced thromboxane metabolite excretion detected in patients with polycythemia vera reflects a change in the biosynthesis of TxA_2 rather than a shift in its metabolic disposition.[10] Enhanced TxA_2 biosynthesis might be a consequence of (1) alterations in platelet biochemistry (e.g., increased substrate availability) or number; (2) abnormal stimuli to platelet activation operating *in vivo;* or (3) increased extraplatelet production of TxA_2. Whereas (1) would be reflected by changes in TxB_2 production detectable *ex vivo,* (2) and (3) would be compatible with perfectly normal biosynthetic capacity of platelets to synthesize and release TxA_2 *in vitro.*

To assess the biosynthetic capacity of polycythemic platelets, we measured TxB_2 production during whole-blood clotting. When corrected for the platelet count, an identical value of TxB_2 production was found in patients and controls. That enhanced TxA_2 biosynthesis found in polycythemia vera did not merely reflect increased platelet numbers was indicated by the lack of any statistically significant correlation between platelet counts and urinary 11-dehydro-TxB_2 or 2,3-dinor-TxB_2 excretion.[10]

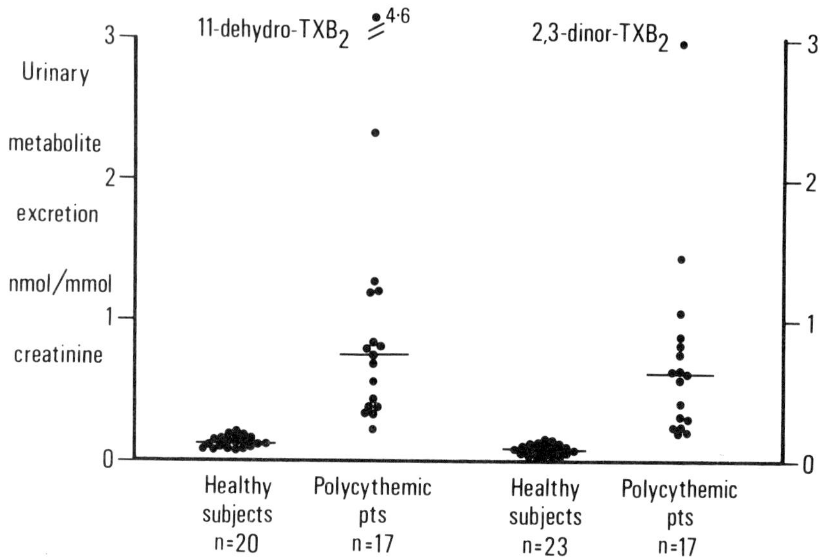

FIGURE 1. Individual rates of excretion of 11-dehydro-TxB_2 and 2,3-dinor-TxB_2 in patients with polycythemia vera and sex- and age-matched controls. (Data from Reference 10.)

Moreover, the finding of unchanged excretion of LTE_4 and of similar LTE_4 production in whole-blood samples of patients and controls (TABLE 1) excludes a generalized abnormality of arachidonate metabolism affecting polycythemic leukocytes, consistent with similar observations of Stenke *et al.*[24]

Urinary metabolites do not necessarily reflect a specific site of eicosanoid biosynthesis.[25] To distinguish between platelet and nonplatelet sites of TxA_2 synthesis, we exploited the capacity of aspirin to acetylate platelet prostaglandin G/H-synthase selectively when it is administered daily in low doses.[26] Other sites of cyclooxygenase activity, such as the kidney, that can be involved in enhanced TxA_2 production under pathophysiologic circumstances[27] are largely unaffected by doses of aspirin in the

TABLE 1. Whole-Blood LTE_4 Production and Urinary LTE_4 Excretion in Patients with Polycythemia Vera and Healthy Controls

Group	LTE_4 (pmol/10^6 WBC)	LTE_4 (pmol/h)
Healthy controls (n = 8)	11.1 ± 7.3	10.9 ± 7.3
PV Patients (n = 10)	10.8 ± 14.3	9.3 ± 6.1

NOTE: All values are mean ± 1 SD. PV, polycythemia vera; WBC, white blood cells.

range of 30 to 50 mg/day.[28] The greater than 80% reduction in TxB_2 metabolite excretion in nine patients after they received 50 mg of aspirin daily for 1 to 2 weeks as well as the slow rate of recovery of TxA_2 production on withdrawal of aspirin administration are consistent with a role for platelets as the major source of enhanced TxA_2 biosynthesis in polycythemia vera. This degree of suppression of TxB_2 metabolite excretion is comparable to that achieved by daily doses of aspirin (40 to 80 mg) in healthy subjects,[29] as well as in patients with episodically[30] or persistently[31,32] enhanced Tx biosynthesis.

Atherosclerotic vascular disease may be associated with accelerated platelet-vascular interactions and provide triggers to platelet activation *in vivo*. However, it should be noted that stable manifestations of coronary[33] and peripheral[30] vascular disease are not associated with detectable changes in TxB_2 metabolite excretion. Because none of our patients had acute manifestations of vascular occlusion at the time of the study, it is unlikely that vascular disease contributed to the observed changes in platelet arachidonate metabolism.

An elevated hematocrit value and increased blood viscosity clearly play a role in the pathogenesis of thrombotic complications in polycythemia vera. As reviewed by Schafer,[1] an increased hematocrit value may permit platelets to achieve more intimate contact with the vessel wall as a function of the axial migration of red blood cells. In our study, TxA_2 biosynthesis was not related to the hematocrit, although in a relatively narrow range of values. Moreover, a 5% to 6% reduction in the hematocrit value associated with phlebotomy did not modify appreciably the rate of TxB_2 metabolite excretion in three patients studied repeatedly. Thus, the mechanisms responsible for enhanced platelet synthesis and release of TxA_2 in polycythemia vera remain to be investigated further. Verification of the pathophysiologic significance of this abnormality would require a controlled trial of low-dose aspirin prophylaxis. The Gruppo Italiano per lo Studio della Policitemia (GISP) is currently testing the feasibility of such a trial by randomly assigning 170 patients with polycythemia vera, for whom the physician is uncertain as to the benefit/risk ratio of antiplatelet therapy, to receive aspirin (40 mg/day) or placebo for 1 year with the aim of assessing the safety of this low-dose regimen in the setting of enhanced hemorrhagic risk.

EICOSANOID BIOSYNTHESIS AND METABOLISM IN ESSENTIAL THROMBOCYTHEMIA

Primary or essential thrombocythemia is a clonal hematologic malignancy characterized by a sustained proliferation of megakaryocytes, which leads to increased numbers of circulating platelets. In addition to platelet counts >600,000/mm^3, this disease has a clinical course punctuated by hemorrhagic and/or thrombotic complications.

In order to test the hypothesis that a common platelet abnormality is shared by different myeloproliferative disorders, we have recently investigated TxB_2 metabolite excretion in patients with essential thrombocythemia.[11] This disease allows exploring TxA_2 biosynthesis over a wider range of platelet numbers than in polycythemia vera and avoiding the *in vivo* stimuli to platelet activation possibly related to the accompanying increase in blood viscosity.

Thrombocythemic patients had significantly higher excretion rates of 11-dehydro-TxB_2 (by 2- to 60-fold) than sex- and age-matched controls. When considering the whole study group of thrombocythemic patients, enhanced metabolite excretion was related neither to the platelet count nor to bleeding time or time from diagnosis. When corrected for the platelet count, whole-blood TxB_2 production was comparable in patients and controls, thus indicating unaltered biosynthetic capacity of thrombocythemic platelets.[11]

Further studies are ongoing in our laboratory in order to define the variables affecting TxA_2 biosynthesis in essential thrombocythemia and to elucidate the mechanism(s) underlying platelet activation in this setting.

CONCLUSIONS

Recent studies of thromboxane biosynthesis and metabolism in polycythemia vera and essential thrombocythemia suggest that myeloproliferative disorders share a common metabolic abnormality, irrespective of hyperviscosity or increased platelet count. TxA_2-mediated amplification of platelet activation may contribute to the increased thrombotic risk associated with these diseases. These studies provide a rationale for assessing the efficacy and safety of a low-dose aspirin regimen in this setting.

ACKNOWLEDGMENT

The authors thank Alessandra Migliavacca for expert editorial assistance.

REFERENCES

1. Schafer, A. I. 1984. Bleeding and thrombosis in the myeloproliferative disorders. Blood **64:** 1.
2. Bolin, R. B., T. Okumara & G. A. Jamieson. 1977. Changes in distribution of platelet membrane glycoproteins in patients with myeloproliferative disorders. Am. J. Hematol. **3:** 63.
3. Kaywin, P., M. McDomough, P. A. Insel & S. J. Shattil. 1978. Platelet function in essential thrombocythemia. Decreased epinephrine responsiveness associated with a deficiency of platelet alpha-adrenergic receptors. N. Engl. J. Med. **299:** 505.
4. Caranobe, C., P. Sie, C. Nouvel, G. Laurent, J. Pris & B. Boneu. 1980. Platelets in myeloproliferative disorders. Serotonin uptake and storage: Correlation with mepacrine labelled dense bodies and with platelet density. Scand. J. Haematol. **25:** 289.
5. Jubelirer, S. J., F. Russel, R. Vaillancourt & D. Deykin. 1980. Platelet arachidonic acid metabolism and platelet function in ten patients with chronic myelogenous leukemia. Blood **56:** 728.
6. Schafer, A. I. 1982. Deficiency of platelet lipoxygenase activity in myeloproliferative disorders. N. Engl. J. Med. **306:** 381.
7. Clezardin, P., J. L. McGregor, M. Dechavanne & K. J. Clemetson. 1985. Platelet membrane glycoprotein abnormalities in patients with myeloproliferative disorders and secondary thrombocytosis. Br. J. Haematol. **60:** 331.

8. MALPASS, T. W., B. SAVAGE, S. R. HANSON, S. J. SLICHTER & L. A. HARKER. 1984. Correlation between prolonged bleeding time and depletion of platelet dense granule ADP in patients with myelodysplastic and myeloproliferative disorders. J. Lab. Clin. Med. **103:** 894.
9. LANDOLFI, R., R. DE CRISTOFARO, M. CASTAGNOLA, E. DE CANDIA, G. D'ONOFRIO, G. LEONE & B. BIZZI. 1988. Increased platelet-fibrinogen affinity in patients with myeloproliferative disorders. Blood **71:** 978.
10. LANDOLFI, R., G. CIABATTONI, P. PATRIGNANI, M. A. L. CASTELLANA, E. POGLIANI, B. BIZZI & C. PATRONO. 1992. Increased thromboxane biosynthesis in patients with polycythemia vera: Evidence for aspirin-suppressible platelet activation in vivo. Blood **80:** 1965.
11. LANDOLFI, R., B. ROCCA, G. CIABATTONI, S. CORTELAZZO, T. BARBUI, R. TARTAGLIONE & C. PATRONO. 1993. Increased thromboxane biosynthesis in essential thrombocythemia. Clin. Res. **41:** 164A.
12. PATRONO, C., G. CIABATTONI, F. PUGLIESE, A. PIERUCCI, I. A. BLAIR & G. A. FITZGERALD. 1986. Estimated rate of thromboxane secretion into the circulation of normal humans. J. Clin. Invest. **77:** 590.
13. PATRONO, C., G. CIABATTONI, P. PATRIGNANI, F. PUGLIESE, P. FILABOZZI, F. CATELLA, G. DAVÍ & L. FORNI. 1985. Clinical pharmacology of platelet cyclooxygenase inhibition. Circulation **72:** 1177.
14. ROBERTS II, L. J., B. J. SWEETMAN & J. A. OATES. 1981. Metabolism of thromboxane B_2 in man: Identification of twenty urinary metabolites. J. Biol. Chem. **256:** 8384.
15. LAWSON, J. A., C. PATRONO, G. CIABATTONI & G. A. FITZGERALD. 1986. Long-lived enzymatic metabolites of thromboxane B_2 in the human circulation Anal. Biochem. **155:** 198.
16. CATELLA, F. & G. A. FITZGERALD. 1987. Paired analysis of urinary thromboxane B_2 metabolites in humans. Thromb. Res. **47:** 647.
17. CIABATTONI, G., F. PUGLIESE, G. DAVÍ, A. PIERUCCI, B. M. SIMONETTI & C. PATRONO. 1989. Fractional conversion of thromboxane B_2 to urinary 11-dehydrothromboxane B_2 in man. Biochim. Biophys. Acta **992:** 66.
18. ORNING, L., L. KAIJSIER & S. HAMMARSTROM. 1985. *In vivo* metabolism of leukotriene C_4 in man: Urinary excretion of leukotriene E_4. Biochem Biophys. Res. Commun. **130:** 214.
19. SALA, A., N. VOELKEL, J. MACLOUF & R. C. MURPHY. 1990. LTE_4 elimination and metabolism in normal human subjects. J. Biol. Chem. **265:** 21771.
20. MACLOUF, J., C. ANTOINE, R. DE CATERINA, R. SICARI, R. C. MURPHY, P. PATRIGNANI, S. LOIZZO & C. PATRONO. 1992. Entry rate and metabolism of leukotriene C_4 into vascular compartment in healthy subjects. Am. J. Physiol. **263:** H244.
21. SINZINGER, H., H. GISSLINGER, H. LINKESCH, W. LUDWIG, R. FLENER & B. A. PESKAR. 1988. Defects in the prostaglandin system. VIII. A pathologic thrombocyte population in myeloproliferative syndrome, which forms no thromboxane from exogenous arachidonic acid. Wien Klin. Wochenschr. **100:** 715.
22. MEHTA, P., J. MEHTA, M. ROSS, N. OSTROWSKI & D. PLAYER. 1985. Decreased platelet aggregation but increased thromboxane generation in polycythemia vera. Arch. Intern. Med. **145:** 1225.
23. BAKER, R. I. & A. MANOHARAN. 1988. Platelet function in myeloproliferative disorders: Characterization and sequential studies show multiple platelet abnormalities, and change with time. Eur. J. Haematol. **40:** 267.
24. STENKE, L., J. SAMUELSSON, J. PALMBLAD, L. DABROWSKI, P. REIZENSTEIN & J. A. LINDGREN. 1990. Elevated white blood cell synthesis of LTC_4 in chronic myelogenous leukaemia but not in polycythemia vera. Br. J. Haematol. **74:** 257.

25. PATRONO, C., G. CIABATTONI & P. PATRIGNANI. 1989. Biochemical indices of arachidonate metabolism by platelets and vascular endothelium in vivo. *In* Platelets and Vascular Occlusion. C. Patrono & G. A. FitzGerald, Eds.: 193. Raven. New York, N.Y.
26. BURCH, J. W., N. STANFORD & P. W. MAJERUS. 1978. Inhibition of platelet prostaglandin synthetase by oral aspirin. J. Clin. Invest. **61:** 314.
27. PATRONO, C., G. CIABATTONI, G. REMUZZI, E. GOTTI, S. BOMBARDIERI, O. DI MUNNO, G. A. TARTARELLI, B. M. SIMONETTI & A. PIERUCCI. 1985. Functional significance of renal prostacyclin and thromboxane A_2 production in patients with systemic lupus erythematosus. J. Clin. Invest. **76:** 1011.
28. PATRIGNANI, P., P. FILABOZZI & C. PATRONO. 1982. Selective cumulative inhibition of platelet thromboxane production by low-dose aspirin in healthy subjects. J. Clin. Invest. **69:** 1366.
29. FITZGERALD, G. A., J. A. OATES, J. HAWIGER, R. L. MAAS, L. J. ROBERTS, J. A. LAWSON & A. R. BRASH. 1983. Endogenous biosynthesis of prostacyclin and thromboxane and platelet function during chronic administration of aspirin in man. J. Clin. Invest. **71:** 676.
30. VEJAR, M., G. FRAGASSO, D. HACKETT, D. P. LIPKIN. A. MASERI, G. V. R. BORN, G. CIABATTONI & C. PATRONO. 1990. Dissociation of platelet activation and spontaneous myocardial ischemia in unstable angina. Thromb. Haemostasis **63:** 163.
31. DAVÍ, G., I. CATALANO, M. AVERNA, A. NOTARBARTOLO, A. STRANO, G. CIABATTONI & C. PATRONO. 1990. Thromboxane biosynthesis and platelet function in type II diabetes mellitus. N. Engl. J. Med. **322:** 1769.
32. DAVÍ, G., M. AVERNA, I. CATALANO, C. BARBAGALLO, A. GANCI, A. NOTARBARTOLO, G. CIABATTONI & C. PATRONO. 1992. Increased thromboxane biosynthesis in type IIa hypercholesterolemia. Circulation **85:** 1792.
33. FITZGERALD, D. J., L. ROY, F. CATELLA & G. A. FITZGERALD. 1986. Platelet activation in unstable coronary disease. N. Engl. J. Med. **315:** 983.

Isoprostanes

Novel Markers of Endogenous Lipid Peroxidation and Potential Mediators of Oxidant Injury[a]

L. JACKSON ROBERTS II AND JASON D. MORROW

Departments of Pharmacology and Medicine
Vanderbilt University
Nashville, Tennessee 37232

INTRODUCTION

In 1990 it was discovered that there are series of prostaglandin (PG)F_2-like compounds, now termed F_2-isoprostanes, that are produced *in vivo* in humans by a noncyclooxygenase mechanism involving free radical-catalyzed peroxidation of arachidonic acid.[1] This was a novel finding in that it had previously been thought that the cyclooxygenase enzyme was obligatory for the formation of prostanoids *in vivo.* The formation of these prostanoids proceeds through bicyclic endoperoxide intermediates that are then reduced to form F-ring isoprostanes. Four regioisomers of the F_2-isoprostanes are produced, each of which can theoretically be composed of eight racemic diastereomers (FIG. 1). Of considerable interest is that, in contradistinction to cyclooxygenase-derived prostaglandins, F_2-isoprostanes have been shown to be initially formed *in situ,* esterified to phospholipids, and subsequently released preformed, presumably by phospholipases.[2]

The fact that prostanoids are produced *in vivo* by a noncyclooxygenase mechanism is of considerable interest from a biochemical perspective. However, this may also be associated with biological ramifications since it has been shown that these compounds are capable of exerting potent biological activity. At present, only one of the F_2-isoprostanes that can be formed, 8-Epi-$PGF_{2\alpha}$, has been available in synthetic form for biological testing. 8-Epi-$PGF_{2\alpha}$ has been found to be an extremely potent renal vasoconstrictor with an EC_{50} in the low nanomolar concentration range.[1] It has also been shown to exert potent vasoconstricting activity in the pulmonary vascular bed.[3,4] Interestingly, SQ29548, a thromboxane A_2 receptor anatagonist, abrogates the vascular effects of 8-epi-$PGF_{2\alpha}$.[5] Curiously, however, 8-epi-$PGF_{2\alpha}$ has been shown to act primarily as an antagonist of thromboxane A_2 receptors in platelets.[6] Insight into this apparent paradox has been obtained, as recent evidence suggests that 8-epi-$PGF_{2\alpha}$ interacts with a receptor on vascular smooth muscle that may be similar to but distinct from the thromboxane A_2 receptor.[7] Because of the potentially important finding that

[a] This work was supported by Grant GM42056 from the National Institutes of Health. J.D.M. is a Howard Hughes Medical Institute Physician Research Fellow and the recipient of a Career Development Award from the International Life Sciences Institute.

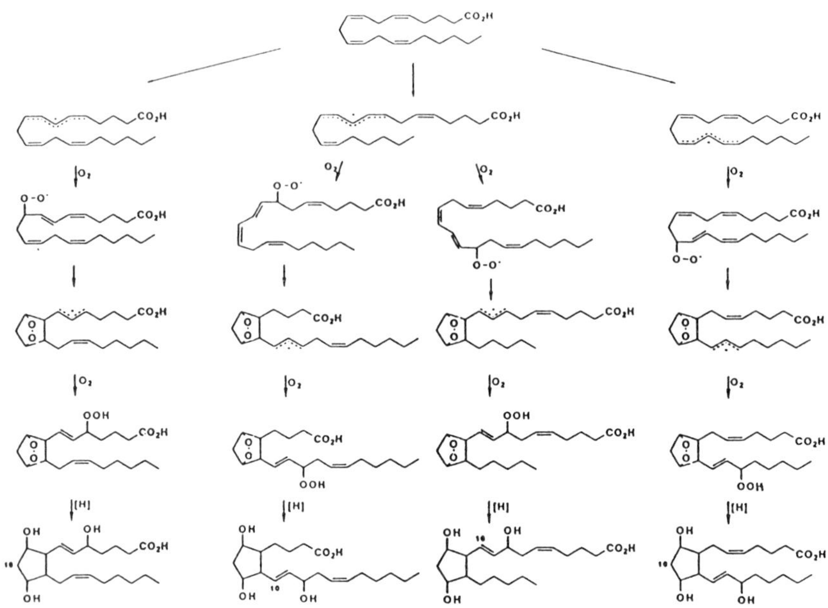

I II III IV

FIGURE 1. Pathway outlining the formation of F_2-isoprostanes.

8-epi-$PGF_{2\alpha}$ exerts potent biological activity, a study was undertaken in an attempt to obtain evidence for its presence *in vivo.* The results obtained indicated that 8-epi-$PGF_{2\alpha}$ is, in fact, an abundant F_2-isoprostane produced *in vivo.*[8]

ISOPROSTANES AS MARKERS OF ENDOGENOUS LIPID PEROXIDATION

It is well recognized that most methods available to assess oxidant stress *in vivo*—in particular, noninvasive methods that can be used for human investigation—are unreliable.[9] However, we have obtained considerable evidence that suggests that quantification of isoprostanes represents an important advance in our ability to assess oxidant status *in vivo.* Sufficient quantities of F_2-isoprostanes can be detected in human biological fluids from normal individuals—for instance, in urine, plasma, and cerebrospinal fluid—to define a normal range. Importantly, these compounds are also very stable molecules. However, precautions must be taken to prevent artifactual generation of F_2-isoprostanes in lipid-containing biological samples—for example, plasma—by autoxidation *ex vivo.* Such precautions include either processing samples immediately after they are obtained or rapidly freezing the samples in liquid nitrogen and storing them at −70 °C. We have found that storage of lipid-containing samples at −70 °C effectively prevents the formation of isoprostanes by autoxidation, whereas storage at −20 °C does not.[10] We have shown in animal models of *in vivo* oxidant injury—for example, by administration of CCl_4 to normal rats and administration of diquat to Se-deficient rats—that free concentrations of F_2-isoprostanes in the circulation increase profoundly by as much as 200-fold.[1] Further, we have shown that measurement of levels of F_2-isoprostanes esterified in tissue lipids can be valuable in localizing endogenous lipid peroxidation to a particular organ or tissue.[11]

Thus, quantification of F_2-isoprostanes can provide a valuable tool to investigate the role of oxidant stress in human disease. In this regard, we have recently obtained data that implicate for the first time oxidant stress in the pathogenesis of the hepatorenal syndrome. The hepatorenal syndrome is defined as the occurrence of renal failure without an identifiable cause in patients with severe liver disease.[12] The etiology of the hepatorenal syndrome remains unknown. The cause of the renal failure in this disease has been attributed to intense renal vasoconstriction. The occurrence of tissue hypoxia and chronic endotoxemia in patients with severe liver disease is an environment conducive for the generation of free radicals.[13,14] Therefore, we tested the hypothesis that oxidant stress may represent the underlying process responsible for the abnormalities observed in the hepatorenal syndrome by measuring F_2-isoprostane production in patients with hepatorenal syndrome and appropriate control groups of patients with severe liver disease and/or chronic renal failure. Circulating concentrations and the urinary excretion of F_2-isoprostanes were markedly increased above normal in patients with hepatorenal syndrome by a mean of 7.8-fold and 23.5-fold, respectively, but were not increased substantially in the control groups of patients. Although enhanced formation of several vasoactive substances—for example, leukotrienes, thromboxane, and endothelins—have been demonstrated in some patients with hepatorenal syndrome, the cause of the renal vasoconstriction that characterizes this disorder remains poorly understood.[15–17] It is of interest that the F_2-isoprostane 8-epi-$PGF_{2\alpha}$, mentioned previously, and the E_2-isoprostane 8-epi-PGE_2 (*vide infra*) are potent renal vasoconstrictors. Thus, isoprostanes may also be contributing to some extent to the renal vasoconstriction in the hepatorenal syndrome. Nonetheless, the demonstration of overproduction of F_2-isoprostanes in the hepatorenal syndrome suggests that oxidant stress may represent the fundamental abnormality involved in the pathogenesis of this disease.

DISCOVERY OF THE FORMATION OF E_2/D_2-ISOPROSTANES

As previously mentioned, the formation of F_2-isoprostanes proceeds through the formation of bicyclic PGH_2-like endoperoxide intermediates, which are reduced to form F_2-isoprostanes. Factors involved in the reduction if the isoprostane endoperoxide intermediates *in vivo* remain to be identified. However, we hypothesized that the endoperoxide intermediates may in part escape reduction and rearrange to form PGE_2-like and PGD_2-like isoprostanes. Using a modified mass spectrometric assay for measurement of cyclooxygenase-derived PGE_2, we initially obtained evidence suggesting that E_2/D_2-isoprostanes can be formed *in vitro* by finding a series of intense peaks representing compounds with similar characteristics to PGE_2 and PGD_2 in plasma that had been allowed to undergo autoxidation. Abundant quantities of what appeared to be the same compounds could also be detected in the circulation and esterified to liver phospholipids following treatment of rats with CCl_4 to induce endogenous lipid peroxidation. Employing a variety of approaches, including the use of deuterated derivatives, catalytic hydrogenation, electron ionization mass spectrometry, and fast atom bombardment mass spectrometry, compelling evidence was obtained that these compounds detected *in vivo* in CCl_4-treated rats represented E_2/D_2-isoprostanes. In untreated rats, levels of E_2/D_2-isoprostanes present esterified to liver lipids were 0.90 ± 0.10 ng/g liver, and levels increased dramatically to 85 ± 33 ng/g liver following administration of CCl_4. E_2/D_2-isoprostanes could not be detected circulating in plasma from untreated rats (lower limit of detection = 5 pg/ml). However, high concentrations were detected in plasma following treatment of rats with CCl_4 (215 ± 90 pg/ml). The relative amounts of E_2/D_2-isoprostanes present esterified to lipids in livers from both untreated and CCl_4-treated rats were approximately one-quarter to one-third of the amount of F_2-isoprostanes present.

We then examined whether 8-epi-PGE_2, the E-ring analog of the F_2-isoprostane 8-epi-$PGF_{2\alpha}$, exerted biological activity. Interestingly, whereas cyclooxygenase-derived PGE_2 is a general vasodilator, 8-epi-PGE_2 was found to be a potent vasoconstrictor of renal vasculature with a potency only slightly less than that of 8-epi-$PGF_{2\alpha}$. Further, we found that the vascular effects of 8-epi-PGE_2 in the kidney could be abrogated by the thromboxane A_2 receptor antagonist SQ29548. Whether 8-epi-PGE_2 exerts these effects by (a) interaction with the thromboxane A_2 receptor, (b) the same receptor that 8-epi-$PGF_{2\alpha}$ interacts with, which is presumably similar to but distinct from the thromboxane A_2 receptor, or (c) another receptor remains to be determined.

The discovery of E_2/D_2-isoprostanes extends our understanding of the biochemistry of the formation of isoprostanes *in vivo*. This discovery also indicates that the variety of novel prostanoids that can be produced by nonenzymatic peroxidation of arachidonic acid *in vivo* is much greater than previously thought. The finding that E_2/D_2-isoprostanes, like F_2-isoprostanes, are capable of exerting biological activity also has potentially important biological implications related to the possibility that isoprostanes may be involved as mediators in the pathophysiology of oxidant injury. Future investigations aimed at understanding factors that may influence the formation of E_2/D_2-isoprostanes relative to F_2-isoprostanes will be of interest and of potential importance in advancing our understanding of the biological consequences of the formation of isoprostanes *in vivo*.

SUMMARY

It was recently discovered that a series of structurally unique prostaglandin F_2-like compounds (F_2-isoprostanes) capable of exerting potent biological activity are produced *in vivo* in humans by a noncyclooxygenase mechanism involving free radical catalyzed peroxidation of arachidonic acid. Considerable evidence has been obtained suggesting that quantification of F_2-isoprostanes represents an important advance in our ability to assess oxidant status *in vivo* in humans. This has allowed us to implicate oxidant stress in the pathogenesis of human disease—for example, the hepatorenal syndrome. In addition to the F_2-isoprostanes, we recently discovered that E-ring and D-ring isoprostanes are also produced in abundance *in vivo* by rearrangement of the isoprostane endoperoxide intermediates. We have also been able to demonstrate that one of the E_2-isoprostanes, 8-epi-PGE_2, is a potent renal vasoconstrictor in the rat. Insights into factors that may influence the formation of E_2/D_2-isoprostanes relative to F_2-isoprostanes should be important in advancing our understanding of the biological consequences of the formation of isoprostanes *in vivo*.

REFERENCES

1. MORROW, J. D., K. E. HILL, R. F. BURK, T. M. NAMMOUR, K. F. BADR & L. J. ROBERTS II. 1990. A series of prostaglandin F_2-like compounds are produced in vivo in humans by a non-cyclooxygenase, free radical catalyzed mechanism. Proc. Natl. Acad. Sci. USA **87:** 9383-9387.
2. MORROW, J. D., J. A. AWAD, H. J. BOSS, I. A. BLAIR & L. J. ROBERTS II. 1992. Non-cyclooxygenase-derived prostanoids (F_2-isoprostanes) are formed in situ on phospholipids. Proc. Natl. Acad. Sci. USA **89:** 10721-10725.
3. BANERJEE, M., K. H. KANG, J. D. MORROW, L. J. ROBERTS II & J. H. NEWMAN. 1992. Effects of a novel non-cyclooxygenase derived prostaglandin, 8-epi-$PGF_{2\alpha}$, in rabbit lung in situ. Am. J. Physiol. **263:** (Heart Circ. Physiol. **32**): H660-H663.
4. KANG, H. K., J. D. MORROW, L. J. ROBERTS II, J. H. NEWMAN & M. BANERJEE. 1993. Airway and vascular effects of 8-epi-prostaglandin $F_{2\alpha}$ in isolated perfused rat lung. J. Appl. Physiol. **74:** 460-465.
5. TAKAHASHI, K., T. M. NAMMOUR, J. EBERT, J. D. MORROW, L. J. ROBERTS II & K. F. BADR. 1992. Glomerular actions of a free radical generated novel prostaglandin, 8-epi-prostaglandin $F_{2\alpha}$, in the rat: Evidence for interaction with thromboxane A_2 receptors. J. Clin. Invest. **90:** 135-141.
6. MORROW, J. D., T. A. MINTON & L. J. ROBERTS II. 1992. The F_2-isoprostane, 8-epi-prostaglandin $F_{2\alpha}$, a potent agonist of the vascular thromboxane/endoperoxide receptor, is a platelet thromboxane/endoperoxide receptor antagonist. Prostaglandins **44:** 155-163.
7. FUKUNAGA, M., N. MAKITA, L. J. ROBERTS II, J. D. MORROW, K. TAKAHASHI & K. F. BADR. 1993. Evidence for the existence of F_2-isoprostane receptors on rat vascular smooth muscle cells. Am. J. Physiol. **263** (Cell Physiol. **33**): C1619-C1624.
8. MORROW, J. D., K. F. BADR & L. J. ROBERTS II. 1994. Evidence for the formation of the F_2-isoprostane, 8-epi-$PGF_{2\alpha}$, in vivo. Biochim. Biophys. Acta **210:** 244-248.
9. HALLIWELL, B. & M. GROOTVELD. 1987. The measurement of free radical reactions in humans. FEBS Lett. **21:** 9-14.
10. MORROW, J. D., T. M. HARRIS & L. J. ROBERTS II. 1990. Non-cyclooxygenase oxidative formation of a series of novel prostaglandins: Analytical ramifications for measurement of eicosanoids. Anal. Biochem. **184:** 1-10.
11. MORROW, J. D., J. A. AWAD, T. KATO, K. TAKAHASHI, K. F. BADR, L. J. ROBERTS II & R. F. BURK. 1992. Formation of novel non-cyclooxygenase derived prostanoids (F_2-isoprostanes) in carbon tetrachloride hepatotoxicity. An animal model of lipid peroxidation. J. Clin. Invest. **90:** 2502-2507.
12. SCHELLING, J. R. & S. L. LINAL. 1990. Hepatorenal syndrome. Semin. Nephrol. **10:** 565-570.
13. HARRISON, P. M., J. A. WENDON, A. E. S. GIMSON, G. J. M. ALEXANDER & R. WILLIAMS. 1991. Improvement by acetylcysteine of hemodynamics and oxygen transport in fulminant hepatic failure. N. Engl. J. Med. **324:** 1852-1857.
14. WILKINSON, S. P., H. MOODIE, J. D. STAMATAKIS, V. V. KAKKAR & R. WILLIAMS. 1976. Endotoxaemia and renal failure in cirrhosis and obstructive jaundice. Br. Med. J. **2:** 1415-1418.

15. Moore, K., P. S. Ward, G. W. Taylor & R. Williams. 1991. Systemic and renal production of thromboxane A_2 and prostacyclin in decompensated liver disease and hepatorenal syndrome. Gastroenterology **100:** 1069-1077.
16. Moore, K. P., G. W. Taylor, N. H. Maltby, D. Siegers, R. Fuller, C. T. Dollery & R. Williams. 1990. Increased production of cysteinyl leukotrienes in hepatorenal syndrome. J. Hepatol. **11:** 263-271.
17. Moore, K., J. Wendon, M. Frazer, J. Karani, R. Williams & K. Badr. 1992. Plasma endothelin immunoreactivity in liver disease and the hepatorenal syndrome. N. Engl. J. Med. **327:** 1774-1778.

Leukotrienes Mediate Delayed Airway Effects of 15-HETE[a]

R. ROBERT SCHELLENBERG[b] AND SAMUEL TSANG

University of British Columbia Pulmonary Research Laboratory
St. Paul's Hospital
Vancouver, British Columbia V6Z 1Y6, Canada

HASSAN SALARI

Jack Bell Research Centre
Vancouver General Hospital
Vancouver, British Columbia, Canada

INTRODUCTION

15-Hydroxyeicosatetraenoic acid (15-HETE) has been demonstrated to be a major arachidonic acid metabolite in human lung[1] and the principal arachidonate metabolite of isolated human airway epithelial cells.[2,3] 15-Lipoxygenase has been cloned from human airway and localized to the epithelium.[4,5] Since the epithelial layer represents the first component of the airway interacting with inhaled stimuli causing bronchoconstriction, the release of 15-HETE may mediate or modulate subsequent responses. It is detected in bronchial alveolar lavage fluid following local provocative challenge with allergen in asthmatics.[6] Although there are no data suggesting that antigen directly interacts with airway epithelial cells to produce 15-HETE, there are a number of potential mechanisms. These include the effects of mediators released by cells adjacent to airway epithelium, which would include inflammatory cells such as eosinophils, which also contain 15-lipoxygenase,[7] dendritic cells, and mast cells. Both mast cells and dendritic cells have high-affinity immunoglobulin (Ig) E receptors that can be activated by antigen-specific IgE. Which of the multiple mediators released by these cells is the primary stimulant of 15-HETE production has not been evaluated, although our previous work has shown that platelet activating factor, a mediator released from both mast cells and eosinophils, generates 15-HETE from isolated human lung epithelial cells.[3] Potential stimuli of 15-HETE production might even include physical stimuli such as alterations in heat and water exchange as implicated in exercise-induced asthma.

A recent study demonstrated that interleukin (IL)-4 can induce 15-lipoxygenase mRNA and protein production in cultured human monocytes. This induction is inhibited by interferon γ.[8] These results are of particular interest regarding allergic asthma since IL-4 enhances IgE production by inducing isotype switching, whereas

[a] This work was supported by the Medical Research Council of Canada.

[b] Address for correspondence: Dr. R. Robert Schellenberg, UBC Pulmonary Research Laboratory, St. Paul's Hospital, 1081 Burrard Street, Vancouver, B.C. V6Z 1Y6, Canada.

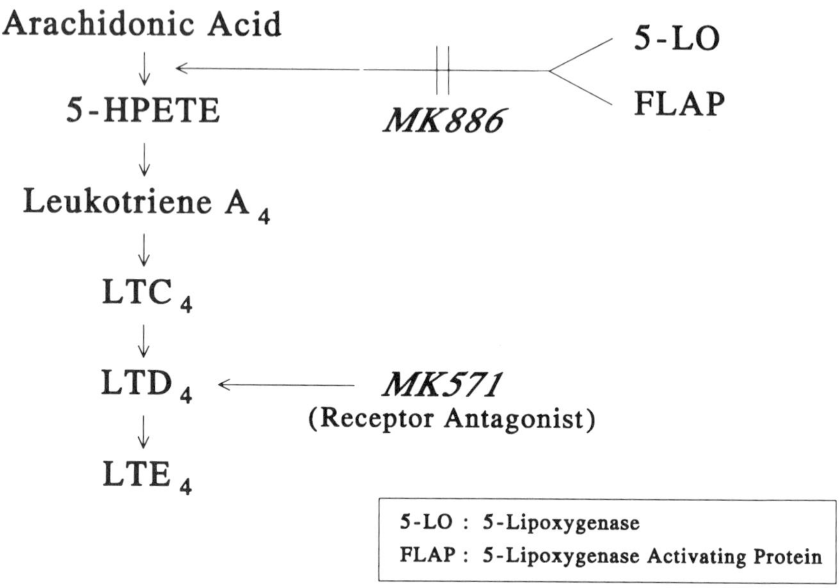

FIGURE 1. Schematic diagram showing the sites of action of the FLAP inhibitor MK886 and the leukotriene D_4 antagonist MK571.

interferon γ inhibits it.[9,10] Whether IL-4 can induce 15-lipoxygenase in airway epithelial cells remains to be determined, as does the effect of other proinflammatory cytokines.

Based on previous findings that 15-HETE induces a delayed contractile response of human airway preparations *in vitro,*[2] we chose to determine whether this contractile response was mediated by the secondary generation of other known contractile agonists of human airway smooth muscle. In this study, we have evaluated the effects of a leukotriene (LT) biosynthesis inhibitor that acts by blocking the binding of the 5-lipoxygenase-activating protein (FLAP),[11,12] MK-886, and the LTD_4 antagonist MK-571,[13] as shown in FIGURE 1. The rationale for choosing leukotrienes as the likely mediators was the sustained contractile response reminiscent of that noted for leukotrienes.[14]

METHODS

Macroscopically normal lobar bronchial tissue was obtained from lobectomy samples within 30 minutes of surgery. In all cases, surgery was performed for removal of carcinoma, and none of the subjects had asthma. Muscle strips approximately 2 × 3 × 10 mm were dissected from airway free of tumor and suspended in 10-ml organ baths containing 37 °C Krebs-Henseleit solution bubbled with 95% O_2-5% CO_2.

Isometric responses to contractile agonists were measured using a Grass FT.03 force displacement transducer connected to a Beckman R511A polygraph. Individual muscle strips were stretched to optimal length as assessed by electrical field stimulation. Where possible, four similar strips were taken from the same preparation so that interventional studies could be performed on matched tissues with control responses.

The 5-lipoxygenase-activating protein inhibitor MK-886 and the LTD_4 antagonist MK-571 were kindly supplied by Merck Frosst Canada (Pt. Claire, Quebec). Drugs were added 10 min prior to the addition of 15-HETE [15(S)-HETE, Sigma Chemicals, St. Louis, MO] after the tissues had equilibrated. Contractile responses of drug-treated tissues to 15-HETE were compared with tissues receiving 15-HETE only. Maximum contractile response of each tissue was assessed at the end of experiment by the addition of carbachol 10^{-3} M.

Measurement of Leukotrienes from Bronchi

Human bronchial tissue was minced, divided into 0.3-g samples, washed twice with Pipes-albumin buffer, pH 7.4, and then incubated with or without 15-HETE 10^{-6} M for 1 h at 37 °C. Incubations were terminated by the addition of equal volumes of 100% methanol. Each 2 ml of sample was centrifuged at 1300 *g* for 20 min at 4 °C and the supernatant removed and acidified to pH 4-5 with 1M HCl. Samples were diluted to 10 ml with 100% methanol and passed through pretreated C_{18} Sep-paks and eluted with serial washings with 10 ml H_2O, 2 ml hexane, and 3 ml 100% methanol. The final eluent was dried under N_2 and resuspended in methanol. Methanol extracts were analyzed by reversed-phase HPLC by the method of Steffenrud and Salari.[15] Sulfidopeptide leukotrienes were detected at 280 nm, 15-HETE at 235 nm, and lipoxins at 310 nm and quantitated using standard curves obtained from known amounts of synthetic compounds. A prostaglandin (PG) B_2 internal standard (50 ng) was added to each sample to evaluate recovery.

RESULTS

Human Bronchial Contraction to 15-HETE

15-HETE 10^{-6} M produced minimal or no immediate contractile responses but caused reproducible delayed responses beginning at approximately 45 min and reaching a plateau at approximately 2 h following addition of 15-HETE. Maximal responses in the absence of drug were 60% of the response obtained with carbachol 10^{-3} M.

Pretreatment with the FLAP inhibitor MK-886 3×10^{-6} M produced significant decreases in 15-HETE-induced contractions as shown in FIGURE 2. The percent inhibition obtained at this concentration was 90.7 ± 3.0% of the control response ($n = 13$).

The LTD_4 antagonist MK-571 at a concentration of 2×10^{-6} M produced 51.5 ± 7.7% inhibition of the 15-HETE-induced contractions ($n = 6$), whereas 2×10^{-5} M produced virtually total inhibition (94.2 ± 3.2%, $n = 17$). Comparative data for both

TABLE 1. Percent Inhibition of 15-Hete-Induced Contraction

Drug	Concentration	n	% Inhibition
MK-886	3×10^{-6} M	13	90.7 ± 3.0
MK-571	2×10^{-6} M	6	51.5 ± 7.7
	2×10^{-5} M	15	94.2 ± 3.3

drugs are shown in TABLE 1. These data suggested that the secondary generation of sulfidopeptide leukotrienes produced the contractile response by activation of LTD_4 receptors present on airway smooth muscle cells. To further validate this conclusion and rule out the possibility that metabolism of 15-HETE to lipoxins produced the contractile response by activation of LTD_4 receptors, arachidonate metabolites were analyzed from airway tissue.

Leukotriene Production from Human Bronchial Tissue

Incubation of minced preparations of fresh human bronchial tissue with 15-HETE 10^{-6} M for 1 h produced measurable quantities of LTC_4 as shown in TABLE 2. A representative chromatogram is shown in FIGURE 3. No control tissues incubated for the same time period had detectable leukotrienes. In additional studies, the extraction

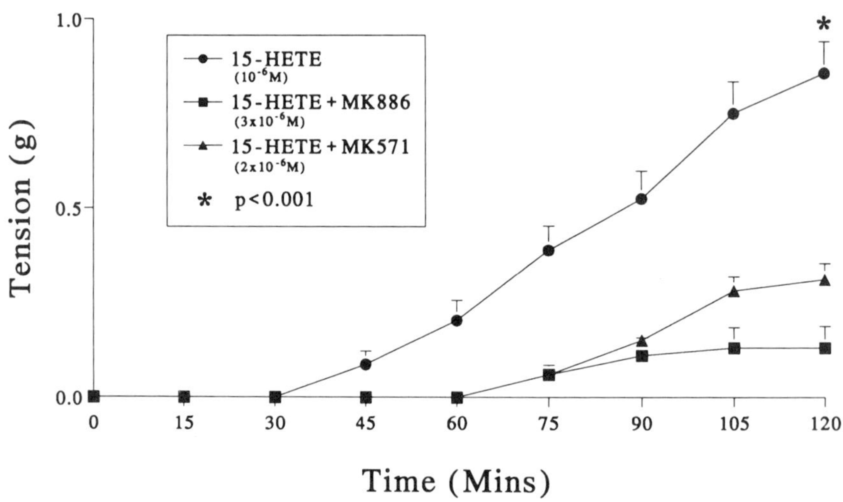

FIGURE 2. Contractile responses of human bronchial strips to 15-HETE in the presence or absence of the leukotriene biosynthesis inhibitor MK886 or the leukotriene D_4 antagonist MK571. *$p < 0.001$. All values of 15-HETE control responses (●) from 60 to 120 minutes were significantly different from the data points for the MK886 or MK571 pretreatment samples.

TABLE 2. LTC_4 Produced by 15-HETE (10^{-6} M)

Sample	Wet Wt. (g)	LTC_4 (ng)	LTC_4 (ng/g Wet Wt)[a]
1	0.34	29.0	85.4
2	0.80	179.2	224.3
3	0.43	160.8	374.3
4	0.40	140.0	350.0
5	0.20	14.0	70.2
6	0.25	39.3	157.2

[a] Mean ± SEM: 210.2 ± 53.1.

of arachidonic metabolites immediately after the addition of 15-HETE vs. 1 h later demonstrated no significant metabolism of the 15-HETE added that was readily detectable by HPLC. Lipoxins were not detected from any tissues. These results confirm that sulfidopeptide leukotrienes were produced following 15-HETE incubation of airway tissue and provided no evidence for lipoxin generation by 15-HETE metabolism.

DISCUSSION

The finding that 15-HETE can cause a delayed contractile response in human airways is of interest and may be relevant to specific stimuli causing bronchoconstriction in asthmatics. We have not seen delayed contractile responses to any other stimuli evaluated, and this unique characteristic suggested to us the production of a secondary mediator capable of producing the contractile response. Delayed responses were highly reproducible in airway tissue and much more pronounced than any acute contractile response to 15-HETE. Acute responses required higher doses of 15-HETE and were small, findings consistent with data of others.[16] The delayed contractions were substantial and of comparable magnitude to those produced by many other contractile mediators or by anti-IgE. The sustained contractile response, once it was produced, was reminiscent of the response to sulfidopeptide leukotrienes, making us pursue the possibility that these mediators were generated following 15-HETE challenge.

The cellular source of leukotrienes in this situation remains to be identified. We believe it unlikely to be mast cells in that there was no evidence for release of other mediators such as histamine. In addition, tissues maintained overnight failed to produce contractile responses to 15-HETE, whereas they still responded nicely to anti-IgE stimulation with leukotriene production and contractions, providing indirect evidence that the cell stimulated by 15-HETE was no longer capable of responding or had lost specific receptors for 15-HETE. We are aware of no evidence suggesting that 15-HETE can induce leukotriene generation from inflammatory cells such as eosinophils or neutrophils, and it would be unlikely that these were present in high numbers in the tissues evaluated that were not asthmatic.

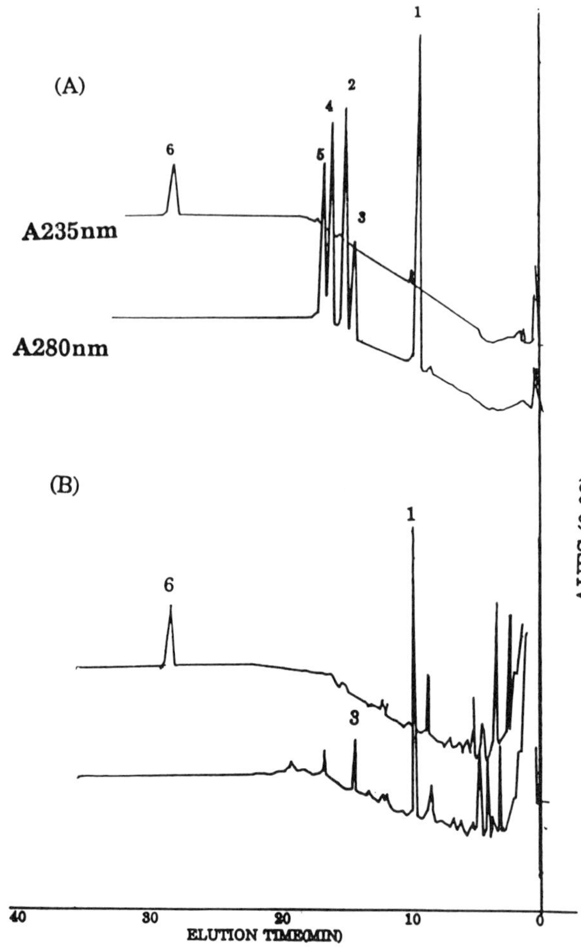

FIGURE 3. Representative chromatography of lipoxygenase products. Chromatograms of leukotriene and 15-HETE standards (**A**) and a sample from human bronchus incubated with 15-HETE 10^{-6}M (**B**) showing LTC_4 formation. **A.** Reverse phase HPLC of 50-ng standards. **B.** Extract of 1-h 15-HETE stimulated human bronchus. 1. PgB_2; 2. LTB_4; 3. LTC_4; 4. LTD_4; 5. LTE_4; 6. 15-HETE.

Although our findings might have been explained on the basis that inhibition of 5-lipoxygenase by MK-886 would inhibit production of lipoxins from 15-HETE and that the LTD_4 antagonist would inhibit the contractile effects of lipoxin A, which has been shown to act via the LTD_4 receptor,[17] the failure to show production of lipoxins or degradation of 15-HETE with incubation of human bronchial tissue, at a time when this stimulus produced LTC_4, dismissed this possibility.

The finding that the major arachidonic metabolite of human airway epithelial cells is capable of causing contraction of human airway smooth muscle raises a number of interesting possibilities. Might this be a mechanism for stimuli that are IgE independent? It is conceivable that alterations in heat and water exchange stimulate epithelial cells to release 15-HETE. Since there is a lack of consensus regarding the role of mast cell mediators in exercise-induced asthma, this remains a possibility, especially in light of the effective role of LTD_4 antagonists in inhibiting exercise-induced asthma.[18] In addition, it is clear that considerable 15-HETE is produced upon antigenic challenge of asthmatic airways *in vivo*,[6] and evidence would suggest that the major source of this is epithelial cells.

Although our studies evaluating contractile responses of human bronchi used 15(S)-HETE as the stimulus, preliminary results with 15(R)-HETE show comparable effects. These findings suggest that stimuli inducing the synthesis of cyclooxygenase-2 (COX-2) could be important triggers of these delayed contractile responses since high levels of this enzyme are found in human lung,[19] cultured ovine tracheal epithelial cells produce 15-HETE via COX-2 following aspirin treatment,[20] and glucocorticoids regulate this inducible enzyme.[21] Further studies to evaluate the chiral specificity of 15-HETE produced with various stimuli and evaluation of COX-2 induction as well as 15-lipoxygenase activation are required.

In conclusion, we have demonstrated that incubation of human bronchial tissue with 15-HETE produces sulfidopeptide leukotrienes that contract human airway smooth muscle. The cellular source of leukotrienes remains to be identified, as does the relative importance of this mechanism in bronchoconstrictor responses of asthma.

ACKNOWLEDGMENTS

We wish to thank Kent Webb for typing the manuscript and Stuart Greene and Randy Thomson for preparation of figures and photography.

REFERENCES

1. Hamburg, M., P. Hedqvist & K. Radegran. 1980. Identification of 15-hydroxy-5,8,11,13-eicosatetraenoic acid (15-HETE) as a major metabolite of arachidonic acid in human lung. Acta Physiol. Scand. **110:** 219-221.
2. Hunter, J. A., W. E. Finkbeiner, J. A. Nadel, E. J. Goetzel & M. J. Holtzman. 1985. Predominant generation of 15-lipoxygenase metabolites of arachidonic acid by epithelial cells from human trachea. Proc. Natl. Acad. Sci. USA **82:** 4633-4637.
3. Salari, H. & R. R. Schellenberg. 1991. Stimulation of human airway epithelial cells by platelet activating factor (PAF) and arachidonic acid produces 15-hydroxyeicosatetraenoic acid (15-HETE) capable of contracting bronchial smooth muscle. Pulmonary Pharmacol. **4:** 1-7.
4. Sigal, E., S. Dicharry, E. Highland & W. E. Finkbeiner. 1992. Cloning of human airway 15-lipoxygenase: Identity to the reticulocyte enzyme and expression in epithelium. Am. J. Physiol. **262:** L392-L398.
5. Sigal, E. & J. A. Nadel. 1991. The airway epithelium and arachidonic acid 15-lipoxygenase. Am. Rev. Respir. Dis. **143:** S71-S74.
6. Murray, J. J., B. Tonnel, A. R. Brash, L. J. Roberts II, P. Gosset, R. Workman, A. Capron & J. A. Oates. 1986. Release of prostaglandin D_2 into human airways during acute antigen challenge. N. Engl. J. Med. **315:** 800-804.

7. TURK, J., R. L. MAAS, A. R. BRASH, L. J. ROBERTS II & J. A. OATES. 1982. Arachidonic acid 15-lipoxygenase products from human eosinophils. J. Biol. Chem. **257:** 7068-7076.
8. CONRAD, D. J., H. KUHN, M. MULKINS, E. HIGHLAND & E. SIGAL. 1992. Specific inflammatory cytokines regulate the expression of human monocyte 15-lipoxygenase. Proc. Natl. Acad. Sci. USA **89:** 217-221.
9. PENE, J., F. ROUSSET, F. BRIERE, I. CHRETIEN, J. Y. BONNEFOY, H. SPITS, T. YOKOTA, J. ARAI, J. BACHEREAU & J. DE VRIES. 1988. IgE production by normal human lymphocytes is induced by interleukin-4 and suppressed by interferon γ and α and prostaglandin E2. Proc. Natl. Acad. Sci. USA **85:** 6880-6884.
10. VERCELLI, D., H. H. JABARA, K. ARAI & R. S. GEHA. 1989. Induction of human IgE synthesis requires interleukin 4 and T/B cell interactions involving T cells receptor/CD3 complex and MHC class II antigens. J. Exp. Med. **169:** 1295-1307.
11. DIXON, R. A. F., R. E. DIEHL, E. OPAS, E. RANDS, P. J. VICKERS, J. F. EVANS, J. W. GILLARD & D. K. MILLER. 1990. Requirement of a 5-lipoxygenase-activating protein for leukotriene synthesis. Nature **343:** 282-284.
12. ROUZER, C. A., A. W. FORD-HUTCHINSON, H. E. MORTON & J. W. GILLARD. 1990. MK886, a potent and specific leukotriene blocks and reverses the membrane association of 5-lipoxygenase in ionophore-challenged leukocytes. J. Biol. Chem. **265:** 1436-1442.
13. JONES, T. R., R. ZAMBONI, M. BELLEY, E. CHAMPION, L. CHARETTE *et al.* 1989. Pharmacology of L-660,711 (MK-571), a novel, potent and selective LTD_4 receptor antagonist. Can. J. Pharmacol. **67:** 17-28.
14. HANNA, C. J., M. K. BACH, P. D. PARE & R. R. SCHELLENBERG. 1981. Slow-reacting substances (leukotrienes) contract human airway and pulmonary vascular smooth muscle in vitro. Nature **290:** 343-344.
15. STEFFENRUD, S. & H. SALARI. 1988. Reversed-phase ion-interaction chromatography of leukotrienes, lipoxins and related compounds. J. Chromatogr. **378:** 35-44.
16. COPAS, J. L., P. BORGEAT & P. J. GARDINER. 1982. The action of 5-, 12- and 15-HETE on tracheobronchial smooth muscle. Prostaglandins Leukotrienes Med. **8:** 105-114.
17. DAHLEN, S. E., L. FRANZEN, J. RAND *et al.* 1988. Actions of lipoxin A_4 and related compounds in smooth muscle preparations and on the microcirculation in vivo. Adv. Exp. Med. Biol. **229:** 107-130.
18. MANNING, P. J., R. M. WATSON, D. J. MARGOLSKEE, V. C. WILLIAMS, J. L. SCHWARTZ & P. M. OBYRNE. 1990. Inhibition of exercise-induced bronchoconstriction by MK-571, a potent leukotriene D_4-receptor antagonist. N. Eng. J. Med. **323:** 1736-1739.
19. O'NEILL, G. P. & A. W. FORD-HUTCHINSON. 1993. Expression of mRNA for cyclooxygenase-1 and cyclooxygenase-2 in human tissues. FEBS Lett. **330:** 156-160.
20. HOLTZMAN, M. J., J. TURK & L. P. SHORNICK. 1992. Identification of a pharmacologically distinct prostaglandin H synthetase in cultured epithelial cells. J. Biol. Chem. **267:** 21438-21445.
21. MASFERRER, J. L., K. SEIBERT, B. ZWEIFEL & P. NEEDLEMAN. 1992. Endogenous glucocorticoids regulate an inducible cyclooxygenase enzyme. Proc. Natl. Acad. Sci. USA **89:** 3917-3921.

Eicosanoids and Asthma[a]

P. M. O'BYRNE[b]

Department of Medicine
Faculty of Health Sciences
McMaster University
1200 Main Street West
Hamilton, Ontario, L8N 3Z5, Canada

INTRODUCTION

Asthma is a disease characterized by symptoms such as dyspnea, wheezing, chest tightness, and cough. These symptoms are usually caused by airflow obstruction, which is characteristically variable. Asthmatics are also known to have airway hyperresponsiveness to a variety of chemical bronchoconstrictor stimuli and physical stimuli such as exercise and hyperventilation of cold, dry air.[1] This means that asthmatics require less of a bronchoconstrictor agonist to induce airway narrowing than do normals. In general, the severity of airway hyperresponsiveness is associated with the severity of asthma[2] and with the amount of treatment needed to optimally control symptoms of asthma.[3] More recently, it has been recognized that asthma symptoms, variable airflow obstruction, and airway hyperresponsiveness occur as a consequence of a characteristic form of cellular inflammation and structural changes in the airway wall.[4] The inflammation consists of the presence of activated eosinophils, lymphocytes, and an increased number of mast cells, which have been described both in bronchoalveolar lavage and airway biopsies from asthmatic subjects,[5–8] as well as characteristic structural changes of the airways.[9,10] The structural changes described in asthmatic airways, which appear to be characteristic of the disease, are likely caused by persisting airway inflammation. These changes include patchy desquamation of the airway epithelium;[10] what was originally described as thickened basement membrane, but is recognized as thickening of the reticular collagen layer of the basement membrane;[11] and airway smooth muscle hypertrophy.[9]

The importance of slow-reacting substance of anaphylaxis (SRS-A) in the events occurring after immunological challenge of the lungs was identified in the early 1940s.[12] Later, Brocklehurst demonstrated that SRS-A was released from lung segments from an asthmatic subject when these segments were exposed to allergen.[13] SRS-A was suggested to be an important mediator in causing symptoms after exposure to allergen in sensitized subjects, mainly because it was a potent airway smooth muscle constrictor with a much longer duration of action than other smooth muscle constrictors, such as histamine. SRS-A is now known to consist of the cysteinyl leukotrienes (LT) C_4, D_4 and E_4.[14] Identifying a role for a mediator in asthma has

[a]From the Asthma Research Group and the Department of Medicine, McMaster University, Hamilton, Ontario, Canada.

[b]Medical Research Council of Canada Scientist.

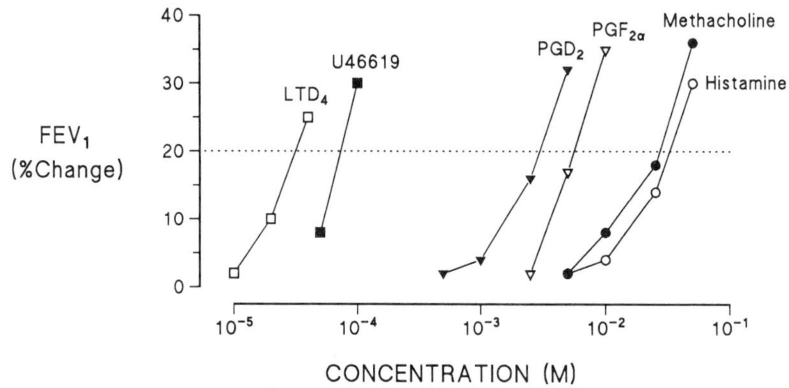

FIGURE 1. Airway responses to a variety of inhaled bronchoconstrictor mediators in one asthmatic subject. The relative potency of the mediators differ, with the cysteinyl leukotrienes and the thromboxane mimetic U46619 being the most potent studied to date in human subjects.

relied on the collection of various types of evidence. Generally, when the structure of the mediator is identified and the mediator is synthesized, the mediator is given (usually by inhalation) to asthmatics to identify whether it can mimic some component of the asthmatic response. Subsequently, when assays for its measurement are available, efforts are made to measure it in biological fluids to determine whether it is released (and excreted) during asthmatic responses. Then, when selective antagonists or synthesis inhibitors are available, efforts are made to block components of the asthmatic responses in clinical models of asthma. The final, and most difficult, hurdle is to determine whether the mediator antagonists or synthetase inhibitors are useful in treating asthmatic patients, thereby proving that the mediator has an important role in its pathogenesis. Each of these steps has been taken with the cysteinyl leukotrienes in asthma; the evidence that they are important in the pathogenesis of asthma will be reviewed.

INHALED CYSTEINYL LEUKOTRIENES IN HUMAN SUBJECTS

The cysteinyl leukotrienes were initially demonstrated to be very potent constrictors of human airway smooth muscle *in vitro.*[15] The first report of the effects of inhaled LTC_4 and LTD_4 *in vivo* was published in 1981 by Holroyde *et al.*[16] who demonstrated that these mediators caused bronchoconstriction in normal human subjects. Subsequently, inhaled LTC_4 and LTD_4 have been demonstrated to be potent bronchoconstrictors in both normal and asthmatic subjects[16–19] (FIG. 1), being up to 10,000 times more potent than methacholine in some normal subjects.[18] The cysteinyl leukotrienes have also been demonstrated to have a longer duration of action than inhaled histamine.[19] The bronchoconstrictor effects of inhaled LTD_4 are generally resolved within 1-2 hours, but are not followed by the development of a late phase bronchoconstrictor response,[20] as occurs after allergen inhalation in many allergic asthmatic subjects.[21]

LTC_4 and LTD_4 appear to have unique characteristics as bronchoconstrictors in asthmatic subjects. Airway hyperresponsiveness in asthmatic subjects is nonspecific. This means that an asthmatic subject who is hyperresponsive to inhaled histamine will be equally hyperresponsive to inhaled methacholine,[22] prostaglandins D_2[23] and F_{2a},[24] and to exercise[25] and isocapnic hyperventilation.[26] However, the relationships between the airway responsiveness to inhaled histamine or metacholine and LTC_4 and LTD_4 are more complex. In normal subjects, LTC_4 and LTD_4 were between 1000-6000 times more potent at causing bronchoconstriction than was histamine.[18] In asthmatic subjects, the concentrations of LTC_4 and LTD_4 needed to cause bronchoconstriction is less than in normals, thereby confirming that asthmatic subjects have airway hyperresponsiveness to the inhaled leukotrienes. However, the degree of the hyperresponsiveness between asthmatics and normals is much less than for other bronchoconstrictors. For example, Griffin *et al.*[27] demonstrated that mild asthmatics required 1/140 the concentration of histamine than normals to induce similar degrees of airway narrowing, while they required only 1/3 the concentration of LTD_4. In addition, Adelroth *et al.*[18] reported that the relative potency of LTC_4 and LTD_4 compared to metacholine subjects was closely related to the level of airway hyperresponsiveness. This is not true for other bronchoconstrictors.[22-24] Therefore, while asthmatics are hyperresponsive to both LTC_4 and D_4, both leukotrienes are less potent bronchoconstrictors than histamine and metacholine in asthmatic compared to normal subjects.

The reasons for the difference in relative potency of LTC_4 and LTD_4 in asthmatic subjects is not yet known. Possible explanations include a different site of action of LTC_4 and LTD_4 on the airways of asthmatics and normal subjects. For example, LTD_4 contracts guinea pig lung parenchyma while having little effect on more central airways.[28] Therefore, more central deposition of LTC_4 or D_4 in asthmatic airways and more peripheral deposition in normal airways may make these bronchoconstrictors appear less potent in asthmatic subjects. Another explanation would be that the presence of leukotrienes in asthmatic airways may be responsible for the airway hyperresponsiveness to methacholine. Alternatively, down-regulation of leukotriene receptors in asthmatic airways may occur, due to the presence of endogenous leukotrienes or to the presence of another substance that specifically antagonizes leukotriene receptors.

We have previously demonstrated that tachyphylaxis exists to the bronchoconstrictor effects of inhaled histamine in stable asthmatic subjects,[29] and to inhaled methacholine in normal but not asthmatic subjects.[30] An important role for inhibitory prostaglandins in causing these effects was initially suggested by studies that demonstrated that histamine-stimulated prostaglandin E_2 release attenuates airway smooth muscle constrictor responses to histamine *in vitro.*[31] Following this observation, tachyphylaxis to other bronchoconstrictor mediators such as histamine in asthmatics[29] and methacholine in normal subjects[30] was shown to be attenuated by the cyclooxygenase inhibitors indomethacin or flurbiprofen. In addition, pretreatment with oral prostaglandin E_1 significantly improves airway responsiveness to both inhaled histamine and methacholine,[32] and inhaled prostaglandin E_2 attenuates allergen-induced early and late asthmatic responses[33] and exercise-induced bronchoconstriction in asthmatics.[34] Tachyphylaxis has been demonstrated to occur to repeated stimulation with inhaled LTD_4 in normal subjects,[35] and we have recently demonstrated that LTD_4 tachyphylaxis also exists in asthmatic subjects and is mediated through LTD_4-stimulated inhibitory prostaglandin release.[36] These studies may also provide an explanation for the reduced relative potency of LTD_4 in asthmatic subjects. Thus, endogenous cysteinyl leukotrienes are released in asthmatic airways, which may result in tachyphylaxis to the effects of exogenous inhaled LTD_4.

The effects of inhaled cysteinyl leukotrienes on airway hyperresponsiveness in asthma have been studied by several investigators. LTC_4 and LTD_4 have not been convincingly demonstrated to increase airway responsiveness to other bronchoconstrictor mediators.[35] However, LTE_4 does appear to increase airway hyperresponsiveness to histamine[37] and also, in a recent study, has been demonstrated to cause influx of eosinophils into asthmatic airways, an effect not seen with another bronchoconstrictor agonist, methacholine.[38] The reasons for this differential effect on airway function in asthmatics of the different cysteinyl leukotrienes, which are thought to act on a common receptor in human airways, is not yet known. However, these results with inhaled LTE_4 suggest that endogenously released cysteinyl leukotrienes may have effects on airway function that persist long after their bronchoconstrictor effects have resolved.

ENDOGENOUS PRODUCTION OF CYSTEINYL LEUKOTRIENES IN ASTHMA

Both allergen inhalation and exercise are important stimuli causing airway narrowing in asthmatics. Both are characterised by acute bronchoconstriction, which resolves usually within 1-1.5 hours if untreated. Allergen inhalation, however, differs from exercise in that in 60-75% of asthmatics who develop allergen-induced acute bronchoconstriction (the early asthmatic response) develop a second phase of bronchoconstriction, usually beginning 2-4 hours after allergen inhalation and progressing over the next 6-12 hours if untreated (the late asthmatic response).[21] The late response is associated with increases in airway responsiveness, which can last days or weeks.

Efforts have been made to measure LTD_4 release and/or LTE_4 excretion in urine after exercise and allergen inhalation and aspirin ingestion in asthmatics. Increases in urinary LTE_4, the metabolite of LTD_4, after exercise have been demonstrated in one,[39] but not another study.[40] By contrast, several investigators have demonstrated increases in urinary LTE_4 after allergen-induced bronchoconstriction.[41,42] The increases in urinary LTE_4 were significantly correlated, in one study, with the magnitude of the bronchoconstriction[41] (FIG. 2). Interestingly, no significant increases in urinary LTE_4 could be demonstrated during the allergen-induced late asthmatic response, even though the magnitude of the bronchoconstriction was similar to the early response.[41] Aspirin ingestion, which causes severe bronchoconstriction in aspirin-sensitive asthmatics, also caused significant increases in urinary LTE_4.[44] These data suggested that the cysteinyl leukotrienes were released during allergen inhalation and aspirin ingestion and were important causes of bronchoconstriction during the allergen-induced early phase, but not of the late phase bronchoconstrictor responses.

Increases in urinary LTE_4 have also been demonstrated to occur during acute spontaneous asthma,[43] which decline as the asthma is treated and the bronchoconstriction resolves. This suggests that the cysteinyl leukotrienes are involved in the development of symptomatic asthma that is occurring outside the clinical research laboratory.

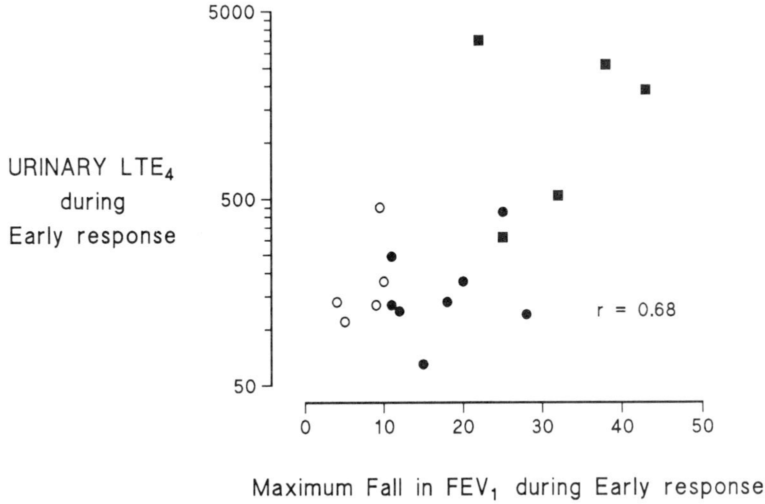

FIGURE 2. A linear correlation exists between the levels of urinary LTE_4 measured between 2-3 hours after inhalation of occupational sensitizing agents or allergens and the maximum fall in the forced expired volume in 1 second (FEV_1) during the early asthmatic response (p = 0.001). The *open circles* represent subjects with isolated late asthmatic responses, the *closed circles* are subjects with dual responses, and the *closed squares* are subjects with isolated early asthmatic responses. (Reprinted with permission from Manning *et al.*[42])

CYSTEINYL LEUKOTRIENES IN CLINICAL MODELS OF ASTHMA

The best evidence for a central role for the cysteinyl leukotrienes in causing exercise-, cold air- and allergen-induced bronchoconstriction is suggested by the observations that a number of different LTD_4 receptor antagonists and leukotriene synthesis inhibitors have been demonstrated to markedly attenuate the bronchoconstrictor responses after exercise (FIG. 3) and exposure to cold air and allergen.[45–48] These studies have indicated that LTD_4 antagonists can inhibit exercise-induced bronchoconstriction by a mean of 50-70%; and, indeed, in some asthmatic subjects they completely inhibit the response.[45] Similarly, both LTD_4 antagonists and leukotriene synthesis inhibitors attenuate allergen-induced early responses by up to 80%,[47,48] and, surprisingly considering the data on urinary LTE_4 excretion during the late response, also attenuate the late response by up to 50%.[47,48] Aspirin-induced bronchoconstriction is also almost completely abolished by treatment with leukotriene synthesis inhibitors.[49,50] These results taken together indicate a pivotal role for the cysteinyl leukotrienes in induced bronchoconstriction in these clinical models of asthma.

Leukotriene receptor antagonists also acutely improve lung function in asthmatic subjects who were selected to have bronchoconstriction prior to beginning treatment with the antagonists.[51,52] The magnitude of the improvement in lung function that was achieved was much less than that achieved with optimal doses of an inhaled β_2-agonist. However, in both studies the magnitude of the improvement with the antago-

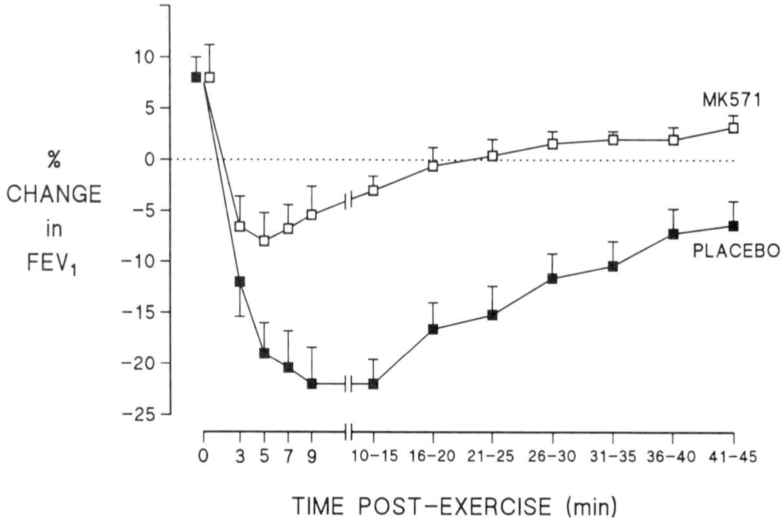

FIGURE 3. The % change in FEV_1 (mean and SEM) over time postexercise following treatment with placebo (*solid squares*) or MK-571 (*open squares*). Measurements were made in all subjects immediately after exercise, then at 3 minutes and every 2 minutes until the FEV_1 began to improve. Treatment with MK-571 significantly reduced the maximal fall in FEV_1 after exercise ($p < 0.001$) and shortened the recovery time ($p < 0.001$). (Reprinted with permission from Manning *et al.*[45])

nist and the β_2-agonist together was significantly greater than that with the β_2-agonist alone. These results suggest that the cysteinyl leukotrienes are, in part, responsible for spontaneous bronchoconstriction in asthma.

CYSTEINYL LEUKOTRIENES IN EXERCISE REFRACTORINESS IN ASTHMA

Repeated exercise challenges in asthmatic subjects, at intervals up to 4 h, usually result in less bronchoconstriction occurring after the second than compared to the initial exercise challenge.[53] This has been termed exercise refractoriness. We[54] and others[55] have demonstrated that this effect is abolished by pretreatment with indomethacin. This again suggested that inhibitor prostaglandins released as a consequence of exercise could modify bronchoconstrictor responses in asthmatics. An initial hypothesis to explain these events in asthmatic airways after exercise was that histamine released following exercise causes bronchoconstriction but also provides partial protection against subsequent exercise-induced bronchoconstriction, through PGE_2 released by stimulation of histamine H_2-receptors.[56] However, several recent studies have suggested that this hypothesis is incorrect. First, the marked attenuation of exercise-induced bronchoconstriction by pretreatment with LTD_4-receptor antagonists[45,57] indicates that LTD_4, rather than histamine, is the mediator mainly responsible for exercise-induced bronchoconstriction. Second, we have been unable to prevent exercise refractoriness by pretreatment with the H_2-receptor antagonists cimetidine or ranitidine, which effectively prevent histamine tachyphylaxis.[58] Therefore, histamine-stimulated inhibitory prostaglandin release does not appear to be the cause of exercise refractoriness. Our studies on leukotriene tachyphylaxis raise the possibility that exercise refractoriness is caused by leukotriene-stimulated inhibitory prostaglandin release. This possibility has recently been tested in a study of asthmatic subjects who develop exercise-induced bronchoconstriction and refractoriness.[59] The study demonstrated that there is an interdependence between the cyclooxygenase and lipoxygenase pathways of arachidonate metabolism in causing exercise bronchoconstriction and refractoriness in asthmatic subjects. This hypothesis is supported by the facts that: (1) exercise refractoriness and LTD_4 tachyphylaxis exist in the same subjects, and the magnitude of the protection afforded by exercise correlates with that afforded by LTD_4; (2) cross-refractoriness exists between exercise and LTD_4; (3) all of these effects are attenuated by cyclooxygenase inhibition, suggesting that the release of inhibitory prostaglandins is the common mechanism affording protection after each of these stimuli.

LEUKOTRIENE ANTAGONISTS AND SYNTHETASE INHIBITORS IN ASTHMA TREATMENT

The success of the leukotriene antagonists and synthetase inhibitors in attenuating bronchoconstriction in the clinical models of asthma suggested that these compounds were likely to be successful in the management of clinical asthma. An early study with a relatively less potent leukotriene receptor antagonist, LY171883, has demonstrated that asthmatic subjects do receive clinical benefit from treatment with this compound.[60] This was demonstrated by improvements in lung function as well as a reduction in the use of inhaled β_2-agonists for symptom relief. Subsequent studies, using both receptor antagonists and synthesis inhibitors, have confirmed that clinical benefit can be achieved with these compounds in patients with moderately severe asthma.[61,62] However, no direct comparisons have yet been made with other drugs, particularly inhaled corticosteroids, known to be very effective in the management of asthma.

CONCLUSIONS

Inhaled LTC_4 and LTD_4 are the most potent bronchoconstrictors yet studied in human subjects. Previously, because of the difficulty in measuring these mediators at their site of action in the airways, the evidence implicating them in this disease process was indirect. More direct evidence became available when potent and specific leukotriene antagonists were used in studies of asthmatic subjects. These studies have supported an important role for this leukotriene in the clinical models of asthma, such as exercise-, allergen-, and aspirin-induced asthma. Also, leukotriene release is in part responsible for spontaneous bronchoconstriction in asthma. Our recent studies have suggested that the cysteinyl leukotrienes can also cause the release of inhibitory prostaglandins in asthmatic airways. This results in the development of LTD_4 tachyphylaxis, which may be the cause of the previously demonstrated reduced relative potency of the cysteinyl leukotrienes in asthmatics when compared to normal subjects. Also, cysteinyl leukotriene-stimulated inhibitory prostaglandin release is also likely the cause of exercise refractoriness in asthmatic subjects. Last, the initial studies of the leukotriene antagonists and synthesis inhibitors have demonstrated clinical efficacy in asthma. This proves that the cysteinyl leukotrienes are involved in the pathogenesis of asthma. However, the eventual usefulness of these compounds in treating asthma awaits studies comparing their efficacy to the drugs currently available to treat asthma.

REFERENCES

1. HARGREAVE, F. E., G. RYAN, N. C. THOMSON, P. M. O'BYRNE, K. LATIMER, E. F. JUNIPER & J. DOLOVICH. 1981. Bronchial responsiveness to histamine or methacholine in asthma: Measurement and clinical significance. J. Allergy Clin. Immunol. **68:** 347-355.
2. COCKCROFT, D. W., D. N. KILLIAN, J. J. A. MELLON & F. E. HARGREAVE. 1977. Bronchial reactivity of inhaled histamine: A method and clinical survey. Clin. Allergy **7:** 235-243.
3. JUNIPER, E. F., P. A. FRITH & F. E. HARGREAVE. 1981. Airway responsiveness to histamine and methacholine: Relationship to minimum treatment to control symptoms of asthma. Thorax **36:** 575-579.
4. DJUKANOVIC, R., W. R. ROCHE, J. W. WILSON, C. R. W. BEASLEY, O. P. TWENTYMAN, P. H. HOWARTH & S. T. HOLGATE. 1990. State of the art: Mucosal inflammation in asthma. Am. Rev. Respir. Dis. **142:** 434-457.
5. KIRBY, J. G., F. E. HARGREAVE, G. J. GLEICH & P. M. O'BYRNE. 1987. Bronchoalveolar cell profiles of asthmatic and nonasthmatic subjects. Am. Rev. Respir. Dis. **136:** 379-383.
6. FLINT, K. C., K. B. P. LEUNG, B. N. HUDSPITH, J. BROSTOFF, F. L. PEARCE & N. M. JOHNSON. 1985. Bronchoalveolar mast cells in extrinsic asthma: A mechanism for the initiation of antigen specific bronchoconstriction. Br. Med. J. **291:** 923-926.
7. BEASLEY, R., W. R. ROCHE, J. A. ROBERTS & S. T. HOLGATE. 1989. Cellular events in the bronchi in mild asthma and after bronchial provocation. Am. Rev. Respir. Dis. **139:** 806-817.
8. JEFFERY, P. K., A. J. WARDLAW, F. C. NELSON, J. V. COLLINS & A. B. KAY. 1989. Bronchial biopsies in asthma. An ultrastructural, quantitative study and correlation with hyperreactivity. Am. Rev. Respir. Dis. **140:** 1745-1753.
9. DUNNILL, M. S., G. R. MASSARELL & J. A. ANDERSON. 1969. A comparison of the quantitive anatomy of the bronchi in normal subjects, in status asthmaticus, in chronic bronchitis and in emphysema. Thorax **24:** 176-179.
10. LAITINEN, L. A., M. HEINO, A. LAITINEN, T. KAVA & T. HAAHTELA. 1985. Damage of the airway epithelium and bronchial reactivity in patients with asthma. Am. Rev. Respir. Dis. **131:** 599-606.
11. ROCHE, W. R., R. BEASLEY, J. H. WILLIAMS & S. T. HOLGATE. 1989. Subepithelial fibrosis in the bronchi of asthmatics. Lancet **i:** 520-523.
12. KELLAWAY, C. H. & E. R. TRETHEWIE. 1940. The liberation of a slow-reacting smooth muscle-stimulating substance in anaphylaxis. Q. J. Exp. Physiol. **30:** 121-145.
13. BROCKLEHURST, W. E. 1960. The release of histamine and formation of a slow-reacting substance (SRS-A) during anaphylactic shock. J. Physiol. **151:** 416-435.
14. LEWIS, R. A. & K. F. AUSTEN. 1984. The biologically active leukotrienes. Biosynthesis, metabolism, receptors, functions and pharmacology. J. Clin. Invest. **73:** 889-897.

15. Dahlen, S. E., P. Hedqvist, S. Hammarstrom & B. Samuelsson. 1980. Leukotrienes are potent constrictors of human bronchi. Nature **288:** 484-486.
16. Holroyde, M. C., R. E. C. Altounyan, M. Cole, M. Dixon & E. V. Elliott. 1981. Bronchoconstriction produced in man by leukotrienes C and D. Lancet **ii:** 17-18.
17. Weiss, J. W., J. M. Drazen, R. McFadden, Jr., P. Weller, E. J. Corey, R. A. Lewis & K. F. Austen. 1983. Airway constriction in normal humans produced by inhalation of leukotriene D. JAMA **249:** 2814-2817.
18. Adelroth, E., M. M. Morris, F. E. Hargreave & P. M. O'Byrne. 1986. Airway responsiveness to leukotrienes C4 and D4 and to methacholine in patients with asthma and normal controls. N. Engl. J. Med. **315:** 480-484.
19. Drazen, J. M. 1986. Inhalation challenge with sulfidopeptide leukotrienes in human subjects. Chest **89:** 414-419.
20. Barnes, N. C., Piper P. J. & Costello, J. F. 1984. Comparative effects of inhaled leukotriene C4, leukotriene D4 and histamine in normal human subjects. Thorax **39:** 500-504.
21. Higgins, D. A. & P. M. O'Byrne. 1987. Inhaled leukotriene D4 does not cause a late response in atopic subjects. J. Allergy Clin. Immunol. **79:** 141A.
22. O'Byrne, P. M., J. Dolovich & F. E. Hargreave. 1987. State of the art: Late asthmatic responses. Am. Rev. Respir. Dis. **136:** 740-751.
23. Juniper, E. F., P. A. Frith, C. Dunnett, D. W. Cockcroft & F. E. Hargreave. 1978. Reproducibility and comparison of responses to inhaled histamine and methacholine. Thorax **33:** 705-710.
24. Hardy, C. C., C. Robinson, A. E. Tattersfield & S. T. Holgate. 1984. The bronchoconstrictor effect of inhaled prostaglandin D2 in normal and asthmatic men. N. Engl. J. Med. **311:** 209-213.
25. Thomson, N. C., R. Roberts, J. Bandouvakis, H. Newball & F. E. Hargreave. 1981. Comparison of bronchial responses to prostaglandin F2a and methacholine. J. Allergy Clin. Immunol. **68:** 392-398.
26. Anderton, R. C., M. T. Cuff, P. A. Frith, D. W. Cockcroft, J. L. C. Morse, N. L. Jones, F. E. Hargreave. 1979. Bronchial responsiveness to inhaled histamine and exercise. J. Allergy Clin. Immunol. **63:** 315-320.
27. O'Byrne, P. M., G. Ryan, M. Morris, D. McCormack, N. L. Jones, J. L. C. Morse & F. E. Hargreave. 1982. Asthma induced by cold air and its relation to nonspecific bronchial responsiveness to methacholine. Am. Rev. Respir. Dis. **125:** 281-285.
28. Griffin, M., J. W. Weiss, A. G. Leitch, E. R. McFadden, E. J. Corey, K. F. Austen & J. M. Drazen. 1983. Effects of leukotriene D4 on the airways in asthma. N. Engl. J. Med. **308:** 436-439.
29. Lewis, R. A., K. F. Austen, J. M. Drazen, D. A. Clark & E. J. Corey. 1980. Slow-reacting substance of anaphylaxis: Identification of leukotriene C-1 and D from human and rat sources. Proc. Natl. Acad. Sci. USA **77:** 3710-3714.
30. Manning, P. J., G. I. Jones & P. M. O'Byrne. 1987. Tachyphylaxis to inhaled histamine in asthmatic subjects. J. Appl. Physiol. **63:** 1572-1577.
31. Stevens, W. H., P. J. Manning & P. M. O'Byrne. 1990. Tachyphylaxis to inhaled methacholine in normal but not asthmatic subjects. J. Appl. Physiol. **69:** 875-879.
32. Anderson, W. H., J. J. Krzanowski, J. B. Polson & A. Szentivanyi. 1979. Increased synthesis of prostaglandin-like material during histamine tachyphylaxis in canine tracheal smooth muscle. Biochem. Pharmacol. **28:** 2223-2226.
33. Manning, P. J., C. G. Lane & P. M. O'Byrne. 1989. The effect of oral prostaglandin E_1 on airway responsiveness in asthmatic subjects. Pulmonary Pharmacol. **2:** 121-124.
34. Pavord, I. D., C. Wong, J. Williams & A. E. Tattersfield. 1993. Effect of inhaled prostaglandin E_2 on allergen-induced asthma. Am. Rev. Respir. Dis. **148:** 87-90.

35. MELILLO, E., K. L. WOOLLEY, P. J. MANNING, R. M. WATSON & P. M. O'BYRNE. Effect of inhaled PGE_2 on exercise-induced bronchoconstriction in asthmatic subjects. Am. Rev. Respir. Dis. In press.
36. KERN, R., L. J. SMITH, R. PATTERSON, R. D. KRELL & P. R. BERNSTEIN. 1986. Characterization of the airway responses to inhaled leukotriene D_4 in normal subjects. Am. Rev. Respir. Dis. **133:** 1124-1126.
37. MANNING, P. J., R. W. WATSON & P. M. O'BYRNE. 1993. Exercise-induced refractoriness in asthmatic subjects involves leukotriene and prostaglandin interdependent mechanisms. Am. Rev. Respir. Dis. **148:** 950-954.
38. LEE, T. H., J. P. ARM & B. W. SPUR. 1987. Leukotriene E_4 (LTE_4) enhances airways histamine responsiveness in asthmatic subjects. J. Allergy Clin. Immunol. **79:** 256A.
39. LAITINEN, L. A., A. LAITINEN, T. HAAHTELA, V. VILKKA, B. W. SPUR & T. H. LEE. 1993. Leukotriene E_4 and granulocytic infiltration into asthmatic airways. Lancet **341:** 989-990.
40. TAYLOR, I. K., R. WELLINGS, G. W. TAYLOR & R. FULLER. 1992. Urinary leukotriene E_4 excretion in exercise-induced bronchoconstriction. Am. Rev. Respir. Dis. **145:** A15.
41. KIKAWA, Y., T. MIYANOMAE, Y. INOUE *et al.* 1992. Urinary leukotriene E_4 after exercise challenge in children with asthma. J. Allergy Clin. Immunol. **89:** 1111-1119.
42. TAYLOR, I. K., R. WELLINGS, G. W. TAYLOR & R. W. FULLER. 1992. Urinary leukotriene E_4 excretion in exercise-induced bronchoconstriction. Am. Rev. Respir. Dis. **145:** A15.
43. MANNING, P. J., J. ROKACH, J. L. MALO, D. ETHIER, A. CARTIER, Y. GIRARD, S. CHARLESON & P. M. O'BYRNE. 1990. Urinary leukotriene E_4 levels during early and late asthmatic responses. J. Allergy Clin. Immunol. **86:** 211-220.
44. TAYLOR, G. W., P. BLACK, N. TURNER, I. TAYLOR, N. H. MALTBY, R. W. FULLER & C. T. DOLLERY. 1989. Urinary leukotriene E_4 after antigen challenge and in acute asthma and allergic rhinitis. Lancet **i:** 584-587.
45. KNAPP, H. R., K. SLADEK & G. A. FITZGERALD. 1992. Increased secretion of leukotriene E_4 during aspirin-induced asthma. J. Lab. Clin. Invest. **119:** 48-51.
46. MANNING, P. J., R. M. WATSON, D. J. MARGOLSKEE, V. WILLIAMS, J. SCHARTZ & P. M. O'BYRNE. 1990. Inhibition of exercise-induced bronchoconstriction by MK-571, a potent leukotriene D_4 receptor antagonist. N. Engl. J. Med. **323:** 1736-1739.
47. ISRAEL, E., R. DERMARKARIAN, M. ROSENBERG *et al.* 1990. The effects of a 5-lipoxygenase inhibitor on asthma induced by cold, dry air. N. Engl. J. Med. **323:** 1740-1744.
48. TAYLOR, I. K., K. M. O'SHAUGHNESSY, R. W. FULLER & C. T. DOLLERY. 1991. Effect of a cysteinyl leukotriene receptor antagonist, ICI 204-219 on allergen-induced bronchoconstriction and airway hyperreactivity in atopic subjects. Lancet **337:** 690-694.
49. O'BYRNE, P. M., R. M. WATSON, H. A. STRONG & G. WYLIE. 1994. The effect of treatment with a 5-lipoxygenase inhibitor, BAYx 1005, on allergen-induced asthmatic responses. Am. J. Respir. Crit. Care. Med. **149:** A532.
50. DAHLEN, B., M. KUMLIN, I. L. JOHANNSON, C. LARSSON, O. ZETTERSTROM, K. MALMSTROM, F. BRUNNER-FERBER, D. M. MARGOLSKEE & S.-E. DAHLEN. 1992. The leukotriene-antagonist, MK-0679, improves baseline pulmonary function and blocks aspirin-induced airway obstruction in aspirin sensitive asthmatics. Am. Rev. Respir. Dis. **145:** A15.
51. ISRAEL, E., A. R. FISCHER, M. A. ROSENBURG, C. M. LILLY, J. C. CALLERY, J. SHAPIRO, J. COHN, R. RUBIN & J. M. DRAZEN. 1993. The pivitol role of 5-lipoxygenase products in the reaction of aspirin-sensitive asthmatics to aspirin. Am. Rev. Respir. Dis. **148:** 1447-1451.
52. GADDY, J., R. K. BUSH, D. MARGOLSKEE, V. C. WILLIAMS & W. BUSSE. 1992. The effects of a leukotriene D_4 (LTD_4) antagonist (MK-571) in mild to moderate asthma. Am. Rev. Respir. Dis. **146:** 358-363.
53. HUI, K. P. & N. C. BARNES. 1991. Lung function improvement in asthma with a cysteinyl-leukotriene receptor antagonist. Lancet **337:** 1062-1063.

54. EDMUNDS, A. T., M. TOOLEY & S. GODFREY. 1978. The refractory period after exercise-induced asthma: Its duration and relation to the severity of exercise. Am. Rev. Respir. Dis. **117:** 247-254.
55. O'BYRNE, P. M. & G. L. JONES. 1986. The effect of indomethacin on exercise-induced bronchoconstriction and refractoriness after exercise. Am. Rev. Respir. Dis. **134:** 69-72.
56. MARGOLSKEE, D. J., B. G. BIGBY & H. A. BOUSHEY. 1988. Indomethacin blocks airway tolerance to repetitive exercise but not to eucapnic hyperpnea in asthmatic subjects. Am. Rev. Respir. Dis. **137:** 842-846.
57. JACKSON, P. A., P. J. MANNING & P. M. O'BYRNE. 1988. A new role for histamine H_2-receptors in asthmatic airways. Am. Rev. Respir. Dis. **138:** 784-788.
58. FINNERTY, J. P., R. WOOD-BAKER, H. THOMSON & S. T. HOLGATE. 1992. Role of leukotrienes in exercise-induced asthma. Am. Rev. Respir. Dis. **145:** 746-749.
59. MANNING, P. J., R. L. WATSON & P. M. O'BYRNE. 1992. The effect of H_2-receptor antagonists on exercise-induced refractoriness in asthma. J. Allergy Clin. Immunol. **88:** 125-126.
60. MANNING, P. J., R. W. WATSON & P. M. O'BYRNE. 1993. Exercise-induced refractoriness in asthmatic subjects involves leukotriene and prostaglandin interdependent mechanisms. Am. Rev. Respir. Dis. **148:** 950-954.
61. CLOUD, M. L., G. C. ENAS, J. KEMP, T. PLATTS-MILLS, L. C. ALTMAN, R. TOWNLEY, D. TINKELMAN, T. KING, E. MIDDLETON, A. L. SHEFFER, E. R. MCFADDEN, JR. & D. S. FARLOW. 1989. A specific LTD_4/LTE_4-receptor antagonist improves pulmonary function in patients with mild, chronic asthma. Am. Rev. Respir. Dis. **140:** 1336-1339.
62. ISRAEL, E., P. RUBIN, J. KEMP *et al.* 1993. The effect of inhibition of 5-lipoxygenase by Zileuton in mild-to-moderate asthma. Ann. Int. Med. **119:** 1059-1066.
63. CHAPMAN, K. R., B. S. FRIEDMAN, S. SHINGO, J. HEYSE, T. REISS & R. SPECTOR. 1994. The efficacy of an oral inhibitor of leukotriene synthesis (MK-0591) in asthmatics treated with inhaled steroids. Am. J. Respir. Crit. Care Med. **149:** A215.

Stereoselective Metabolism of the 5-Lipoxygenase Inhibitor A-78773

GEORGE W. CARTER, RANDY L. BELL,
KENNAN MARSH, CARMINE LANNI,
WALID M. AWNI, JENNIFER BOUSKA,
ANDREW O. STEWART, ROBERT HANSEN,
LOUISE DUBÉ, AND DEE W. BROOKS

Abbott Laboratories
Abbott Park, Illinois 60064

INTRODUCTION

Recent clinical results with selective 5-lipoxygenase (5-LO) inhibitors and leukotriene antagonists have added considerable support to the hypothesis that leukotrienes play an important mediation role in human asthma.[1–21] One of the most advanced compounds undergoing clinical investigation is zileuton, a selective, safe, orally effective 5-lipoxygenase inhibitor,[22,23] currently in phase III trials. Encouraging results were obtained with this compound in asthma as well as other inflammatory conditions.[12,24–27] In a 28-day study in mild to moderate asthmatics, oral dosing with zileuton caused a rapid (within 30 to 60 minutes) and sustained improvement in lung function, decreased the usage of inhaled β agonists, and reduced disease symptoms. These favorable clinical responses were accompanied by a significant reduction in the urinary excretion of leukotriene E_4.[1]

As a result of our ongoing research to identify 5-lipoxygenase inhibitors with improved properties compared to currently available compounds, we recently reported on the discovery of A-78773, a potent, selective, long-acting inhibitor.[28] This compound has a single chiral center. The present report highlights some of the *in vitro* and *in vivo* properties of the R (+), A-79175 and S (–), A-79176 enantiomers of this new inhibitor. While the two stereoisomers of A-78773 display similar *in vitro* inhibitory activity, there are significant differences in the susceptibility of these enantiomers to metabolism. The major route of metabolism of these *N*-hydroxyureas is by glucuronidation of the *N*-hydroxy moiety. Both *in vitro* and *in vivo* studies in several species confirmed that the R (+) enantiomer is more resistant to conjugation compared to the S (–) stereoisomer. Accordingly, the long duration of leukotriene biosynthesis inhibition seen in experimental animals and man following oral administration of A-78773 is mainly due to the R (+) enantiomer, A-79175.

METHODS

RBL Cell Lysate 5-Lipoxygenase

Adherent rat basophilic leukemia (RBL) cells (2H3 subline) were harvested by trypsinization, suspended at 3.0×10^7 cells/ml in pH 6.8 buffer, and lysed by sonication.

The lysate was centrifuged at 20,000 *g* for 20 minutes and the supernatant containing 5-lipoxygenase activity stored frozen until used. Enzyme assays were performed as described by Carter *et al.*[22]

Human Neutrophil Leukotriene Biosynthesis

The ability of test compounds to prevent human neutrophil leukotriene biosynthesis was evaluated using a modification of the method described by Carter *et al.*[22] Heparinized blood samples were collected from healthy volunteers, layered over Ficoll-Hypaque Mono-Poly Resolving Media (Flow Laboratories, McLean, VA), and centrifuged at 300 *g* for 30 minutes. The polymorphonuclear leukocyte (PMNL) layer was removed, subjected to brief hypotonic lysis to remove RBC contamination, and washed three times. Aliquots of the PMNL suspension were incubated with test agents and leukotriene biosynthesis stimulated by addition of calcium ionophore A23187 (final concentration 7 μM). Incubations were extracted with four volumes of methanol, and the LTB_4 content assayed by enzyme immunoassay (EIA; Cayman, Ann Arbor, MI).

Metabolism by Liver Microsomes

The susceptibility of *N*-hydroxyurea inhibitors to glucuronidation was initially assessed *in vitro* with liver microsomal preparations from various species using a modification of the procedure described by Sweeney *et al.*[29] Livers were homogenized and centrifuged at 10,000 *g* for 20 minutes. The supernatants were centrifuged at 100,000 *g* for one hour and the microsomal pellet stored at −80 °C until used.

Compounds (100 μM) were incubated with microsomes (1 mg protein/ml) and uridine 5′-diphosphoglucuronic acid at 37 °C. Aliquots were removed at various times (0, 5, 10, 20, 30, and 60 minutes) and added to acetonitrile to stop the reactions. Following the removal of precipitated proteins by centrifugation, samples were analyzed by high-performance liquid chromatography (HPLC) for the appearance of glucuronide conjugates and unmetabolized compounds.

Inhibition of Ex Vivo *Leukotriene B_4 Biosynthesis*

A-78773 was suspended in 0.2% hydroxypropyl methylcellulose (HPMC) and administered by oral gavage to experimental animals. A polyethylene glycol : alcohol solution of A-78773 was orally administered to healthy human volunteers. Heparinized blood samples were drawn at various times after dosing and leukotriene synthesis triggered by addition of 50 μM calcium ionophore A23187. After a 30-minute incubation at 37°C, samples were centrifuged and plasma aliquots extracted with four volumes of methanol. The levels of LTB_4 in extracted plasma samples were measured by EIA (Cayman, Ann Arbor, MI).

TABLE 1. The 5-Lipoxygenase Inhibitory Properties of A-78773 and Its Enantiomers

Compound	RBL Lysate[a] IC_{50} (nM)[c]	Human Neutrophils[b] IC_{50} (nM)[c]
A-78773	300	17
A-79175R(+)	250	25
A-79176 S(−)	230	10
Zileuton	500	600

[a] Rat basophilic leukemia cells were disrupted by sonication, the 20,000 × *g* supernatant was incubated with ^{14}C-arachidonic acid, and the production of 5-HETE was measured.
[b] Human peripheral blood neutrophils were stimulated with A23187 and LTB_4 generation in the presence and absence of inhibitor measured by EIA.
[c] IC_{50} values were calculated using linear regression analysis. Data are the means of duplicate determinations.

Determination of Drug Concentration in Plasma

Blood samples were collected at various times following compound administration, centrifuged, and the plasma removed and stored frozen until assayed. Plasma samples were thawed and plasma proteins precipitated by the addition of two volumes of methanol. Precipitated materials were removed by centrifugation and the supernatants assayed by HPLC for A-78773. Plasma extracts were injected directly onto a C18 reversed-phase column (Adsorbosphere 3 micron; Alltech, Deerfield, IL) and chromatographed using a mobile phase composed of acetonitrile : tetrahydrofuran : triethylamine acetate buffer at a flow rate of 0.5 ml/minute. Compound elution was monitored at 260 nm and the concentration determined by comparison to an external calibration curve.

A chiral HPLC procedure was used to measure R (+) and S (−) enantiomers in plasma samples following A-78773 administration. Plasma was extracted with methylene chloride and the organic extract evaporated to dryness. The reconstituted samples were chromatographed on an acid glycoprotein (AGP) column using an acetonitrile : phosphate buffer mobile phase with UV quantitation at 260 nm.

RESULTS

TABLE 1 shows that A-78773 is a potent inhibitor of 5-lipoxygenase activity when tested against a crude cellular lysate or intact cells. The chirality of the methyl branch (FIG. 1) on the carbon adjacent to the *N*-hydroxy group does not appear to be important

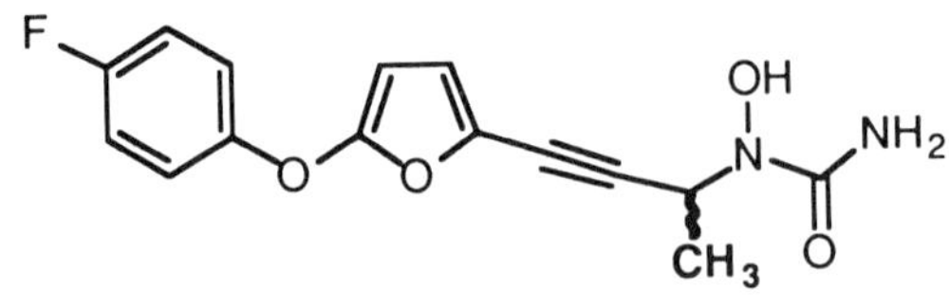

FIGURE 1. Structure of A-78773 (Racemate). The chiral center and methyl branch is in *bold.* Individual enantiomers are referred to as: R (+), A-79175 and S (−), A-79176.

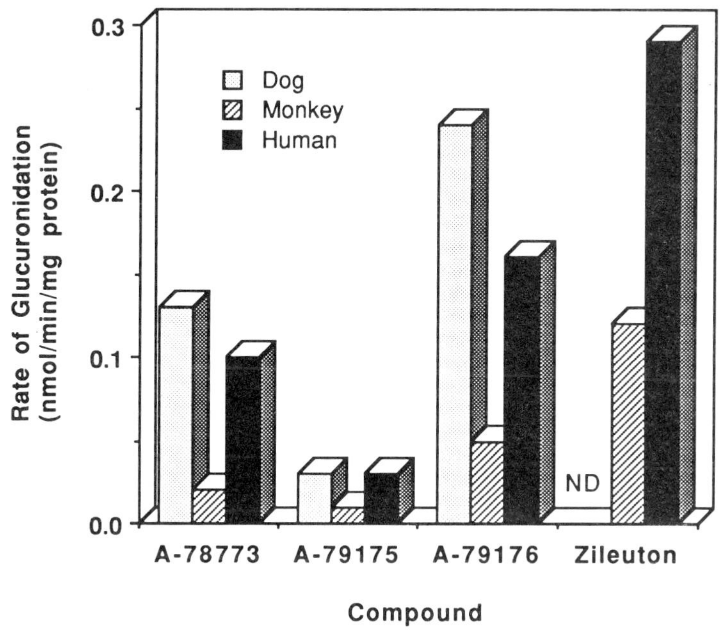

FIGURE 2. Glucuronidation of *N*-hydroxyurea inhibitors by dog, monkey, and human liver microsomes. Test compounds (A-78773, racemate; R (+), A-79175; S (–), A-79176; and zileuton) were incubated at 100 μM with microsomes and glucuronidation estimated by HPLC. Both the appearance of glucuronide conjugates and the disappearance of unmetabolized compound were monitored. ND = not determined. Plotted are means of two to nine experiments done in duplicate.

for inhibitory activity since the R (+) and S (–) stereoisomers display similar potencies in these two assays. The racemate and its enantiomers were approximately twice as potent as zileuton in the RBL assay and as much as 60-fold more potent in preventing LTB_4 synthesis by ionophore-stimulated human neutrophils.

We and others have shown that 5-lipoxygenase inhibitors of the acetohydroxamate and *N*-hydroxyurea classes are susceptible to metabolic inactivation by glucuronidation of the *N*-hydroxy group.[29–31] To facilitate the identification of inhibitors with long durations of *in vivo* action, a simple *in vitro* glucuronidation assay was established using microsomes from various species, including man. Compounds that were identified to be resistant to rapid conjugation in this simple assay were then selected for further testing. FIGURE 2 shows that the racemate A-78773 was stereoselectively metabolized by microsomal preparations from dog, cynomolgus monkey, and human livers. In all three species, the R (+) enantiomer was found to be the most resistant to glucuronidation. Compared to the S (–) enantiomer, the rate of conjugation of R (+) stereoisomer A-79175 was approximately five-fold less in all species tested. The rate of metabolism of A-79175 in human microsomes is about one-tenth that of zileuton. If glucuronidation proves to be the major route of metabolism of A-79175

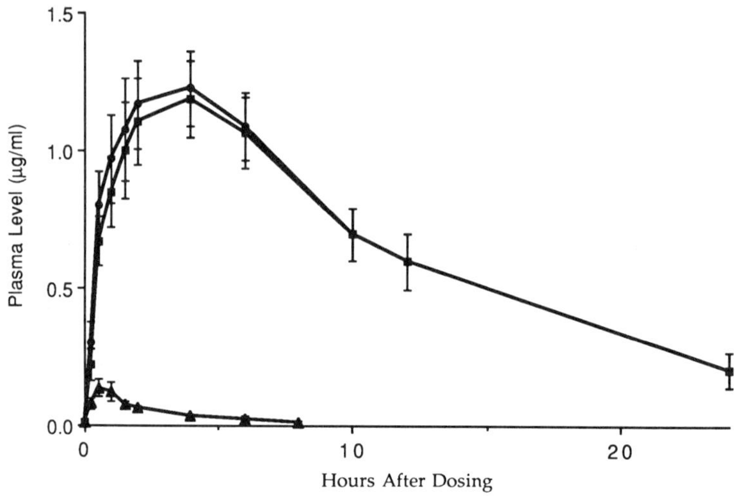

FIGURE 3. Enantiomeric distribution of the A-78773 after oral dosing in the dog. Symbols represent mean ± SEM for four animals. *Triangles* = S (–), A-79716; *squares* = A-79175, R (+); and *circles* = A-78773, racemate. Plasma concentrations measured as described in METHODS section.

in man, then this greater metabolic stability would suggest that A-79175 has the potential for being longer acting than zileuton in man.

When A-78773 was administered to experimental animals and plasma levels of the R (+) and S (–) enantiomers were analyzed by chiral HPLC, significant stereoselective metabolism was observed. In fact, the same trends that were seen in the *in vitro* microsomal assays were also observed in the intact animals. For example in the dog following a 4-mg/kg oral dose, plasma concentrations of the R (+) enantiomer increased rapidly to a peak concentration of 1.25 μg/ml at 4 hours and then declined slowly, with an estimated elimination half-life of 6 hours (FIG. 3). On the other hand, the S (–) enantiomer reached a peak concentration of only 0.125 μg/ml within 30 minutes after dosing, approximately one-tenth the maximum concentration (C_{max}) of the R (+) enantiomer. The elimination half-life of the S (–) enantiomer was rather short at 0.3 hours, and plasma concentrations of this enantiomer were not detectable 8 hours after dosing. At these later times, only the R (+) enantiomer was found circulating in the plasma.

The *in vivo* stereoselective metabolism of A-78773 in various species was reflected in the enantiomeric distribution of the area under the plasma concentration curves (AUC) as shown in FIGURE 4. It can be seen that in the rat, dog, and monkey the major circulating compound in plasma following oral administration of A-78773 was the R (+) enantiomer. In the dog, the R (+) enantiomer accounted for 93% of the total circulating drug. Similar but smaller differences between the circulating levels of the R (+) and S (–) enantiomers were seen in other species. In the rat 61% and in the monkey 78% of the AUC was accounted for by the R (+) enantiomer.

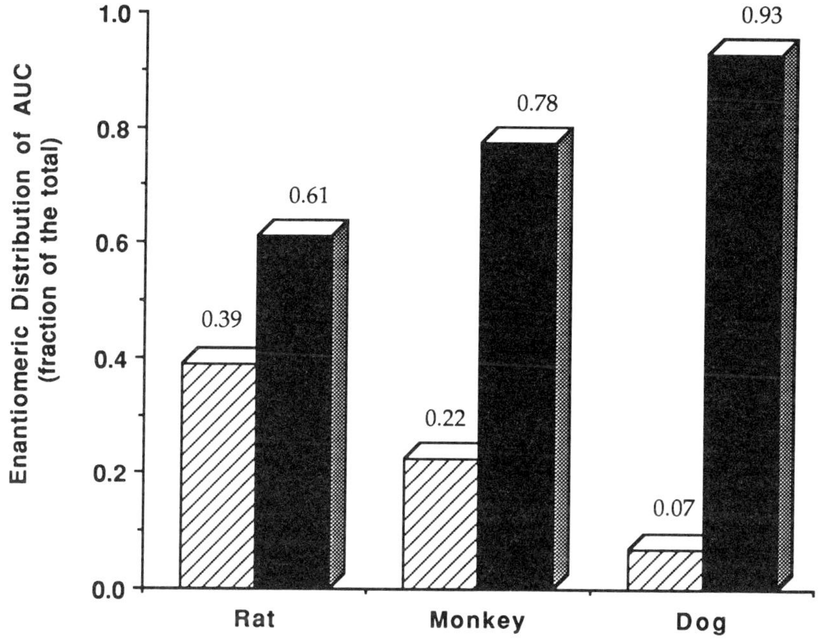

FIGURE 4. Enantiomeric distribution of the A-78773 expressed as area under the plasma concentration curve (AUC) after oral dosing in rat, dog, and monkey. *Cross-hatched bars* indicate mean fraction of S (–) enantiomer, A-79176, and the *solid bars* represent R (+) enantiomer, A-79175. Racemate, A-78773 was administered orally to rats (n = 4), dogs (n = 4), and monkeys (n = 2) at doses of 50, 4 mg/kg and 2 mg/kg, respectively. Plasma concentrations of enantiomers were measured as described in METHODS section.

In a phase I study, increasing single oral doses of A-78773 were administered to healthy male volunteers; blood samples were collected at various times and concentrations of A-78773 and its component stereoisomers measured. The capacity of blood samples to synthesize leukotrienes was also assessed by challenging samples with the calcium ionophore A23187 and measuring LTB_4 levels by EIA. As shown in FIGURE 5, A-78773 was stereoselectively metabolized in man. Consistent with *in vitro* observations in liver microsomes, the S (–) enantiomer was more rapidly eliminated compared to the R (+) enantiomer. The peak concentrations of the S (–) enantiomer were reached within 4 hours compared to 5.3 hours with the R (+) enantiomer. The elimination half-life of the S (–) enantiomer was 2.4 hours compared to 6.5 hours for the R (+) enantiomer. The AUC for the R (+) enantiomer was approximately three- to fourfold greater than that for the S (–) stereoisomer. Following a 400-mg dose of A-78773, leukotriene biosynthesis was essentially eliminated for 16 hours. Even 24 hours after dosing, blood leukotriene production continued to be suppressed by nearly 70%. Clearly at these later times, the R (+) enantiomer was mainly responsible for the observed inhibition of leukotriene formation.

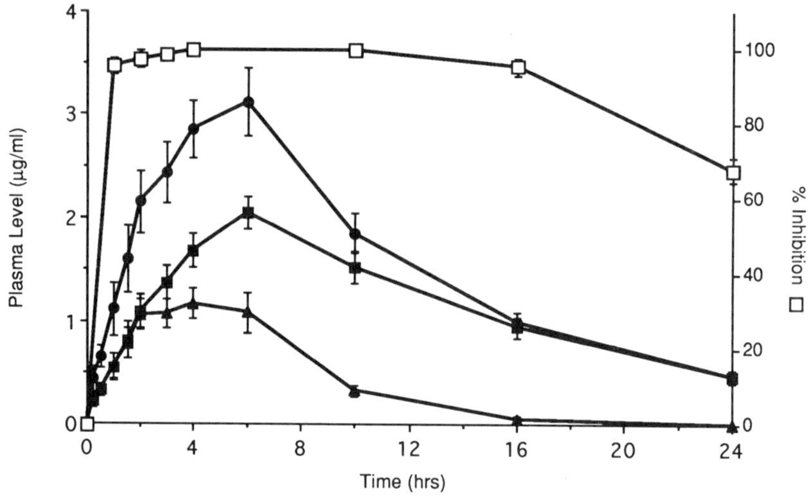

FIGURE 5. Plasma levels of A-78773 and its enantiomers with LTB_4 inhibition after a single 400-mg oral dose of A-78773 in man. *Solid symbols* represent mean ± SEM for eight subjects: *triangles* = S (−), A-79716; *squares* = R (+), A-79175; and *circles* = racemate, A-78773. *Open boxes* represent mean ± SEM inhibition of LTB_4 biosynthesis. Plasma concentrations and inhibition were measured as described in METHODS section.

DISCUSSION

Since the discovery of the *N*-hydroxyurea zileuton, this laboratory has focused on the identification of other promising 5-LO inhibitors of this class. A main emphasis of this research was to find more potent and longer-acting compounds. Using zileuton as a template, a large number of analogs were prepared in which the benzothiophene ring was replaced by other heteroaryl moieties. A-69412 was one interesting compound to emerge from these efforts. Compared to zileuton, this simple water-soluble compound was a somewhat less potent 5-lipoxygenase inhibitor *in vitro* but was a much longer-lived inhibitor *in vivo,* particularly in the monkey.[32] Another focus of inhibitor modification was the nature of the structural linker between the heteroaryl group and the *N*-hydroxyurea that influenced the spatial orientation of these two components. Incorporating a butynyl linker between the heteroaryl system and the *N*-hydroxyurea led to a series of potent, long-acting inhibitors. We recently disclosed a promising inhibitor, A-78773, which was both more potent and longer acting that zileuton.[28] *In vitro* this compound was up to 60 times more potent than zileuton in preventing leukotriene biosynthesis by ionophore-stimulated human neutrophils. The plasma half-life of this compound was also significantly longer than that of zileuton. In the cynomolgus monkey, a single oral dose of 2 mg/kg totally attenuated *ex vivo* whole-blood leukotriene biosynthesis through 24 hours.

Since A-78773 contains a chiral center created by the methyl-substituted carbon adjacent to the *N*-hydroxy group, we investigated the properties of the individual enantiomers of this compound. The R (+) and S (−) enantiomers of A-78773 were found to exhibit similar 5-LO inhibitory activity when tested in a number of *in vitro* test systems. Against RBL supernatant 5-LO there was no significant difference in potency between the two isomers; in a human neutrophil assay, there was only a slight difference in potency in favor of the S (–) enantiomer (IC_{50}s for the R (+) and S (–) forms were 25 and 10 nM, respectively). This relative lack of stereoselective inhibition by these enantiomers of A-78773 is not surprising in light of the fact that the methyl branch can be eliminated entirely without affecting 5-lipoxygenase inhibitory activity. It would appear that when A-78773 inhibits 5-lipoxygenase product formation, the stereochemical orientation of the methyl branch is not important for inhibitory activity. Similar results concerning the lack of stereoselective inhibition by a series of related acetohydroxamic acid inhibitors have been reported by Salmon *et al.*[31] However, as discussed later, this chiral center does influence the metabolic lability and therefore the *in vivo* pharmacokinetic behavior of these molecules.

A major route of metabolism of zileuton[29] as well as other compounds in this class[32] is glucuronidation of the *N*-hydroxy group. In man and cynomolgus monkey conjugation of the *N*-hydroxy group followed by urinary excretion of the conjugate is the dominant route of metabolism and elimination.[23] Therefore, as part of a strategy to find longer-acting inhibitors, compounds with interesting potencies against 5-lipoxygenase were tested for their susceptibility to conjugation using microsomes prepared from livers of selected species. Compounds that showed appropriately slow rates of metabolism in *in vitro* glucuronidation assays were chosen as candidates for more extensive pharmacokinetic studies in animals. Compared to zileuton, the racemate A-78773 was more slowly conjugated by microsomal preparations from dog, monkey, and human. For example, human microsomes metabolized A-78773 about three times more slowly than zileuton. Realizing the possibility for stereoselective metabolism of A-78773, the individual enantiomers were synthesized and tested for their *in vitro* metabolic lability. In microsomes from all three species, the S (–) enantiomer was more rapidly conjugated than the R (+) stereoisomer. The magnitude of the difference varied somewhat between species. In the dog there was an eightfold difference in the rate of metabolism between the R (+) and S (–) enantiomers, whereas in monkey and human the difference was approximately fivefold. The possibility of interconversion of S (–) to the R (+) enantiomer was dismissed because administration of the S (–) enantiomer failed to generate detectable levels of the R (+) stereoisomer and vice versa (unpublished observations). If these findings have predictive value for the *in vivo* pharmacokinetic behaviors of these compounds, then the R (+) enantiomer would be predicted to be a longer-lived species than the S (–) stereoisomer.

Validation of the utility of the *in vitro* liver microsomal metabolism results came from studies in which the racemate A-78773 was orally administered to various animals and blood samples collected and assayed by chiral HPLC for levels of both R (+) and S (–) enantiomers. Consistent with the *in vitro* results, the S (–) and R (+) enantiomers were metabolized differently in the intact animals. Oral administration of the racemate A-78773 to the dog resulted in the rapid appearance of both enantiomers; however, the peak concentration achieved by the R (+) enantiomer was some 7 to 14 times higher than the maximum levels reached by the S (–) enantiomer. Comparing elimination half-lives of the two enantiomers in the dog, the R (+) was found to last approximately 20 times longer than the S (–). Similar results were also seen in the other species. A comparison of the plasma AUC highlights differences in the pharmacokinetic behavior of the enantiomers. In the rat, dog, and monkey, the R (+) enantiomer was the major circulating compound following oral dosing with A-78773. In the dog, 93% of the total circulating drug was as the R (+) enantiomer. Similarly, this enantiomer accounted for 78% and 61% of the drug exposure in the rat and cynomolgus monkey. These data demonstrate that the R (+) enantiomer is more resistant to metabolism and elimination compared to the S (−) enantiomer.

If these preclinical results are predictive of what will happen in man, then the R (+) enantiomer would be expected to have superior pharmacokinetic properties compared with the more metabolically labile S-stereoisomer. Phase I studies with A-78773 verified these predictions. Administration of this racemate at a 400-mg dose to healthy human volunteers demonstrated that the major circulating form of the compound was, as predicted, the R (+) enantiomer. Plasma levels of this enantiomer peaked at approximately 2 μg/ml at around 6 hours and then gradually declined. The estimated elimination half-life for the R (+) enantiomer was 6.5 hours, whereas the S (−) enantiomer peaked at about half the level of the R (+) enantiomer and was eliminated with a half-life of around 2.4 hours. The AUC of the R (+) enantiomer accounted for approximately 78% of the total drug concentration. Following this dose of A-78773, blood leukotriene biosynthesis was essentially 100% inhibited through 16 hours. These results confirm the effectiveness of A-78773 as a 5-LO inhibitor and suggest that once- or twice-daily dosing would be adequate to strongly suppress leukotriene formation throughout the day. The favorable pharmacokinetic properties and equivalent 5-lipoxygenase inhibitory potency of the R (+) enantiomer compared to A-78773 and the S (−) enantiomer make the R (+), A-79175 enantiomer an interesting compound for further development.

SUMMARY

Based on the knowledge that glucuronidation was a major route of metabolism of the *N*-hydroxyurea class of 5-lipoxygenase inhibitors, a simple *in vitro* glucuronidation assay was established using liver microsomes from various species, including man. Compounds that were potent inhibitors of 5-LO and showed a reduced metabolic liability *in vitro* were then characterized more extensively in experimental animals. This prudent usage of *in vitro* glucuronidation proved to be highly efficient and was indispensable in the identification of A-78773, a potent new long-acting 5-LO inhibitor. Further studies with liver microsomes revealed that glucuronidation of A-78773 was stereoselective and that the R (+) enantiomer was considerably more resistant to conjugation than the S (−) stereoisomer. Pharmacokinetic studies in experimental animals and humans confirmed the greater metabolic stability of the R (+) enantiomer. A single 400-mg oral dose of A-78773 inhibited *ex vivo* leukotriene biosynthesis for more than 24 hours. Since 78% of the drug plasma AUC following A-78773 administration was accounted for by the R (+) enantiomer, it is reasonable to assume that the majority of the leukotriene inhibition caused by the racemate is attributable to the R (+) enantiomer, A-79175, particularly at the later times. The equivalent 5-lipoxygenase inhibitory potency coupled with the superior pharmacokinetic profile of the R (+) enantiomer, A79175, compared to the S (−) enantiomer, A-79176, indicate that the development of this compound may be preferable to the racemate A-78773.

ACKNOWLEDGMENTS

The authors acknowledge the technical support of E. Otis, S. Majest, C. Goodfellow, and W. Hinz. We also thank A. Niemi for assistance in preparing the manuscript.

REFERENCES

1. ISRAEL, E., P. RUBIN, J. P. KEMP, J. GROSSMAN, W. PIERSON, S. C. SIEGEL, D. TINKELMAN, J. J. MURRAY, W. BUSSE, A. T. SEGAL, J. FISH, H. B. KAISER, D. LEDFORD, S. WENZEL, R. ROSENTHAL, J. COHN, C. LANNI, H. PEARLMAN, P. KARABALLOS & J. DRAZEN. 1993. The effect of inhibition of 5-lipoxygenase by zileuton in mild-to-moderate asthma. Ann. Intern. Med. **119:** In press.
2. ISRAEL, E., R. DERMARKARIAN, M. ROSENBERG, R. SPERLING, G. TAYLOR, P. RUBIN & J. M. DRAZEN. 1990. The effects of a 5-lipoxygenase inhibitor on asthma induced by cold, dry air [see comments]. N. Engl. J. Med. **323:** 1740-1744.
3. ISRAEL, E., E. F. JUNIPER, J. T. CALLAGHAN, P. N. MATHUR, M. M. MORRIS, A. R. DOWELL, G. G. ENAS, F. E. HARGREAVE & J. M. DRAZEN. 1989. Effect of a leukotriene antagonist, LY171883, on cold air-induced bronchoconstriction in asthmatics. Am. Rev. Respir. Dis. **140:** 1348-1353.
4. TAYLOR, I. K., K. M. O'SHAUGHNESSY, R. W. FULLER & C. T. DOLLERY. 1991. Effect of cysteinyl-leukotriene receptor antagonist ICI 204.219 on allergen-induced bronchoconstriction and airway hyperreactivity in atopic subjects [see comments]. Lancet **337:** 690-4.
5. MANNING, P. J., R. M. WATSON, D. J. MARGOLSKEE, V. C. WILLIAMS, J. I. SCHWARTZ & P. M. O'BYRNE. 1990. Inhibition of exercise-induced bronchoconstriction by MK-571, a potent leukotriene D4-receptor antagonist [see comments]. N. Engl. J. Med. **323:** 1736-1739.
6. MAKKER, H. K., L. C. LAU, H. W. THOMSON, S. M. BINKS & S. T. HOLGATE. 1993. The protective effect of inhaled leukotriene D4 receptor antagonist ICI 204,219 against exercise-induced asthma. Am. Rev. Respir. Dis. **147:** 1413-1418.
7. RASMUSSEN, J. B., L. O. ERIKSSON, P. TAGARI, E. G. STAHL & K. E. ANDERSSON. 1992. Reduced nonspecific bronchial reactivity and decreased airway response to antigen challenge in atopic asthmatic patients treated with the inhaled leukotriene D4 antagonist, L-648,051. Allergy **47:** 604-609.
8. RASMUSSEN, J. B., L. O. ERIKSSON, D. J. MARGOLSKEE, P. TAGARI, V. C. WILLIAMS & K. E. ANDERSSON. 1992. Leukotriene D4 receptor blockade inhibits the immediate and late bronchoconstrictor responses to inhaled antigen in patients with asthma. J. Allergy Clin. Immunol. **90:** 193-201.
9. HUI, K. P. & N. C. BARNES. 1991. Lung function improvement in asthma with a cysteinyl-leukotriene receptor antagonist. Lancet **337:** 1062-1063.
10. DAHLÉN, S. E., B. DAHLÉN, E. ELIASSON, H. JOHANSSON, T. BJORCK, M. KUMLIN, K. BOO, J. WHITNEY, S. BINKS, B. KING *et al.* 1991. Inhibition of allergic bronchoconstriction in asthmatics by the leukotriene-antagonist ICI-204,219. Adv. Prostaglandin Thromboxane Leukotriene Res. **21a:** 461.
11. DAHLÉN, B., M. KUMLIN, D. J. MARGOLSKEE, C. LARSSON, H. BLOMQVIST, V. C. WILLIAMS, O. ZETTERSTROM & S. E. DAHLÉN. 1993. The leukotriene-receptor antagonist MK-0679 blocks airway obstruction induced by inhaled lysine-aspirin in aspirin-sensitive asthmatics. Eur. Respir. J. **6:** 1018-1026.
12. BEL, E. H., M. C. TIMMERS, J. H. DIJKMAN, E. G. STAHL & P. J. STERK. 1990. The effect of an inhaled leukotriene antagonist, L-648,051, on early and late asthmatic reactions

and subsequent increase in airway responsiveness in man. J. Allergy Clin. Immunol. **85:** 1067-1075.

13. BRITTON, J., S. HANLEY & A. TATTERSFIELD. 1987. Effect of an oral leukotriene D4 antagonist L-649,923 on the response to inhaled antigen in asthma. Adv. Prostaglandin Thromboxane Leukotriene Res. **17b:** 1003.
14. FUJIMURA, M., S. SAKAMOTO, Y. KAMIO & T. MATSUDA. 1993. Effect of a leukotriene antagonist, ONO-1078, on bronchial hyperresponsiveness in patients with asthma. Respir. Med. **87:** 133-138.
15. GADDY, J. N., D. J. MARGOLSKEE, R. K. BUSH, V. C. WILLIAMS & W. W. BUSSE. 1992. Bronchodilation with a potent and selective leukotriene D4 (LTD4) receptor antagonist (MK-571) in patients with asthma. Am. Rev. Respir. Dis. **146:** 358-363.
16. LAMMERS, J. W., D. P. VAN, D. E. F. VAN, M. DECRAMER, A. BUNTINX, L. I. DE & B. FRIEDMAN. 1992. Bronchodilator properties of an inhaled leukotriene D4 antagonist (verlukast—MK-0679) in asthmatic patients. Pulmonary Pharmacol. **5:** 121-125.
17. FINDLAY, S. R., J. M. BARDEN, C. B. EASLEY & M. GLASS. 1992. Effect of the oral leukotriene antagonist, ICI 204,219, on antigen-induced bronchoconstriction in subjects with asthma. J. Allergy Clin. Immunol. **89:** 1040-1045.
18. FINNERTY, J. P., B. R. WOOD, H. THOMSON & S. T. HOLGATE. 1992. Role of leukotrienes in exercise-induced asthma. Inhibitor effect of ICI 204219, a potent leukotriene D4 receptor antagonist. Am. Rev. Respir. Dis. **145:** 746-749.
19. CHRISTIE, P. E., C. M. SMITH & T. H. LEE. 1991. The potent and selective sulfidopeptide leukotriene antagonist, SK&F 104353, inhibits aspirin-induced asthma. Am. Rev. Respir. Dis. **144:** 957-958.
20. RASMUSSEN, J. B., L. O. ERIKSSON & K. E. ANDERSSON. 1991. Reversal and prevention of airway response to antigen challenge by the inhaled leukotriene D4 antagonist (L-648,051) in patients with atopic asthma. Allergy **46:** 266-273.
21. CLOUD, M. L., G. C. ENAS, J. KEMP, M. T. PLATTS, L. C. ALTMAN, R. TOWNLEY, D. TINKELMAN, T. J. KING, E. MIDDLETON, A. L. SHEFFER *et al.* 1989. A specific LTD4/LTE4-receptor antagonist improves pulmonary function in patients with mild, chronic asthma. Am. Rev. Respir. Dis. **140:** 1336-1339.
22. CARTER, G. W., P. R. YOUNG, D. H. ALBERT, J. BOUSKA, R. DYER, R. L. BELL, J. B. SUMMERS & D. W. BROOKS. 1991. 5-Lipoxygenase inhibitory activity of zileuton. J. Pharmacol. Exp. Ther. **256:** 929-937.
23. RUBIN, P., L. DUBE, R. BRAECKMAN, L. SWANSON, R. HANSEN, D. ALBERT & G. CARTER. 1991. Pharmacokinetics, safety, and ability to diminish leukotriene synthesis by zileuton, an inhibitor of 5-lipoxygenase. Agents Actions Suppl. **35:** 103-116.
24. KNAPP, H. R. 1990. Reduced allergen-induced nasal congestion and leukotriene synthesis with an orally active 5-lipoxygenase inhibitor [see comments]. N. Engl. J. Med. **323:** 1745-1748.
25. LAURSEN, L. S., J. NAESDAL, K. BUKHAVE, K. LAURITSEN & M. J. RASK. 1990. Selective 5-lipoxygenase inhibition in ulcerative colitis. Lancet **335:** 683-685.
26. RASK, M. J., K. BUKHAVE, L. S. LAURSEN & K. LAURITSEN. 1992. 5-Lipoxygenase inhibitors for the treatment of inflammatory bowel disease. Agents Actions Spec. No. pC37.
27. WEINBLATT, M. E., J. M. KREMER, J. S. COBLYN, S. HELFGOTT, A. L. MAIER, G. PETRILLO, B. HENSON, P. RUBIN & R. SPERLING. 1992. Zileuton, a 5-lipoxygenase inhibitor in rheumatoid arthritis. J. Rheumatol. **19:** 1537-1541.
28. BELL, R. L., D. W. BROOKS, P. R. YOUNG, C. LANNI, A. O. STEWART, J. BOUSKA, P. E. MALO & G. W. CARTER. 1993. A-78773: A selective, potent 5-lipoxygenase inhibitor. J. Lipid Mediat. **6:** 259-264.
29. SWEENEY, D. J., J. BOUSKA & J. MACHINIST. 1992. Glucuronidation of zileuton (A-64077) in human hepatic microsomes. Drug Metab. Disp. **20:** 328-329.

30. Sweeney, D. J. & H. Nellans. 1992. Enantiomeric activation of glucuronidation in dog hepatic microsomes. J. Biol. Chem. **267:** 13171-13174.
31. Salmon, J. A. & L. G. Garland. 1991. Leukotriene antagonists and inhibitors of leukotriene biosynthesis as potential therapeutic agents. Prog. Drug Res. **37:** 9-90.
32. Bell, R. L., J. Bouska, P. R. Young, C. Lanni, J. Machinist, P. E. Malo, J. B. Summers, D. W. Brooks & G. W. Carter. 1993. The properties of A-69412: A small hydrophilic 5-lipoxygenase inhibitor. Agents Actions **38:** 178-187.

Studies of the Combination of Ro 24-5913, a Peptidoleukotriene Antagonist, and Ro 24-4736, a PAF Antagonist, in Guinea Pig and Rat Models of Lung Inflammation

A. F. WELTON,[a] M. O'DONNELL,[a] L. RENZETTI,[a]
B. SIMKO,[a] J. TOCKER,[a] C. CASHIN,[b] P. NEWBOLD,[b]
AND M. A. WASSERMAN[a]

[a]*Department of Bronchopulmonary Research*
Roche Research Center
Nutley, New Jersey 07110

[b]*Biology Department*
Roche Products Limited
Welwyn Garden City
Herts AL7 3AY, United Kingdom

INTRODUCTION

A variety of substances including lipids, biogenic amines, and cytokines have been identified as potential mediators of various aspects of complex inflammatory diseases. In particular we have been interested, for a number of years, in the role of lipid mediators, such as peptidoleukotrienes (LTD_4) and platelet-activating factor (PAF), in respiratory diseases with a large inflammatory component (e.g., asthma, rhinitis, and adult respiratory distress syndrome). Two drugs, Ro 24-5913, a peptidoleukotriene antagonist,[1] and Ro 24-4736, a PAF antagonist,[2] have been identified that will be useful in probing the role of these mediators in these human diseases. The pharmacological profiles of Ro 24-5913 and Ro 24-4736 are summarized in TABLES 1 and 2, respectively. It is possible, however, that for optimum therapy of the diseases cited above a combination of these mediator antagonists may be most useful. To investigate this hypothesis more thoroughly, we have studied the activity of Ro 24-5913 and Ro 24-4736, alone and in combination, in several animal models of lung inflammation in guinea pigs and rats.

METHODS

Ovalbumin Induced Anaphylaxis in Guinea Pigs

Male (Hartley) guinea pigs (200-300 g) were sensitized by intraperitoneal (ip) injection with ovalbumin (OVA, 10 mg) in one ml of saline solution on days 1 and

TABLE 1. Pharmacological Profile of the LTD_4 Antagonist Ro 24-5913

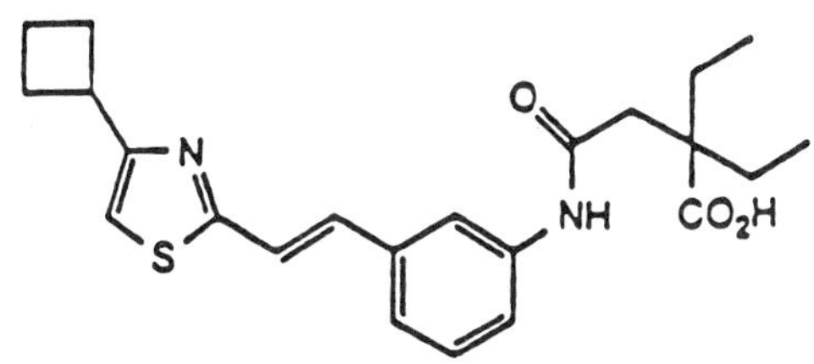

In vitro activity	
^{3}H-LTD_4 binding (guinea pig lung; IC_{50})	6.4 nM
pA_2 (guinea pig trachea)	9.6
pK_B (human bronchi)	9.3
In vivo activity	
LTD_4 bronchoconstriction (guinea pig; ID_{50})	0.05 mg/kg iv and oral
OVA bronchoconstriction (guinea pig; ID_{50})	0.02 mg/kg iv and oral
LTD_4 vascular permeability (ID_{50})	1.3 mg/kg iv 4.8 mg/kg oral

3 and used for study 20-30 days later. For the measurement of cardiopulmonary parameters the animals were anesthetized with urethane (2 g/kg, ip), and the jugular vein and right carotid artery were cannulated for intravenous (iv) drug administration and for systemic blood pressure measurements (Model 50 pressure transducer, Statham Instruments), respectively. The trachea was also cannulated for positive pressure ventilation with a Harvard pump set at a 4.5-ml stroke volume at a rate of 40 breaths per minute. Respiratory muscles were paralyzed by periodic injections of succinylcholine (1.2 mg/kg, iv).

To measure transpulmonary pressure, a 12-gauge hypodermic needle attached via a short length of Tygon tubing to one side of a Statham differential pressure transducer (Model P131TC) was introduced through the fifth or sixth intercostal space into the intrapleural space near the sternum, leaving a pneumothorax. The other side of this transducer was connected to the side arm of the tracheal cannula by a second Tygon tubing of identical length and bore as the first. Airflow was measured with a Fleisch pneumotachograph head (type #0000, Instrumentation Associates, New York, NY) connected to a tracheal cannula and a second Statham differential pressure transducer (Model PM 15). Electrical integration of the flow signal provided a recording of tidal volume.

Pulmonary resistance and dynamic lung compliance were calculated manually from the transpulmonary pressure, airflow, and volume signals as described by Amdur and Mead,[3] and the results were used to calibrate an on-line analog pulmonary computer (Buxco Electronics, Sharon, CT). Blood pressure was recorded with the cannula inserted into the carotid artery. Heart rate was calculated by an on-line analog cardiovascular computer (Buxco Electronics, Sharon, CT). All six parameters (airflow, transpulmonary pressure, tidal volume, resistance, compliance, and blood pressure)

TABLE 2. Pharmacological Profile of the PAF Antagonist Ro 24-4736

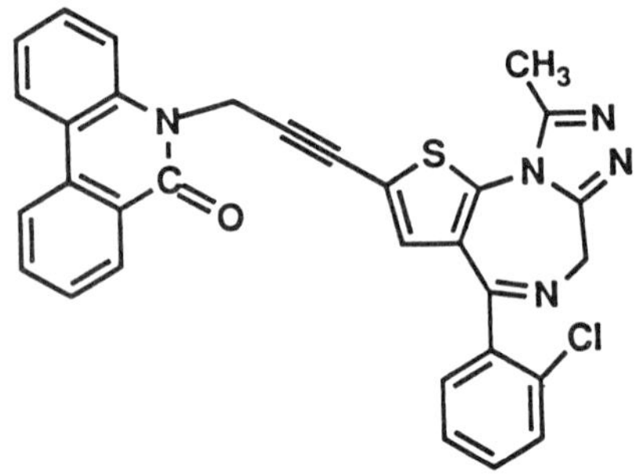

In vitro activity	
^{3}H-PAF binding (dog platelets; IC_{50})	9.5 nM
PAF aggregation (dog platelets; IC_{50})	6 nM
PAF aggregation (guinea pig platelets; IC_{50})	6 nM
PAF aggregation (human platelets; IC_{50})	160 nM
In vivo activity	
PAF-induced bronchoconstriction (ID_{50})	0.02 mg/kg, iv
	0.01 mg/kg, po
PAF-induced vascular permeability (ID_{50})	0.05 mg/kg, iv
	0.05 mg/kg, po

were recorded simultaneously on a polygraph recorder (Model 7758A, Hewlett Packard, Paramus, NJ), and these parameters, plus heart rate and respiratory rate, were printed out on an electronic data terminal (Texas Instruments, Houston, TX). After completion of surgery, a period of no less than 5 min was allowed for cardiopulmonary functions to stabilize. During this period, control values were recorded continuously.

Following surgical preparation, the animals were pretreated with intravenous indomethacin (10 mg/kg) 22 min before OVA (antigen) challenge. Spontaneous breathing was arrested with succinylcholine 7 min before antigen challenge. Treatment with succinylcholine resulted in a constant baseline ventilatory pressure of 6-8 cm H_2O. The bronchoconstrictor effects of histamine were inhibited with pyrilamine (2 mg/kg, iv), administered 6 min before antigen challenge. To potentiate the bronchoconstrictor effect of antigen, propranolol (0.1 mg/kg, iv) was administered 5 min before antigen challenge.

Ro 25-5913, Ro 24-4736, the combination of these two drugs, or control vehicles were administered intravenously through a cannula in the jugular vein 5 min before the animals were challenged with a maximum constrictory dose of OVA (1 mg/kg, iv). The percent change in pulmonary resistance, dynamic lung compliance, and systemic blood pressure and survival time were monitored for 120 min.

OVA-Induced Plasma Extravasation in Guinea Pig Airways

Male (Hartley) guinea pigs (200-300 g) were sensitized subcutaneously with OVA [10 μg plus 1 mg $Al(OH)_3$ in 0.5 ml saline] on days 1 and 14 and used for study 21-35 days later.

On the day of the experiment the animals were anesthetized with urethane (2 g/kg, ip), the jugular vein was cannulated, and the animals were allowed to stabilize for 15 min. Pyrilamine (2 mg/kg, iv) and indomethacin (10 mg/kg, iv) then were administered followed, 14 min later, by Evans blue dye (30 mg/kg, iv).

Airway microvascular leakage was induced by a single injection of OVA (1 mg/kg, iv) given 1 min after the Evans blue dye. After 10 min the thorax was opened, and a catheter (PE-50 tubing) was inserted through the left ventricle into the aorta. The right atria was cut, and blood was expelled by perfusing the animal with approximately 100 ml of saline using a cartridge pump (Masterflex, Cole-Parmer, Chicago, IL) with a rate of 100 ml/min. The pulmonary circulation then was perfused with an additional 50 ml of saline by inserting the catheter into the pulmonary artery and cutting the left atria. The lungs were removed and the surrounding parenchyma and vasculature were dissected away. The airways were divided into the trachea (distal 5 mm), main bronchi, and the intrapulmonary airways. All tissues were blotted between filter paper, and weighed. Evans blue dye was extracted in formamide (37 °C, 24 h) and quantified by measuring the absorbance at 620 nm with a spectrophotometer (Model DU-64, Beckman Instruments, Somerset, NJ). Tissue dye content was interpolated from a standard curve of Evans blue dye concentrations (0.5-10 μg/ml) and expressed in ng/mg of tissue wet weight.

Guinea pigs were dosed orally with Ro 24-5913, Ro 24-4736, a combination of these antagonists, or vehicle 1 h prior to OVA challenge. Baseline or background plasma leakage was measured in a group of sensitized animals that were given only saline, intravenously. The mean ± standard error (SEM) were calculated for each group, and statistical differences were determined at the 0.05 level using a two-tailed Student's *t*-test for unpaired observations.

OVA-Challenged Perfused Guinea Pig Lung

Male Hartley strain guinea pigs (300-350 g) were sensitized to OVA as described above for the OVA-induced plasma extravasation model. The methodology for removing and preparing this perfused lung system has been described previously.[4]

For challenge, OVA (30 μg in 30 μl of saline) was injected directly into the pulmonary artery of perfused lungs that had been pretreated with perfusion buffer (phosphate-buffered Ringer's Solution containing 0.5% bovine serum albumin and 137 mM NaCl, 1.8 mM $CaCl_2$, 1.05 mM $MgCl_2$, 2.68 mM KCl, 0.06 mM $NaHCO_3$, 0.130 mM NaH_2PO_4, 0.869 mM Na_2PO_4, and 5.55 mM dextrose) or perfusion buffer containing 1 μM Ro 24-5913, 1 μM Ro 24-4736, or a combination of 1 μM Ro 24-5913 and 1 μM Ro 24-4736. Changes in intratracheal pressure, pulmonary artery pressure, and lung weight were monitored for 60 min. Data are expressed as means ± SEM. Statistical significance was determined by analysis of variance followed by multiple comparison testing using Dunnett's testing procedure.

Sephadex-Induced Lung Inflammation in Rats

Male Alderley Park strain-1 rats (150-250 g), bred at Roche-Welwyn, were used in these studies. Sephadex G100 SF, suspended in 0.9% saline (1.5 mg/ml) and allowed to swell for 48 h before use (mean particle size approximately 100 μm), was administered intravenously to the animals (7.5 mg of Sephadex/kg). At the time of termination of an experiment, the rats were anesthetized with Sagatal (60 mg/kg, ip), the trachea cannulated, and the lungs gently lavaged twice with 10 ml/kg of phosphate-buffered saline (PBS). The bronchoalveolar lavage (BAL) fluid was analyzed for total leukocyte count (Coulter counter) and eosinophil content (50 μl diluted into 450 μl of Discombes Fluid for Fuchs-Rosenthal chamber count). Total protein content was determined by transferring a 1.5-ml aliquot of BAL fluid to Eppendorf tubes, centrifuging at 10,000 rpm for 15 min, and determining protein using the Bradford method on the CoBAS-BIO. The mean ± SEM for each measurement was determined for each group, and the significance of the difference from controls was calculated by an unpaired Student's *t*-test (2-tail).

For measurement of drug effects, animals were dosed with drug 3 h and 1 h prior to the iv injection of Sephadex and, then, at 24-h intervals for the next two days. On the third day animals were sacrificed, and BAL was isolated and analyzed as described above.

RESULTS

OVA-Induced Guinea Pig Anaphylaxis

Ro 24-5913 and Ro 24-4736 were initially evaluated, alone and in combination, in a model of OVA-induced anaphylaxis in actively sensitized guinea pigs for effects on antigen-induced airway resistance, dynamic lung compliance, blood pressure changes, and mortality. The animals used in these studies were pretreated with (1) indomethacin to inhibit prostaglandin production and accentuate metabolism through the 5-lipoxygenase pathway, (2) pyrilamine maleate to eliminate histamine-mediated effects, and (3) propranolol to prevent effects mediated by endogenously released catecholamines. Under these conditions OVA challenge induced a 550% increase in pulmonary resistance, an 83% decrease in dynamic lung compliance, a decrease in systemic blood pressure of 50 mm Hg, and 100% mortality within 40 min (FIG. 1).

Ro 24-5913 and Ro 24-4736 were studied in this OVA-anaphylaxis model at intravenous doses of 1 mg/kg (similar effects were also seen subsequently when the drugs were tested orally at these doses). The doses studied were previously shown to completely inhibit bronchoconstriction in guinea pigs induced by exogenous LTD_4[1] or PAF,[2] respectively. OVA-induced pulmonary resistance increases in this guinea pig model seem to be primarily mediated by LTD_4, since Ro 24-5913 alone inhibited the airway resistance change by approximately 90% (FIG. 1A). Ro 24-4736, as a single agent or in combination with Ro 24-5913, had little effect on resistance changes, thereby suggesting that PAF has little role in mediating this physiological parameter. Interestingly, also shown in FIGURE 1A, the resistance measurements for those animals challenged with OVA alone, or OVA with Ro 24-5913 or Ro 24-4736, as single treatments, all terminated between 30 and 55 min because the animals died. When Ro 24-5913 and Ro 24-4736 were studied in combination, however, the animals lived until the experiment was finished. This suggested that, although the drug combination had little effect on pulmonary resistance, the combination likely was having a beneficial effect on other physiological processes that affected the animals' survival. For example, the combination of Ro 24-5913 and Ro 24-4736 was more effective than either drug alone when the parameters of OVA-induced dynamic compliance (FIG. 1B) and systemic blood pressure (FIG. 1C) were evaluated. FIGURE 1D more directly depicts the survival rate of the animals treated with Ro 24-5913 or Ro 24-4736, alone or in combination, and further demonstrates the synergistic protective activity of the drugs in combination.

OVA-Induced Plasma Extravasation in Guinea Pig Airways

Another closely related model where we have evaluated the activity of a combination of our LTD_4 antagonist and PAF antagonist is a model of OVA-induced plasma extravasation into the airways of actively sensitized guinea pigs.

In this model, animals were also pretreated with indomethacin and pyrilamine maleate, as described above for the OVA-anaphylaxis model; but plasma extravasation into the trachea, main bronchi, and intrapulmonary airways was the experimental endpoint measured. In control animals, this endpoint was measured by injecting Evans Blue dye (30 mg/kg, iv) into the animals 1 min prior to antigen challenge and observing the migration of the dye into the airway tissues. As shown in FIGURE 2, OVA challenge induced a substantial migration of Evans Blue into the airway tissues over the 10 min after OVA challenge (that this extravasation was monitored). Both Ro 24-5913 and Ro 24-4736 blocked this effect in a dose-related fashion (FIGS. 2A and B), but the maximal inhibition seen with each was different. Ro 24-5913 (1 mg/kg, po) inhibited ($p < 0.05$) Evans Blue extravasation by 73%, 67%, and 53% in the trachea, main bronchus, and intrapulmonary airways, respectively; whereas the inhibition ($p < 0.05$) for Ro 24-4736 (1 mg/kg, po) was less–46%, 40%, and 44%, respectively. Most notably, when maximal doses of these drugs (1 mg/kg, po) were studied in this model, the level of inhibition obtained (94%, 92%, and 90% in the trachea, main bronchus and intrapulmonary airway, respectively) was greater than seen with either drug alone (FIG. 2C). This observation supports the hypothesis that combinations of mediator antagonists such as Ro 24-5913 and Ro 24-4736 may offer improved therapy in preventing plasma extravasation in certain respiratory inflammatory diseases.

OVA-Challenged Perfused Guinea Pig Lung

To more directly evaluate the effect of Ro 24-5913 and Ro 24-4736, alone or in combination, on lung physiology, a perfused guinea pig lung model was utilized. Drug effects (1 μM in the perfusing medium perfusate) on OVA-induced intratracheal pressure, pulmonary artery pressure, and lung weight were monitored. As described in FIGURE 3, neither drug effectively inhibited OVA-induced changes in intratracheal pressure. Both were effective, however, in inhibiting changes in pulmonary artery

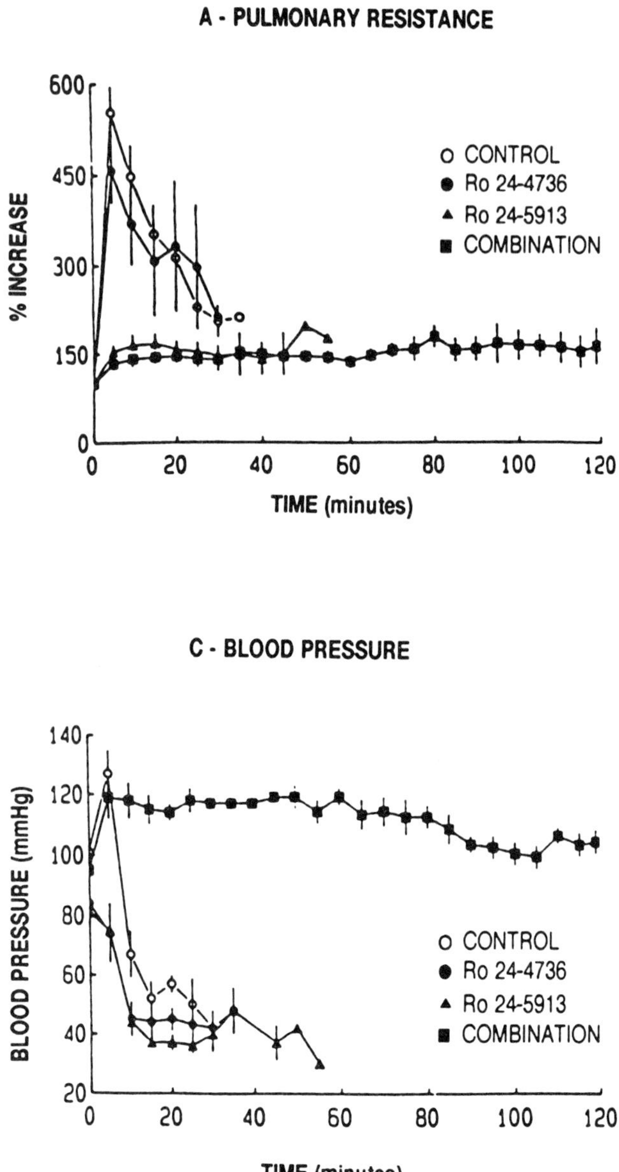

FIGURE 1. Protective effects of intravenous control vehicle (○), Ro 24-5913 (1 mg/kg, iv, ▲), Ro 24-4736 (1 mg/kg, iv, ●), and the combination of Ro 24-5913 and Ro 24-4736 (■) against OVA-induced changes in guinea pigs in (**A**) pulmonary resistance, (**B**) dynamic lung compliance, and (**C**) blood pressure. (**D**) Protective effects of Ro 24-5913, Ro 24-4736, and the combination against OVA-induced mortality. Each *point* represents the mean ± SEM for determinations made on six animals.

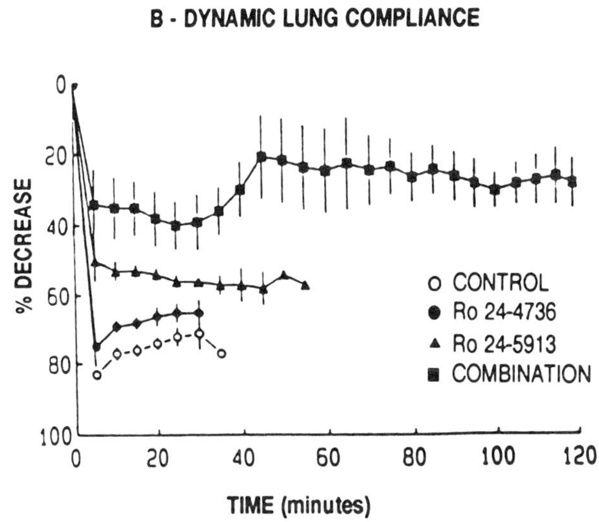
B - DYNAMIC LUNG COMPLIANCE
% DECREASE
0
20
40
60
80
100
0
20
40
60
80
100
120
TIME (minutes)
CONTROL
Ro 24-4736
Ro 24-5913
COMBINATION

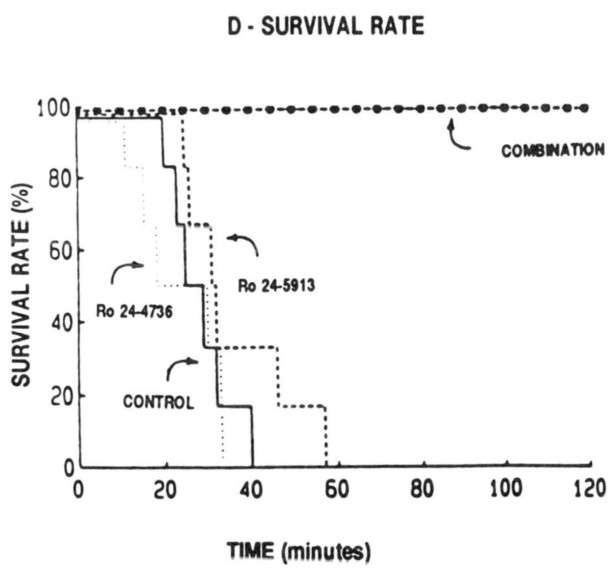
D - SURVIVAL RATE
SURVIVAL RATE (%)
100
80
60
40
20
0
0
20
40
60
80
100
120
TIME (minutes)
COMBINATION
Ro 24-5913
Ro 24-4736
CONTROL

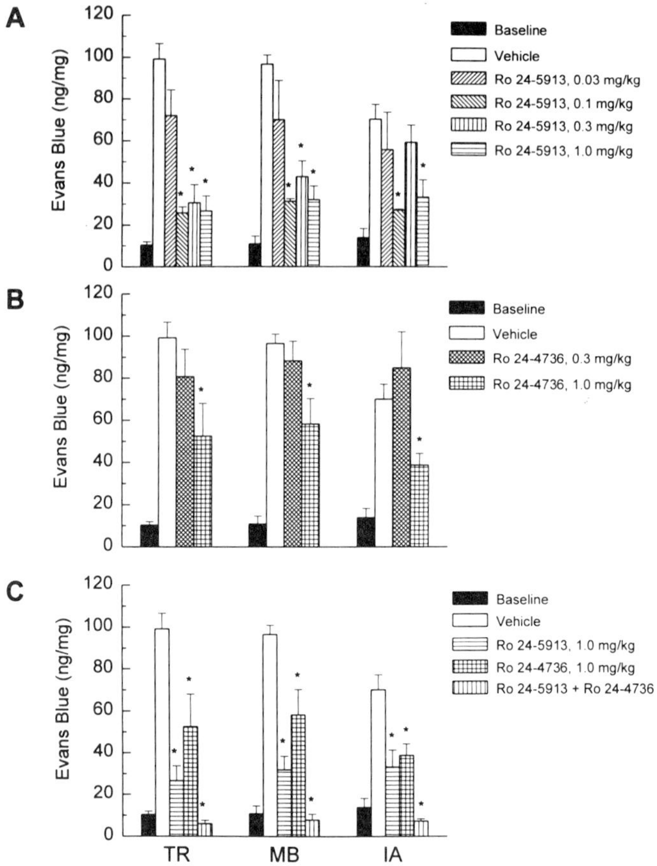

FIGURE 2. Effect of Ro 24-5913, Ro 24-4736, or the combination on antigen (OVA)-induced plasma extravasation in guinea pig airways. Sensitized animals were dosed orally one hour prior to OVA (1.0 mg/kg, iv) challenge. (**A**) Treatment with Ro 24-5913 alone at 0.03, 0.1, 0.3, or 1 mg/kg; (**B**) treatment with Ro 24-4736 alone at 0.3 and 1 mg/kg; and (**C**) treatment with Ro 24-5913 and Ro 24-4736 (both at 1 mg/kg). Baseline values were determined in sensitized animals that received iv saline. All animals from each group were pretreated with pyrilamine (2.0 mg/kg) and indomethacin (10.0 mg/kg) as described in METHODS. Each *bar* represents the mean ± SEM for determinations made on at least $n = 4$ animals per group. *Asterisks* indicate significant ($p < 0.005$) difference from vehicle-treated OVA-challenged animals.

pressure (80-85% for Ro 24-4736 and 60-70% for Ro 24-5913). Ro 24-4736 also provided statistically significant inhibition of OVA-induced changes in lung weight (65-67%). When the drugs were studied in combination, an improvement was observed in the ability to inhibit intratracheal pressure (FIG. 4). However, the combination did not improve the ability to inhibit OVA-induced changes in pulmonary arterial pressure or lung weight (data not shown).

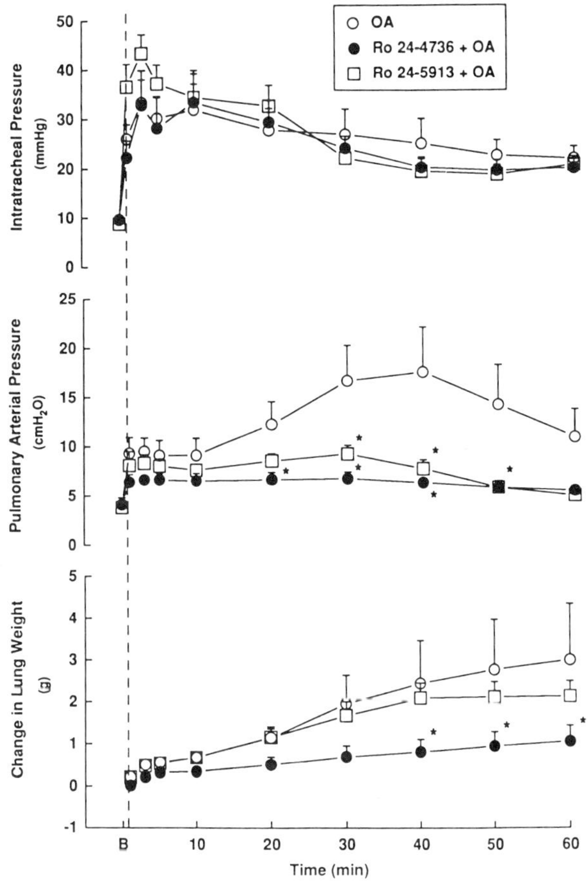

FIGURE 3. Effect of intraarterial ovalbumin (OA; 30 μg) challenge on intratracheal pressure (**top**), pulmonary arterial pressure (**middle**), and change in lung weight (**bottom**) in sensitized guinea pig lungs pretreated with Ro 24-4736 or Ro 24-5913 (1 μM each; n = 8). OA was administered after baseline measurement (indicated by *dashed line*). OA-challenged lungs (30 μg OA; n = 8) provided for comparison. Values are mean ± SEM. *Asterisks* indicate significant ($p < 0.05$) difference of drug-treated from OA-challenged control group at corresponding time.

Sephadex-Induced Lung Inflammation in Rats

In the final model in which a combination of Ro 24-5913 and Ro 24-4736 was evaluated, rats were challenged intravenously with an injection of hydrated Sephadex G-100 SF beads. The particles embolize in the lung vasculature, leading to thrombus and granuloma formation in the lungs, accompanied by pulmonary inflammation, which is reflected by increases of inflammatory cells, especially eosinophils, and protein in the BAL fluid. These cellular and protein changes in BAL were maximal at 2 to 3 days after Sephadex injection (FIG. 5). Therefore, the effects of Ro 24-5913

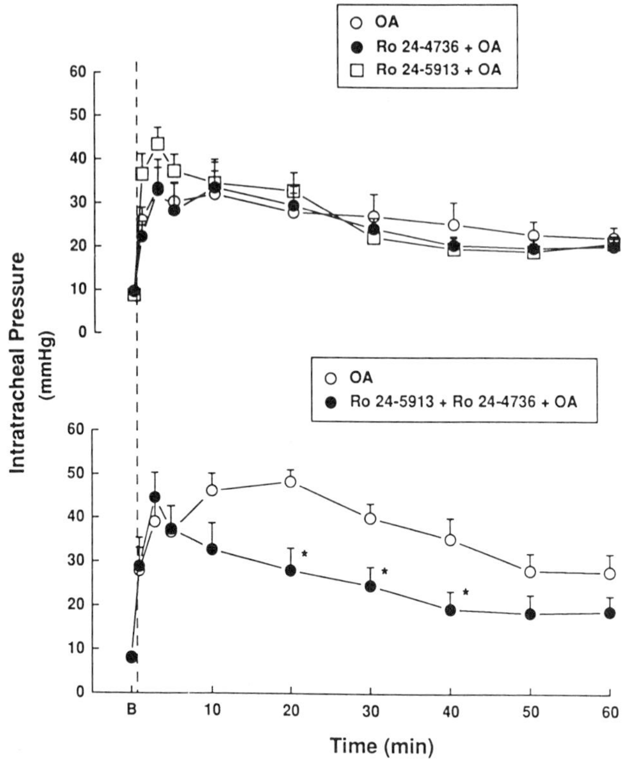

FIGURE 4. Effect of intraarterial ovalbumin (OA; 30 μg) challenge on intratracheal pressure in sensitized guinea pig lungs pretreated with: (**top**) Ro 24-4736 (1 μM; $n = 7$) or Ro 24-5913 (1 μM; $n = 7$); or (**bottom**) Ro 24-5913 and Ro 24-4736 (1 μM each, $n = 8$). OA was administered after baseline measurement (indicated by *dashed line*). OA-challenged lungs (30 μg OA, $n = 8$ for **top**; $n = 7$, for **bottom**) provided for comparison. Values are mean ± SEM. *Asterisks* indicate significant ($p < 0.05$) difference of drug-treated from OA-challenged control group at corresponding time.

(5 mg/kg) and Ro 24-4736 (1 mg/kg), alone and in combination, on the BAL content were evaluated 3 days after Sephadex challenge (FIG. 6). The data in FIGURE 6 demonstrate that the combination of Ro 24-5913 and Ro 24-4736 yielded better inhibition of Sephadex-induced total cell infiltration and eosinophil infiltration, and increased protein content in the BAL than either drug alone, although only the effect on eosinophil infiltration (59% inhibition, $p < 0.05$ compared to vehicle control) reached statistical significance.

DISCUSSION

The studies reported herein describe situations in which a combination of multiple mediator antagonists afforded improved and, in some cases, synergistic protection

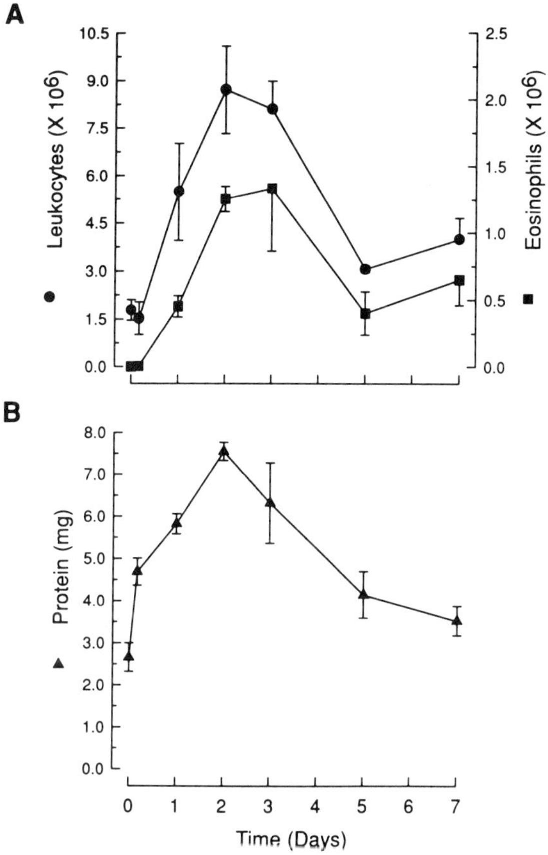

FIGURE 5. Time course of (**A**) cellular changes and (**B**) protein changes in BAL fluid from Sephadex-treated rats. Each *point* represents the mean ± SEM for determinations made on three animals.

against the development of inflammatory reactions in lung tissue. Although the animal models where this was demonstrated do not directly mimic human diseases, the implications of these studies for the optimum treatment of respiratory diseases are obvious. Intuitively, the fact that numerous mediators have been implicated in chronic inflammatory reactions should make the notion that combinations of multiple mediator antagonists may offer optimum therapy relatively easy to defend.

In a previously described study of asthmatics, it has been demonstrated that a combination of the peptidoleukotriene antagonist SK&F 104, 353 and the antihistamine terfenadine afforded additive protection against the early-phase bronchoconstriction induced by antigen.[5] Our studies suggest that the addition of a PAF antagonist to this therapeutic regimen may yield additional improvement, perhaps in the late-phase reaction.

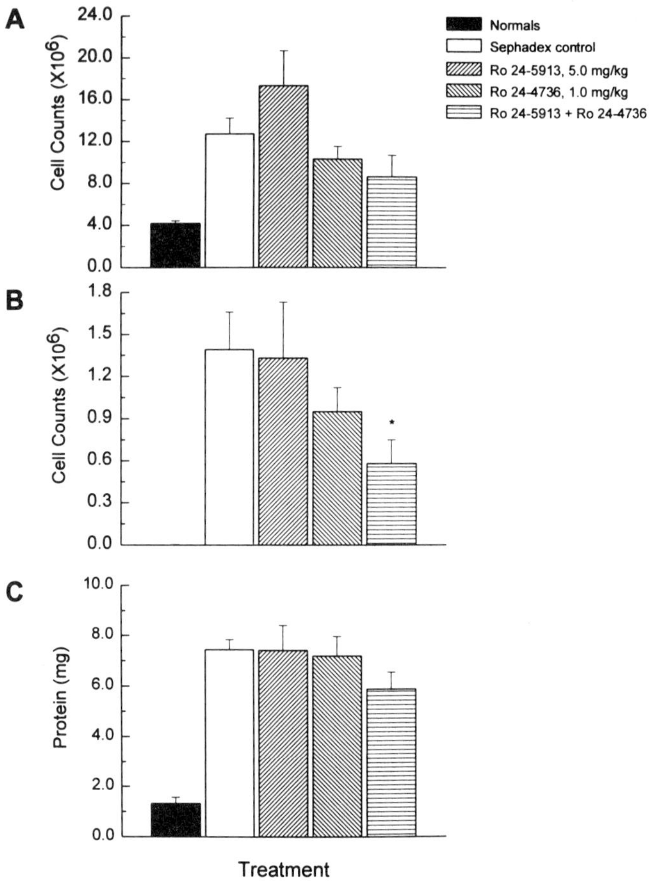

FIGURE 6. The effect of Ro 24-5913 (5 mg/kg, po), Ro 24-4736 (1 mg/kg, po), and the combination on BAL composition from Sephadex-G100SF-challenged rats. (**A**) total leukocytic number, (**B**) eosinophil number, and (**C**) protein composition. Animals were pretreated with the indicated drugs 3 hours and 1 hour prior to Sephadex challenge and at 24 and 48 hours postchallenge. Analysis of BAL content (mean ± SEM) was performed 72 hours after challenge. The group sizes were: normal treatment, 4 animals; Sephadex-treated, 8 animals; drug-treated, 6 animals. *Asterisk* indicates significant ($p < 0.05$) difference from vehicle-treated Sephadex-100SF-challenged animals.

Several PAF antagonists including WEB2086,[6] MK287,[7] and UK74,505[8] have not demonstrated significant activity by themselves in clinical studies of antigen-challenge allergic asthmatics. Our studies would suggest that these observations should not be a deterrent to the further study of PAF antagonists in combination with other mediator antagonists, since we have observed in our studies in the guinea pig OVA-anaphylaxis model and the Sephadex-lung inflammation model in rats situations where a PAF antagonist showed no activity by itself and yet, when combined with a peptidoleukotriene antagonist, demonstrated synergistic improvements. PAF antagonists are likely to supplement efficacy by preventing inflammatory symptoms associated with increased vascular permeability (e.g., edema, plasma extravasation) and eosinophil migration and activation.

Although the concept of clinically evaluating a combination of a peptidoleukotriene antagonist and a PAF antagonist is interesting, there are likely to be a number of complex regulatory hurdles that may need to be overcome. These hurdles include the need to do combination toxicology with a variety of ratios of the two drugs to evaluate the safety of the combination at all possible clinical doses that might be administered in the study. Furthermore, many regulatory agencies may be hesitant to allow the simultaneous study of two experimental drugs in humans at once. This may mean that it will be necessary to have one of the drugs registered first, before combination studies can commence. Based on promising clinical information that is now appearing for several peptidoleukotriene antagonists,[9-12] it is possible that drugs of this type may be approved for use in the next several years. If this is the case, then some of the difficulty in evaluating the combination of a peptidoleukotriene antagonist and a PAF antagonist will have been circumvented.

In summary, we have demonstrated in several animal models, that the combination of a peptidoleukotriene antagonist, Ro 24-5913, and a PAF antagonist, Ro 24-4736, affords better protection against the development of inflammatory symptoms with lung tissue than either drug alone. This suggests the clinical potential of this combination for the treatment of respiratory diseases such as asthma, rhinitis, and adult respiratory distress syndrome. Combination therapy with multiple mediator antagonists may become the norm in the future, as selective antagonists for a variety of different types of substances are identified and profiled in human diseases.

REFERENCES

1. O'Donnell, M., H. J. Crowley, B. Yaremko, N. O'Neill & A. F. Welton. 1991. Pharmacological actions of Ro 24-5913, a novel antagonist of leukotriene D_4. J. Pharmacol. Exp. Ther. **259:** 751-758.
2. Crowley, H. J., B. Yaremko, W. M. Selig, D. Janero, C. Burghardt, A. F. Welton & M. O'Donnell. 1991. Pharmacology of a potent platelet-activating factor antagonist: Ro 24-4736. J. Pharmacol. Exp. Ther. **259:** 78-85.
3. Amdur, M. & J. Mead. 1958. Mechanics of respiration in anesthetized guinea pigs. Am. J. Physiol. **192:** 364-368.
4. Selig, W. M., S. K. Durham & A. F. Welton. 1989. Pulmonary responses to phospholipase A_2 in the perfused guinea-pig lung. J. Appl. Physiol. **67**(6): 2495-2503.
5. Eiser, N. M., W. Denman, M. Hayhurst, P. Ancie & J. Rodrigues. 1989. Role of histamine and leukotrienes in the early response to inhaled antigen. Thorax **44:** 334.
6. Freitag, A., R. M. Watson, G. Matsos, C. Eastwood & P. M. O'Bryne. 1991. The effect of treatment with a oral platelet-activating factor antagonist (WEB2086) in allergen-induced asthmatic responses in human subjects. Am. Rev. Respir. Dis. **143:** A157.
7. Bel, E. H., M. De Smut, T. H. Rossing, M. C. Timrus, J. H. Dijkman & P. J. Sterzk. 1991. The effect of a specific oral PAF antagonist MK-287, on antigen-induced early and late asthmatic reactions. Am. Rev. Respir. Dis. **143:** A811.

8. KUITERT, L. M., K. P. HUI, S. UTHAYARKUMAR, W. BURKE, A. C. NEWLAND, S. UDEN & N. C. BARNES. 1992. Effect of a platelet-activating antagonist, UK74,505, on allergen-induced early and late responses. Am. Rev. Respir. Dis. **145:** A292.
9. CLOUD, M. L., G. C. ENAS, J. KEMP, T. PLATTS-MILLS, L. C. ALTMAN, R. TOWNLEY, D. TINKELMAN, T. KING, E. MIDDLETON, A. L. SHEFFER, E. R. MCFADDEN & D. S. FARLOW. 1989. A specific LTD_4/LTE_4-receptor antagonist improves pulmonary function in patients with mild, chronic asthma. Am. Rev. Respir. Dis. **140:** 1336–1339.
10. TAYLOR, I. K., K. M. O'SHAIGHNESSY, R. W. FULLER & C. T. DOLLERY. 1991. Effect of cysteinyl-leukotriene receptor antagonist IC1204,219 on allergen-induced bronchoconstriction and airway hyperreactivity in atopic subjects. Lancet **337:** 690–694.
11. MARGOLSKEE, D., S. BODMAN, R. DOCKHORN, E. ISRAEL, J. KEMP, H. MANSMANN, D. A. MINOTTI, S. SPECTOR, W. STRICKER, D. TINKELMAN, R. TOWNLEY, J. WINDER & V. WILLIAMS. 1991. The therapeutic effect of MK571, a potent and selective leukotriene D_4 receptor antagonist in patients with chronic asthma. J. Allergy Clin. Immunol. **87:** 309.
12. SPECTOR, S. L., M. GLASS, M. C. MINKWITZ & ICI ASTHMA TRIAL GROUP. 1991. The effect of six weeks therapy with oral doses of ICI 204,219 in asthmatics. Am. Rev. Respir. Dis. **145:** A16.

Preclinical Exploration of the Potential Antiinflammatory Properties of the Peptide Leukotriene Antagonist ICI 204,219 (Accolate™)

ROBERT D. KRELL,[a] CHRISTOPHER J. DEHAAS, DAVID J. LENGEL, EDWARD J. KUSNER, JOSEPH C. WILLIAMS, AND CARL K. BUCKNER

ZENECA Pharmaceuticals Group
Department of Pharmacology
Wilmington, Delaware 19897

INTRODUCTION

The search for antagonists of the leukotrienes or inhibitors of their production by the enzyme 5-lipoxygenase began in earnest over a decade ago following the disclosure of the structures of the leukotrienes LTB_4, LTC_4, LTD_4, and LTE_4 by Samuelsson *et al.*[1] During the past few years, several highly potent and selective receptor antagonists—e.g., Accolate™,[2] SKF 104,353,[3,4] MK571[5]—and synthesis inhibitors—e.g., D2138[6] and MK886[7]—have been discovered, with several entering clinical trials. Clinical findings reported to date indicate that these various compounds are capable of blocking or modulating bronchoconstriction induced by antigen, cold air, exercise, etc.[8–16] While seductive, indeed, these clinical paradigms are not necessarily accurate predictors of efficacy in the "at large" asthma population. Thus, it has been heartening that early results reported from phase IIb and III trials also appear to demonstrate the efficacy of modulators of the leukotrienes. At the present time, it appears that the validity of the hypothesis that peptide leukotrienes are involved in the pathophysiology of allergic disease, specifically, asthma, is going to be verified. As a consequence, it became of interest to explore in greater depth the biological mechanism(s) responsible for this clinical activity. Preclinical evaluation of the potential antiinflammatory properties of the peptide leukotriene antagonist Accolate™ was undertaken as a first step toward this goal.

MATERIALS AND METHODS

Male albino Hartley guinea pigs weighing 400 to 800 g were used in all studies. They were anesthetized with 1.5 g/kg (ip) urethane as appropriate.

[a]Current address: Biofor, Inc., P.O. Box 629, Waverly, Pennsylvania 18471.

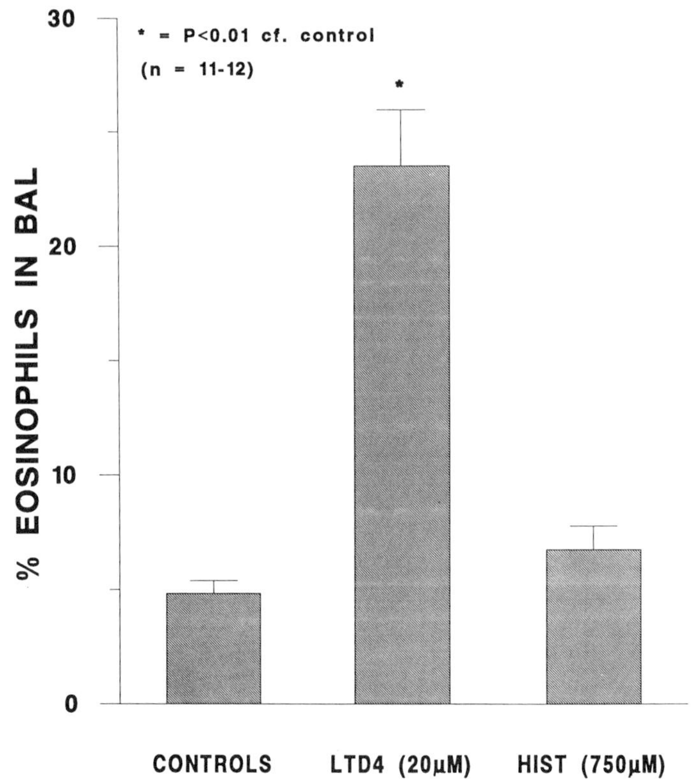

FIGURE 1. Lung eosinophilia induced in guinea pigs by aerosol administration of LTD_4, but not histamine or phosphate buffered saline (controls). Conscious guinea pigs were exposed to aerosol LTD_4, histamine, or vehicle for 65 s, killed 18 h later, their lungs lavaged, and eosinophils counted as percent of total lavagable white cells. Values are mean ± SEM of *n* animals.

Eosinophil Migration

Evaluation of eosinophil immigration into guinea pig lung in response to aerosol administration of LTD_4 was accomplished using methods similar to those described previously.[18] In brief, conscious guinea pigs were placed in a Plexiglas chamber that allowed head-only exposure. After pretreatment with the appropriate pharmacologic agent(s), they were exposed to an aerosol of LTD_4. At varying time intervals following exposure, the animals were killed, and the lungs and airways were lavaged with phosphate-buffered saline (2 ml × 5). The lavage was centrifuged (700 × g × 10 min) and the pellet resuspended in 200 µl of buffer. Slides were prepared in a Hemaspinner 0107, and eosinophils were stained with eosin Y and methylene blue. Eosinophils were counted as a percentage of total white cells recovered.

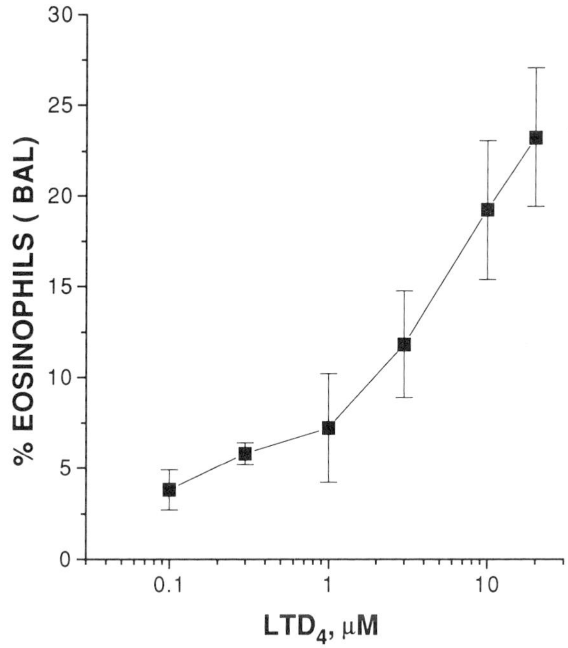

FIGURE 2. Eosinophilia induced in guinea pig lungs by aerosol administration of various concentrations of LTD_4. LTD_4 was administered for 65 s, the animals killed 24 h later, their lungs lavaged, and eosinophils counted as percent of total lavagable white cells. Basal levels of eosinophils in this study were 3.6% of lavagable white cells. Values are mean ± SEM of 3-6 animals.

Measurement of Airway Vascular Permeability

Guinea pigs were anesthetized with urethane (1.5 g/kg, ip) and the right jugular vein cannulated for iv drug and dye administration. Animals were pretreated iv with indomethacin (10 mg/kg) and propranolol (0.5 mg/kg) 20 and 15 min, respectively, before challenge with leukotriene D_4 (LTD_4), in order to decrease modulation of the response by cyclooxygenase products and activation of β-adrenergic receptors. Accolate was administered iv 10 min before LTD_4 or vehicle.

Each animal was given a single bolus injection of LTD_4 or vehicle 1 min after administration of Evans blue dye (30 mg/kg) and perfused with 0.9% sodium chloride for 2 min via the aorta beginning 5 min after challenge. The distal trachea (about 1 cm length above the bifurcation) was removed, weighed, and incubated at 60 ° in 2 ml formamide for 18 h. The concentration of Evans blue in the formamide was quantified using a Biomek 1000 spectrophotometer. Values for Evans blue were calculated as ng/mg of tissue, and those obtained after LTD_4 challenge were normalized by expressing them as multiples of those obtained after sham (LTD_4 vehicle) challenge.

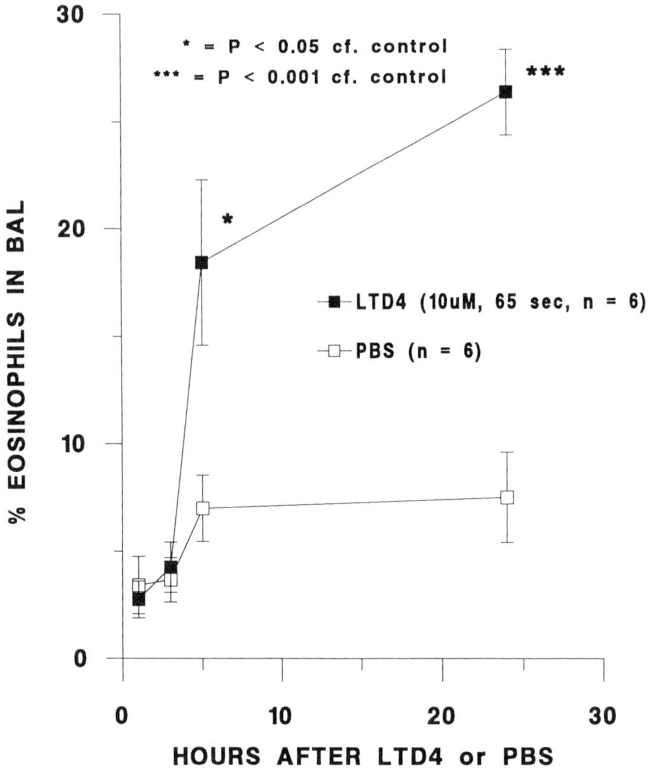

FIGURE 3. Time course of eosinophil influx into guinea pig lungs following aerosol administration of LTD_4 or vehicle (phosphate-buffered saline). See legend to FIG. 1 for additional details. Values represent mean ± SEM for *n* animals.

RESULTS

Previous studies have indicated that LTD_4 is chemotactic for eosinophils *in vivo* when injected into the eye of a guinea pig[17] or administered by aerosol to guinea pig airways.[18] Accordingly, the ability of aerosolized LTD_4 to induce a pulmonary eosinophilia was confirmed in this species. As illustrated in FIGURE 1, aerosol administration of a 20-μM solution of LTD_4 to conscious guinea pigs produced a substantial influx of eosinophils into the lung as assessed by lavage. In contrast, aerosol administration of the bronchoconstrictor histamine produced no increase in lung eosinophils. It is noteworthy that only modest, inconsistent increases in lavagable eosinophils were apparent when platelet activating factor (0.1 μM-6 mM) or LTB_4 (0.01-30 μM) were aerosolized into the lung (data not shown).

FIGURE 2 illustrates that eosinophilia in response to aerosol administration of LTD_4 appeared to be concentration dependent. One of the difficulties with these experiments is the variability of the response to LTD_4 and the highly variable numbers

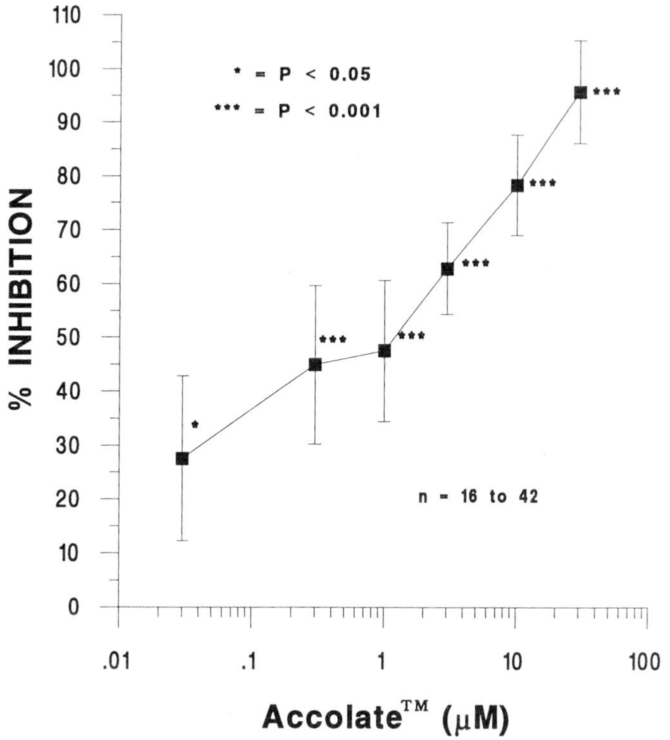

FIGURE 4. Inhibition of aerosol LTD_4-induced lung eosinophilia in guinea pigs by pretreatment with aerosol Accolate™. Various concentrations of aerosolized Accolate were administered for 5 min, 10 prior to LTD_4 administration. See legend to FIG. 1 for additional details.

of eosinophils that exist in the lungs of control animals. Basal numbers of eosinophils vary from 0% to >15% of lavagable white cells. Aerosol LTD_4 consistently increased the number of lavagable eosinophils above basal; however, because of the variability in basal, the fold-increase ranged from 2 to 6. It should also be noted that from experiment to experiment the concentrations of LTD_4 required to induce eosinophilia varied by as much as 10-fold.

FIGURE 3 illustrates the temporal response to LTD_4. A 10-μM solution of LTD_4 was aerosolized for 65 s and animals lavaged at varying time intervals for up to 24 h after exposure. As is apparent, eosinophils appeared to begin moving into the lung between 3 and 5 h following administration. However, they appeared to continue to accumulate for up to 24 h, the latest time point measured. The response to aerosol administration of the vehicle, phosphate-buffered saline, was much smaller than to LTD_4.

To explore potential pharmacologic mechanisms involved in the response to LTD_4, the histamine H_1-receptor antagonist pyrilamine (1 mg/kg, ip, 30 min prior), the cyclooxygenase inhibitor indomethacin (10 mg/kg, iv, 30 min prior), the PAF

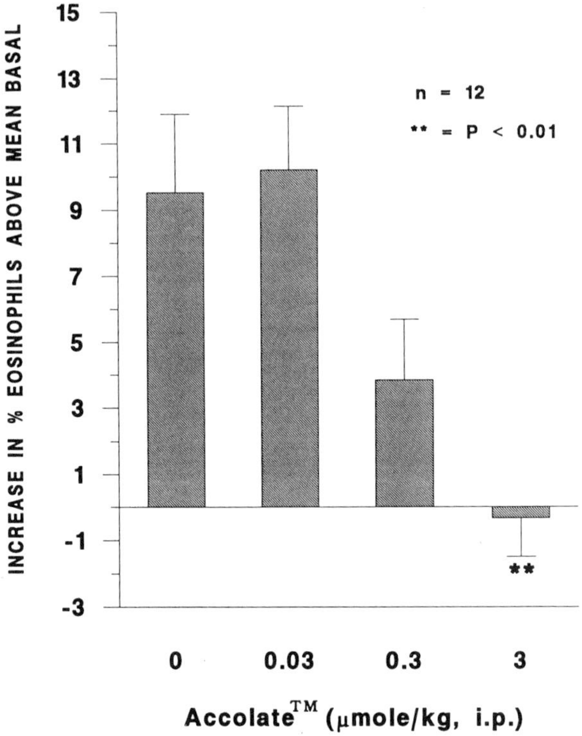

FIGURE 5. Inhibition of aerosol LTD_4-induced eosinophilia in guinea pig lungs by pretreatment with Accolate™ (ip). Accolate was administered 5 min prior to exposure to LTD_4. See legend to FIG. 1 for additional details.

antagonist WEB 2086 (10 μmole/kg, iv, 30 min prior), and dexamethasone (0.3 mg/ml, 5 min aerosol administered 24, 18, 1 h prior to and 1 h after LTD_4), and the LTB_4 receptor antagonist LY 255,283 (30 mg/kg, po) were examined for their capacity to inhibit LTD_4-induced eosinophil influx. None of these agents altered significantly the influx of eosinophils in response to LTD_4.

The ability of Accolate administered prior to exposure to LTD_4 to inhibit this response was evaluated. As illustrated in FIGURE 4, aerosol administration of Accolate 5 min prior to LTD_4 provided a concentration-dependent inhibition of the eosinophilia. Similarly, when administered by the intraperitoneal route, Accolate also produced an apparent dose-dependent inhibition of the response (FIG. 5). Finally, the ability of LTC_4 and LTE_4, administered topically, to induce eosinophil immigration into guinea pig lungs and the ability of Accolate to inhibit were evaluated. As illustrated in FIGURE 6, both LTC_4 and LTE_4 (20 μM, 65-s aerosol) produced an eosinophil recruitment to the lung. Prior aerosol administration of a 30-μM solution of Accolate completely inhibited the response to both LTC_4 and LTE_4.

As peptide leukotrienes are well-known edemagenic agents, it was also of interest to determine if Accolate was capable of inhibiting LTD_4-induced edema in guinea

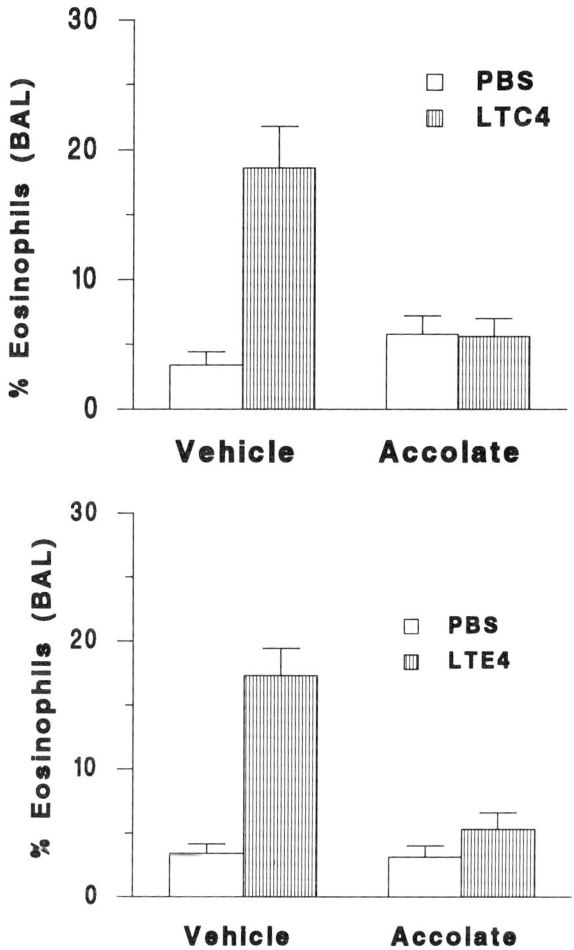

FIGURE 6. Inhibition of LTC_4- and LTE_4-induced eosinophil influx into guinea pig lung by Accolate™. LTC_4 (**upper panel**) and LTE_4 (**lower panel**) (both 20 μM) were aerosolized for 65 s following aerosol administration of Accolate (30 μM, 5 min) or vehicle. See legend to FIG. 1 for additional details.

pig trachea. In the presence of indomethacin and propranolol, iv LTD_4 produced a dose-dependent increase in the content of Evan's blue dye of tracheal tissue. (FIG. 7). This response was inhibited in a dose-dependent manner by prior iv administration of Accolate.

DISCUSSION

The steady increase in positive clinical phase II and III studies with modulators of the 5-lipoxygenase pathway of arachidonic acid metabolism, namely, leukotriene

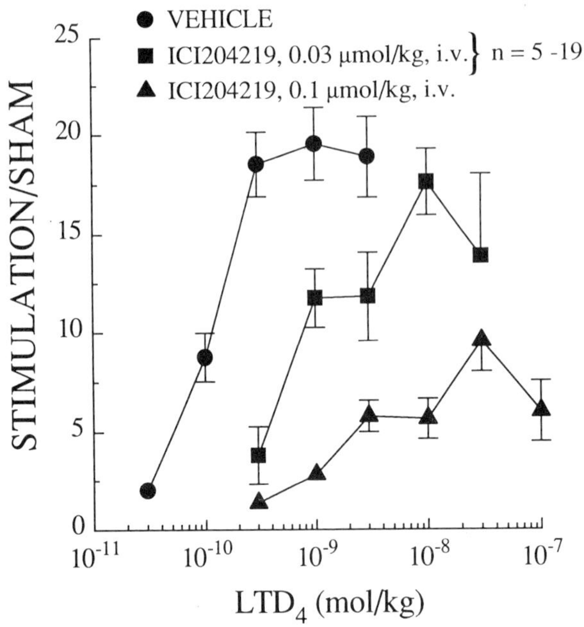

FIGURE 7. Inhibition of LTD_4-induced edema in guinea pig trachea by Accolate™. Guinea pigs received pretreatment with 0.5 mg/kg, iv, propranolol, and 10 mg/kg, iv, indomethacin and either Accolate™ or vehicle 30, 25, and 20 min, respectively, prior to initiating iv dose-response curves with LTD_4. Symbols with *vertical lines* represent the mean ± SEM of 5-19 animals.

synthesis inhibitors or receptor antagonists, argues strongly for the involvement of these lipid mediators in the allergic disease process. Consequently, it is of interest to further explore the preclinical biology of these modulators to better define their mechanism(s) of action, which may ultimately lead to a better understanding of pathophysiologic processes.

Previous reports indicated that LTD_4 was chemotactic for guinea pig eosinophils *in vivo.*[17,18] Moreover, there are similar reports of lung eosinophilia in humans following aerosol administration of LTE_4.[19] The present findings confirm that LTD_4 is chemotactic for guinea pig eosinophils. LTC_4 and LTE_4 also appear to possess similar properties. The chemotactic response to all three peptide leukotrienes can be inhibited by prior administration of Accolate. Similarly Accolate antagonized the ability of LTD_4 to produce edema. These findings reaffirm the proinflammatory properties of peptide leukotrienes and indicate that modulators of products of the 5-lipoxygenase pathway may be capable of exerting an antiinflammatory activity. A pivotal next step is to determine whether peptide leukotrienes contribute to antigen-induced eosinophilia and airway edema in the guinea pig model. A positive outcome in these studies would establish a reasonable background for conducting similar studies in the clinic.

ACKNOWLEDGMENTS

The authors express their appreciation to Ms. Sharon Bailor and Ms. Terri Bryden for their assistance in the preparation of this manuscript.

REFERENCES

1. MURPHY, R. C., S. HAMMARSTROM & B. SAMUELSSON. 1979. Leukotriene C: A slow reacting substance from murine mastocytoma cells. Proc. Natl. Acad. Sci. USA **76:** 4275-4279.
2. KRELL, R. D., D. AHARONY, C. K. BUCKNER, R. A. KEITH, E. J. KUSNER, D. W. SNYDER, P. R. BERNSTEIN, V. G. MATASSA, Y. K. YEE, F. J. BROWN, B. HESP & R. E. GILES. 1990. The preclinical pharmacology of ICI 204,219: A peptide leukotriene antagonist. Am. Rev. Respir. Dis. **141:** 978-987.
3. WEICHMAN, B. M., M. A. WASSERMAN, D. A. HOLDEN, R. R. OSBORN, D, F. WOODWARD, T. W. KU & J. G. GLEASON. 1983. Antagonism of the pulmonary effects of peptidoleukotrienes by a leukotriene D_4 analog. J. Pharmacol. Exp. Ther. **227:** 700-705.
4. MONG, S., H-L. WU, J. MILLER, R. F. HALL, J. G. GLEASON & S. T. COOKE. 1987. SKF 104353, a high affinity antagonist for human and guinea pig lung leukotriene D_4 receptor, blocked phosphatidylinositol metabolism and thromboxane synthesis induced by leukotriene D_4. Mol. Pharmacol. **32:** 223-229.
5. JONES, T. R., R. ZAMBONI, M. BELLEY *et al.* 1989. Pharmacology of L-660,711 (MK571): A novel, potent and selective leukotriene D_4 receptor antagonist. Can. J. Physiol. Pharmacol. **67:** 17-28.
6. MCMILLAN, R. M., K. E. SPRUCE, G. C. CRAWLEY, E. R. H. WALKER & S. J. FOSTER. 1992. Preclinical pharmacology of ICI D2138, a potent orally-active non-redox inhibitor of 5-lipoxygenase. Br. J. Pharmacol. **107:** 1042-1047.
7. GILLARD, J., A. W., FORD-HUTCHINSON, C. CHAN, S. CHARLESON, D. DENIS, A. FOSTER, R. FARTIN, S. LEGER, C. S. MCFARLANE, H. MARTIN, H. PRECHUTA, D. RIENDEAU, C. A. ROUZER, R. YOUNG, D. E. MACINTYRE, L. PETERSON, T. BACH, G. EIRMANN, S. HOPPLE, J. HUMES, L. HOPE, S. LUELL, J. METZGER, R. MEUER, D. K. MILLER, E. OPAS, S. PACHOLOK. 1989. L-663,556 (MK-886), a novel orally active leukotriene biosynthesis inhibitor. Can. J. Physiol. Pharmacol. **67:** 456-464.
8. MANNING, P. J., R. M. WATSON, D. J. MARGOLSKEE, V. C. WILLIAMS, J. I. SCHWARTZ & P. M. O'BRYNE. 1990. Inhibition of exercise-induced bronchoconstriction by MK-571, a potent leukotriene D_4 receptor antagonist. N. Engl. J. Med. **323:** 1736-1739.
9. ISRAEL, E., R. DERMARKARIAN, M. ROSENBERG, R. SPERLING, G. TAYLOR, P. RUBIN & J. M. DRAZEN. 1990. The effects of a 5-lipoxygenase inhibitor on asthma induced by cold, dry air. N. Engl. J. Med. **323:** 1740-1744.
10. TAYLOR, I. K., K. M. O'SHAUGHNESSY, R. W. FULLER & C. T. DOLLERY. 1991. Effect of cysteinyl-leukotriene receptor antagonist ICI 204,219 on allergen-induced bronchoconstriction and airway hyperreactivity in atopic subjects. Lancet **337:** 690-694.
11. FINNERTY, J. P., R. WOOD-BAKER, H. THOMSON & S. T. HOLGATE. 1992. Role of leukotrienes in exercise-induced asthma: Inhibitory effect of ICI 204,219, a potent leukotriene D_4 receptor antagonist. Am. Rev. Respir. Dis. **145:** 746-749.
12. FINDLAY, S. R., J. M. BARDEN, C. B. EASLEY & M. GLASS. 1992. Effect of the oral leukotriene antagonist, ICI 204,219, on antigen-induced bronchoconstriction in subjects with asthma. J. Allergy Clin. Immunol. **89:** 1040-1045.
13. MAKKER, H. K., L. C. LAU, H. W. THOMSON, S. M. BINKS & S. T. HOLGATE. 1993. The protective effect of inhaled leukotriene D_4 receptor antagonist ICI 204,219 against exercise-induced asthma. Am. Rev. Respir. Dis. **147:** 1413-1418.

14. O'Shaughnessy, K. M., I. K. Taylor, B. O'Connor, F. O'Connell, H. Thomson & C. T. Dollery. 1993. Potent leukotriene D_4 receptor antagonist ICI 204,219 given by the inhaled route inhibits the early but not the late phase of allergen-induced bronchoconstriction. Am. Rev. Respir. Dis. **147:** 1431-1435.
15. Friedman, B. S., E. H. Bel, A. Buntinx, W. Tanaka, Y-H. R. Han, S. Shingo, R. Spector & P. Sterk. 1993. Oral leukotriene inhibitor (MK-886) blocks allergen-induced airway responses. Am. Rev. Respir. Dis. **147:** 839-844.
16. Christie, P. E., B. W. Spur & T. H. Lee. 1991. The effect of inhalation of the leukotriene receptor antagonist, SK&F 104353, on leukotriene C_4- and leukotriene E_4-induced bronchoconstriction in subjects with asthma. J. Allergy Clin. Immunol. **88:** 193-198.
17. Chan, C-C., K. McKee, P. Tagari, P. Chee & A. Ford-Hutchinson. 1990. Eosinophil-eicosanoid interactions: Inhibition of eosinophil chemotaxis in vivo by an LTD_4 receptor antagonist. Eur. J. Pharmacol. **191:** 273-280.
18. Chan, C. C. & A. Foster. 1991. Peptide leukotriene involvement in pulmonary eosinophil migration upon antigen challenge in the actively sensitized guinea pig. Int. Arch. Allergy Appl. Immunol. **96:** 279-284.
19. Laitinen, L. A., A. Laitinen, T. Haahtela, V. Vilkka, B. W. Spur & T. H. Lee. 1993. Leukotriene E_4 and granulocytic infiltration into asthmatic airways. Lancet **341:** 989-990.

Modulation of Eicosanoid Biosynthesis by Novel Pyridinylpyrimidines

WALTER C. PICKETT, CHRISTINE SILVERSTRIM, CATHERINE DONDERO, AND LAWRENCE TORLEY

Oncology Section
Lederle Laboratories
American Cyanamid
Pearl River, New York 10965

A simple but effective strategy for drug development has been controlling those molecules that cause disease either through specific antagonists or synthesis inhibitors. Accordingly, structural elucidation of key mediators profoundly affects this approach by providing clear targets. The 1911 discovery of histamine is a classic example, eventually leading to a number of specific antagonists that remain therapeutically useful today. More recently, the discovery of the slow-reacting substance of anaphylaxis (SRS-A) as the sulfidopeptidyl leukotrienes (LT),[1] and the rabbit aorta-constricting substance as a mixture of thromboxane A_2 (TXA_2), prostaglandin H_2 (PGH_2), and PGG_2,[2] for example, have led to the synthesis of a number of specific inhibitors and antagonists with therapeutic potential. Although specific inhibition or antagonism is advantageous with respect to toxicity by minimally affecting other homeostatic processes, it is often insufficient due to redundancy of mediators of diverse origin and structure such as histamine and LTC. Because of this, an alternative to specific inhibitors and antagonists is the design of cell "stabilizing" agents or mediator release inhibitors. These agents prevent the release of both preformed mediators and the biosynthesis of other mediators such as eicosanoids.

The demonstration that chromolyn sodium—a useful drug for the management of asthma—inhibits mediator release from antigen-stimulated mast cells[3] greatly bolstered the mast cell-based search for mediator release inhibitors, despite clear evidence that this was chromolyn's mechanism of action in asthma. A new generation of mediator release inhibitors containing a pyridinylpyrimidine structure is described below,[4] but unlike chromolyn, the pyridinylpyrimidines also inhibit basophils and phagocytes. This series of compounds inhibits eicosanoid biosynthesis primarily at levels proximal to the phospholipase A_2 (PLA_2), but to a lesser extent at distal sites such as the thromboxane synthetase. From *in vivo* and *in vitro* data, the pyridinylpyrimidines appear to be a promising new class of compounds for development as therapy for allergic and inflammatory diseases where eicosanoids and granule release are essential components of the etiology.

STRUCTURE-FUNCTION RELATIONSHIPS

The 2-, 3-, and 4-pyridinylpyrimidines are potent inhibitors of antigen-induced histamine release from *in vitro* preparations of human leukocyte containing basophils.[4]

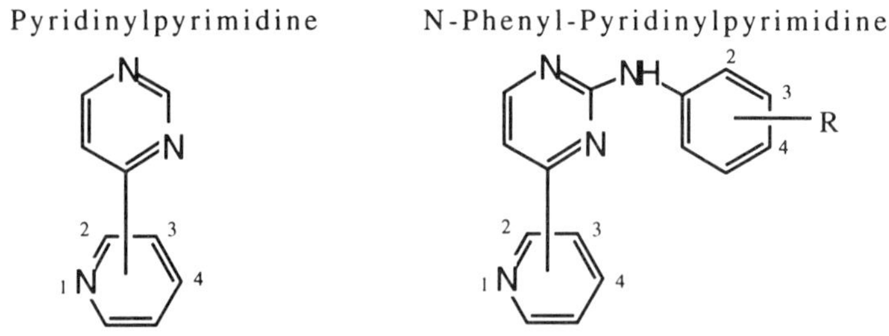

FIGURE 1. Substituted pyridinylpyrimidines.

Although various nonpyridinyl derivatives such as 2-furyl, 4-thienyl, and miscellaneous heterocycles are also efficacious, they are less thoroughly characterized and, therefore, will not be included in this discussion. With respect to structure-function relationships, activity is significantly improved when phenyl substituents or substituted phenyl substituents are attached to the 2-amino group (see general structures in FIG. 1 and activity in TABLE 1). For the 2-pyridinyl series, the unsubstituted N-phenyl is inactive, while a variety of substituted phenyls irrespective of electronic

TABLE 1. Summary of Structure-Activity Relationships

Compound	Pyridinyl Position[a]	R[a]	Baso EC_{50} (μM)[b]
A	2	H	>5
B	2	4-F	>5
C	2	3-F	1.0
D	2	4-ethyl	2.1
E	2	3-ethyl	1.4
F	2	4-*N*-dimethyl	3.3
G	2	3-*N*-dimethyl	0.4
H	2	4-(1-imidazole)	1.4
I(**1**)[c]	2	3-(1-imidazole)	0.3
J(**2**)[c]	3	H	0.9
K	4	H	0.9
L	4	4-F	1.9
M	4	3-F	0.7
N	4	4-methyl	>5
O	4	3-methyl	8.8
P	4	4-(1-imidazole)	7.6
Q	4	3-(1-imidazole)	0.6
R	4	4-amino, 3-methyl	0.9

[a] See FIGURE 1.

[b] Micromolarity of compound that inhibits 50% of the antigen-induced histamine release in mixed human leukocytes.[4,5]

[c] Compounds I and J are referred to as **1** and **2** within the text.

1 **2**

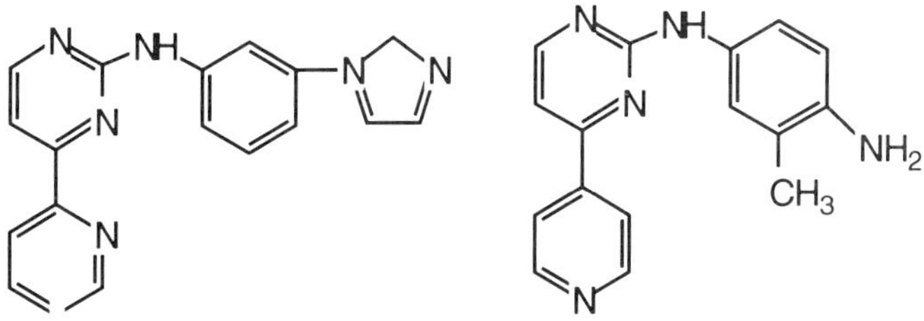

FIGURE 2. *In vivo* leukotriene biosynthesis inhibitors.

effects consistently improve activity. Of these, the meta substituents yield greater inhibition than ortho or para (TABLE 1, see A through I). Similarly, the meta substituents increase potency for the 4-pyridinyl series (TABLE 1, see L through Q), but not for the 3-pyridinyl series, where there are many exceptions to this rule (data not shown). In contrast to the 2-pyridinyl series, the 3- and 4-pyridinylpyrimidines do not require the N-phenyl substitution for activity (TABLE 1, see J and H).

SCREENING FOR *IN VIVO* INHIBITION OF LEUKOTRIENE BIOSYNTHESIS

Because of the important role eicosanoids play in the inflammatory and allergic responses, we chose to identify *in vivo* leukotriene inhibitors among this new series of compound—a selected few are listed in TABLE 1. These compounds were selected for their ability to inhibit leukotriene biosynthesis after oral administration. Although this approach is cumbersome and complicates structure-function relationships due to additional considerations such as bioavailability and toxicity, it was chosen because of ease of further development. Compounds defined by this criterion lack gross toxicity, are orally bioavailable, as well as inhibit leukotriene biosynthesis. Specifically, the assay involves treating mice orally with compound or placebo, then challenging them with an intraperitoneal (ip) injection of zymosan (100 mg/kg). After peritoneal lavage and extraction, LTE is measured by HPLC using non-end capped octadecyl silane columns. During the zymosan challenge, peritoneal peptidases efficiently convert nascent LTC to LTE. Although inhibitors of these peptidases could be wrongly interpreted as inhibitors of more proximal enzymes, studies with exogenous LTC indicate that, in general, the pyridinylpyrimidines do not affect peptidase activity.

Two active compounds of this series are shown in FIGURE 2: the first, **1,** is a 2-pyridinylpyrimidine containing an imidazole-substituted phenyl; and the second, **2,** is a 4-pyridinylpyrimidine containing a methyl, amino-substituted phenyl. In a dose-dependent fashion, compounds **1** and **2** inhibit zymosan-induced rat peritoneal LTE

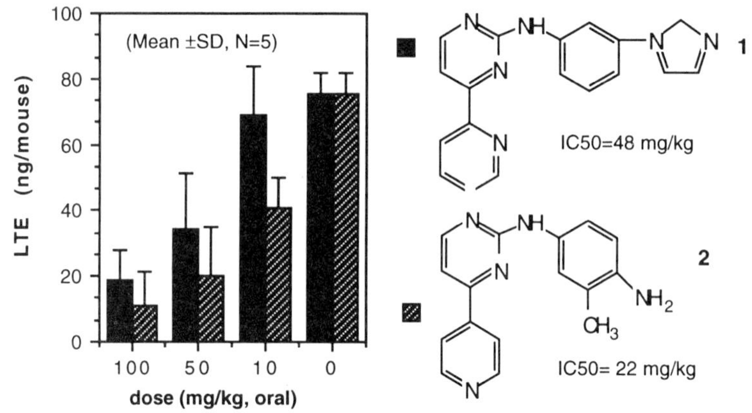

FIGURE 3. Effect of pyridinylpyrimidines (compounds **1** and **2**) on zymosan-induced LTE biosynthesis in the mouse peritoneum. Shown on the Y-axis is the amount of LTE produced per mouse challenged with zymosan (ip). Shown on the X-axis is the oral dose of the indicated compound. Each *column* represents the mean and standard deviation of five mice. See text for additional details.

biosynthesis with IC_{50}s of 49 and 18 mg/kg, respectively (FIG. 3). Note in TABLE 1 and FIGURE 3 the reversal in relative potencies. For *in vitro* histamine release, **1** is three times more potent than **2,** while **2** is twice as potent as **1** in the *in vivo* assay. This may be related to the fact that histamine release is basophil and IgE dependent, while LTE biosynthesis, as measured by the mouse peritoneal assay, is complement and phagocyte dependent. Alternatively, this difference may reflect subtle differences in bioavailability of **1** and **2.** Concerning bioavailability, **1** and **2** are efficacious by many routes including oral, ip, and subcutaneous (sc). Protonation (0.05 M HCl) is required for ip and sc efficacy, while gastric acids are sufficient to protonate the free base after oral administration.

EFFECT OF COMPOUNDS 1 AND 2 ON PLATELET EICOSANOIDS

We tested the effect of **1** and **2** on eicosanoid biosynthesis in platelets and PMNs. In both of these systems, we specifically examined the effect of **1** and **2** on the release of endogenous and the oxygenation of exogenous arachidonic acid (20:4) such that drug effects either distal or proximal to the PLA_2 could be detected. For the platelet system, the release of endogenous 20:4 after A23187 (10 μM) treatment served as a measure of the compounds' effect or lack of effect on the PLA_2, while the inhibition of eicosanoid biosynthesis derived from incubations with exogenous 20:4 served as a measure of effects distal to PLA_2. Endogenous 20:4 release, measured by gas chromatography, was unaffected by either **1** or **2** at concentrations up to 50 μg/ml (data not shown). Furthermore, neither **1** or **2** affected thrombin-induced 20:4 release, suggesting that **1** or **2** does not directly affect PLA_2 or proximal events in platelet signal transduction.

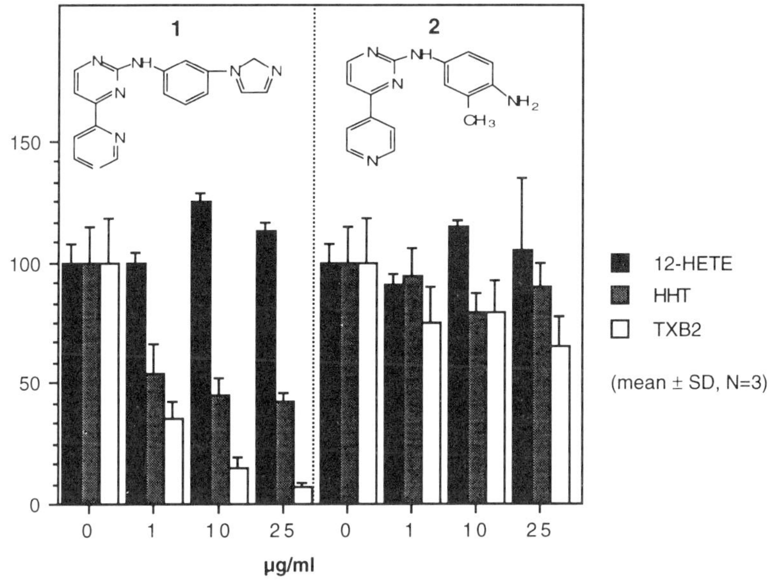

FIGURE 4. Effect of compounds **1** and **2** on platelet eicosanoids derived from exogenous arachidonic acid. Shown on the Y- and X-axes are the % of control and oral dose of either **1** and **2,** respectively. Each *column* represents the mean and standard deviation of triplicate measurements from platelets incubated 20 min with 50 μM 20 : 4.

In contrast, **1** and **2** differentially affected the oxygenation of exogenous 20:4. While 12(S)-hydroxyeicosatetraenoic acid (12-HETE), a 12-lipoxygenase product, was unaffected by either **1** or **2,** the cyclooxygenase-dependent products 12-hydroxy-heptadecatrienoic acid (HHT) and TXB_2 were significantly inhibited by **1**—with respective IC_{50}s of 1 and 0.5 μg/ml—but uninhibited by **2** (FIG. 4). Because HHT can be derived from both the cyclooxygenase-dependent endoperoxides PGG_2/PGH_2 and TXA_2,[6] it is difficult to determine whether **1** specifically inhibits the thromboxane synthetase or whether both the cyclooxygenase and the thromboxane synthetase are inhibited. The dose response data (FIG. 4), however, suggests the former, since TXB_2 is more completely inhibited (90%) at higher doses than HHT (45%). In general, cyclooxygenase inhibitors equally inhibit HHT and TXB_2 biosynthesis. Finally, the thromboxane inhibition may be due to the presence of the imidazole moiety, a known TXB_2 synthetase inhibitor,[7] in **1** but not in **2.** Surprisingly, **1** inhibits TXB_2 synthesis more potently than free imidazole.

EFFECT OF COMPOUNDS 1 AND 2 ON PMN EICOSANOIDS

We tested the effect of **1** and **2** on eicosanoids derived from caseinate-elicited rat peritoneal PMNs, analogous to the platelet system. Release of 20:4 after A23187 treatment served as a measure of the compounds' direct effect on the PLA_2, while effects on LTB_4 and 5-HETE derived from exogenous 20:4 served as a measure of enzyme inhibition distal to the PLA_2, such as the 5-lipoxygenase. In these experiments, PMNs were treated with 2.5 μM A23187 in the presence of 0.25% fatty acid-free albumin before measuring the endogenous 20:4 by gas chromatography. To measure the effects of **1** and **2** on products derived from exogenous 20:4, the PMNs were treated with 10 μM A23187 and 50 μM 20:4 in calcium-containing buffer. The combination of A23187 and 20:4 was required because PMNs, unlike the platelet, poorly oxygenate 20:4. A23187 greatly facilitates the oxygenation of exogenous 20:4 by activating the 5-lipoxygenase[8] with minimal (<1%) release of endogenous 20:4, which was determined from arachidonate labeling experiments.

In A23187-treated cells, neither **1** nor **2** affected the release of endogenous 20:4 when tested at concentrations up to 50 μg/ml (data not shown). Similarly, **1** or **2** did not significantly affect the conversion of exogenous 20:4 to 5-HETE or LTB, indicating a lack of effect on the 5-lipoxygenase, the 5-HPETE reductase, or the 5-lipoxygenase-activating protein. Furthermore, there was no significant change in the ratio of the two nonenzymatic hydrolysis products of LTA_4, 6-t-LTB_4 or 12-epi-6-t-LTB_4, to LTB_4, additionally suggesting that **1** and **2** do not affect LTA hydrolysis or synthesis.

CONCLUSIONS

Pyridinylpyrimidines **1** and **2** potently inhibit preformed mediator release and eicosanoid biosynthesis in a number of *in vivo* and *in vitro* models. Although the primary site of action is not clearly defined, it may be common to those events required for histamine release and PLA_2 activation; furthermore, the inhibitory effects are greater for IgE-dependent cells and phagocytes. Neither **1** nor **2** have any effect on agonist (thrombin)- or ionophore-induced 20:4 release from platelets. Only **1** inhibits platelet eicosanoids (thromboxane) derived from exogenous 20:4, which is probably related to the imidazole substituents. Unexpectedly, the effect of **1** on thromboxane synthesis inhibition is greater than that of imidazole alone.[7]

Neither **1** or **2** affect the oxygenation of exogenous 20:4 or the A23187 induced release of endogenous 20:4 in PMNs, which was similar to what was found in the platelet system. However, as shown in FIGURE 3, both **1** and **2** inhibit LTE biosynthesis in agonist-stimulated peritoneal cells. Accordingly, this data forms the working hypothesis that **1** and **2** are mediator release inhibitors with preferential effects on IgE-containing (TABLE 1) and phagocytic cells (FIG. 3), respectively.

Finally, since these compounds are orally active with no overt signs of toxicity, **1** and **2** are both excellent candidates for development as new treatment for asthma and inflammatory diseases where eicosanoids are part of the etiology.

ACKNOWLEDGMENTS

The authors acknowledge Florence Nocera for her excellent technical assistance.

REFERENCES

1. MURPHY, R. C. *et al.* 1979. Proc. Natl Acad. Sci. USA **76:** 4275-4279.
2. HAMBERG, M. *et al.* 1975. Proc. Nat. Acad. Sci. USA **72:** 2994-2998.

3. SHAPIRO, G. G. & P. KONIG. 1985. Pharmacotherapy **5:** 156-170.
4. PAUL, R. *et al.* 1993. J. Medicinal Chem. **36:** 2716-2725.
5. SIRAGANIAN, R. P. *et al.* 1977. J. Immunol. **119:** 2078-2083.
6. VINCENT, J. *et al.* 1980. *In* Advances in Prostaglandin and Thromboxane Research, Vol. 6. B. Samuelsson, P. Ramwell & R. Paoletti, Eds.: 463-466.
7. NEEDLEMAN, P. *et al.* 1975. Proc. Natl. Acad. Sci. USA **74:** 1716-1720.
8. BORGEAT, P. & B. SAMUELSSON. 1979. Proc. Natl. Acad. Sci. USA **76:** 2148-2152.

Effects of Saturated and Polyunsaturated Fatty Acids and Their 3-Thia Fatty Acid Analogues on MCF-7 Breast Cancer Cell Growth

F. ABDI-DEZFULI, R. K. BERGE,[a] M. RASMUSSEN, T. THORSEN, AND A. AAKVAAG

Division of Endocrinology
[a]Division of Biochemistry
Department of Clinical Biology
University of Bergen
N-5021 Bergen, Norway

Epidemiological[1] and experimental[2] studies have revealed an association between dietary fat and the incidence of breast cancer. ω-6 polyunsaturated fatty acids (PUFAs) appear to enhance tumor growth and proliferation, while ω-3 PUFAs seem to have an inhibitory influence on these processes. In addition, recent studies have shown that a non-β-oxidizable 3-thia analogue of palmitic acid (3-thia heptadecanoic acid; THDA) can markedly inhibit cancer cell growth *in vitro*.[3] This study has investigated the possible mechanisms by which these fatty acids may modulate mammary neoplasia.

A range of concentrations of palmitic acid, eicosapentaenoic acid (EPA), docosahexaenoic acid (DHA), and two 3-thia PUFAs were studied for their effects on the growth of MCF-7 breast cancer cells in culture. At the highest concentration of fatty acid used (64 μM) THDA was found to be the most effective of all fatty acids tested in inhibiting cell growth, while palmitate and DHA had no significant effect on cell growth (FIG. 1). In all cases the inhibitory effect of the fatty acid was found to be reversible. The inhibition brought about by THDA was not overcome by increasing concentrations of insulin-like growth factor 1 (IGF-1), insulin or estradiol (FIG. 2). Thus, THDA appears to inhibit MCF-7 cell growth via a pathway that is independent of the established means by which these growth stimulatory factors facilitate the growth of this cell line.

The growth of human mammary carcinomas maintained in immunodeficient mice has been found to be significantly curtailed when the animals are fed a diet containing high levels of fish oil.[4] This inhibitory effect was attributed to an increased incorporation of fish oil EPA and DHA into tumor cell membranes, which, it was suggested, increased their vulnerability to peroxidative damage and led, ultimately, to cell death. However, the results obtained in this study and by others[5] do not support such an assertion, since in the present investigation EPA inhibited cell growth to a small extent (FIG. 1e), while DHA, which has a higher degree of unsaturation compared

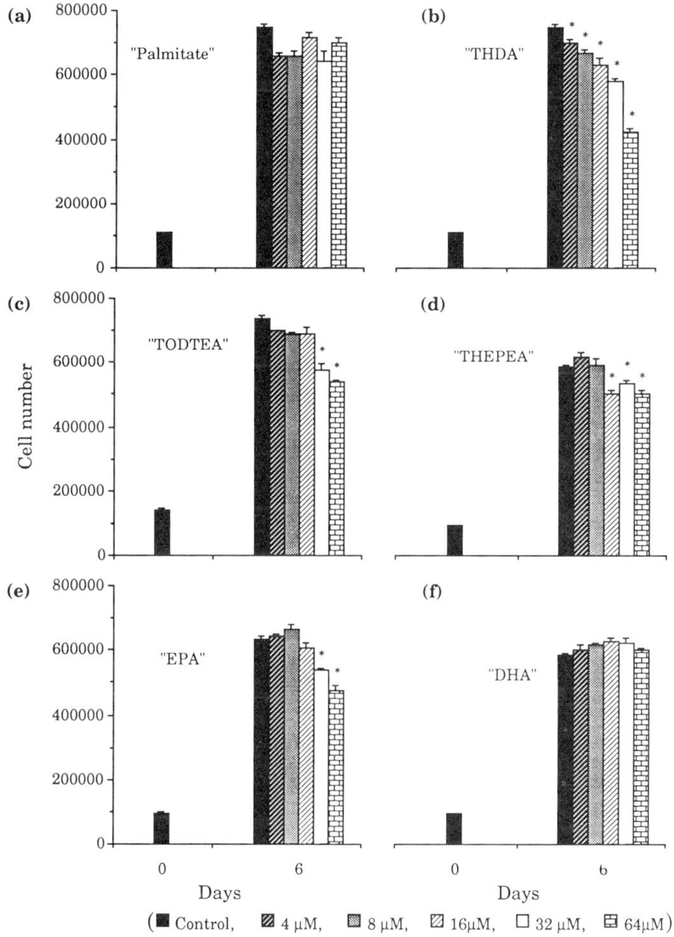

FIGURE 1. Effects of increasing concentrations of (**a**) palmitic acid (Palmitate); (**b**) 3-thia heptadecanoic acid (THDA); (**c**) 3-thia octadeca-6,9,12,15-tetraenoic acid (TODTEA); (**d**) 3-thia heneicosa-6,9,12,15,18-pentaenoic acid (THEPEA); (**e**) eicosapentaenoic acid (EPA); and (**f**) docosahexaenoic acid (DHA) on the growth of the MCF-7 breast cancer cell line. Cells were ''plated'' by incubating with Dulbecco's minimum essential medium (DMEM) containing 5% estradiol-lacking (charcoal-stripped) fetal calf serum, insulin (5 nM), and estradiol (10 pM) for 72 h (seeding density ~40000 cells/0.5 ml); and were subsequently cultured for 24 h with serum-free DMEM containing insulin (5 nM) and estradiol (10 pM), before being cultured (Day 0) in serum-free DMEM containing insulin (50 nM) and estradiol (0.5 nM) for a further six days. Each *bar* represents the mean (± SEM) number of cells present in six separate wells. *Asterisk,* significantly ($p < 0.01$) lower than control; determined using Student's unpaired *t*-test.

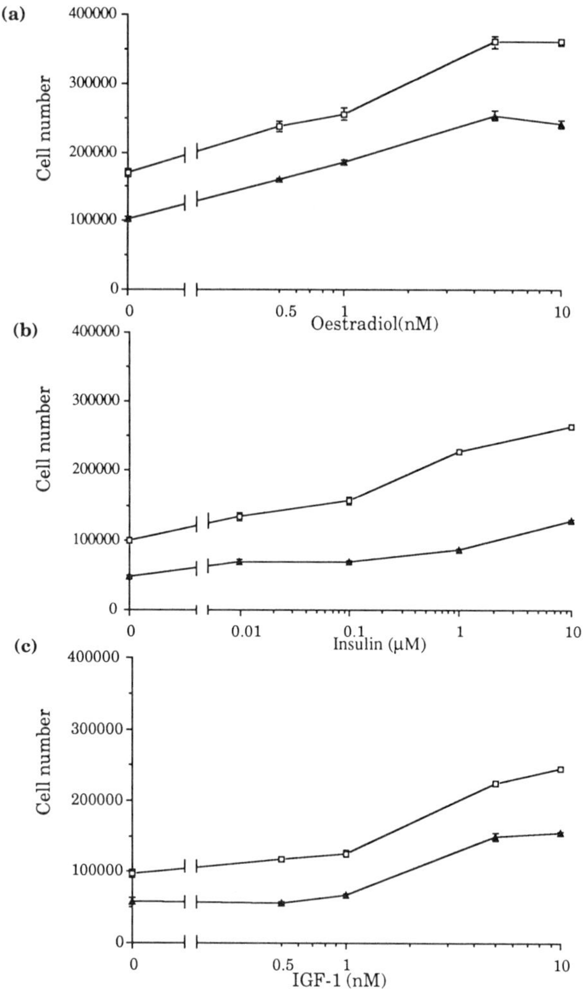

FIGURE 2. Effect of increasing concentrations of (**a**) estradiol, (**b**) insulin, and (**c**) insulin-like growth factor (IGF-1) on MCF-7 cells cultured in the absence (*squares*) and presence (*triangles*) of THDA (64 μM). Cells were "plated" by incubating with Dulbecco's minimum essential medium (DMEM) containing 5% estradiol-lacking fetal calf serum, insulin (5 nM) and estradiol (10 pM) for 24 h (seeding density ~40000 cells/0.5 ml), and were subsequently cultured for 24 h with serum-free DMEM containing insulin (5 nm) and estradiol (10 pM), before being cultured (Day 0) in serum-free DMEM containing insulin (50 nM) and estradiol (0.5 nM) for a further six days. Each *point* represents the mean (± SEM) number of cells present in six separate wells.

to EPA, had no effect on cell growth (FIG 1f). It is possible that these PUFAs inhibit cell growth by altering the profile of eicosanoids involved in cell growth, since both EPA and DHA block arachidonate metabolism. Previous studies carried out in this laboratory revealed that indomethacin was capable of inhibiting the growth of MCF-7 cells in a dose-dependent manner. Furthermore, in the present investigation, arachidonic acid (64 μM) was found to produce a marked inhibition of cell growth (data not shown). THDA, which is a potent peroxisome proliferator, has also been shown to increase hepatic H_2O_2 and lipid peroxide content.[6] It was, therefore, hypothesized that THDA may stem cell growth by increasing the level of oxidative stress within the cells. However, neither tocopherol, cysteamine (which raises glutathione levels), nor 1-buthionine-(S,R)-sulphoximine (BSO; a glutathione depleting agent) influenced the inhibitory effect of THDA on cell growth.

Overall, these results suggest that manipulation of the arachidonic acid cascade can affect the growth of these cells. Thus, whether or not THDA inhibits cell growth by interfering with growth mechanisms that involve arachidonic acid metabolism is currently being investigated in our laboratory.

REFERENCES

1. LANDS, W. E. M., T. HAMAZAKI, T. YAMAZAKI, H. OKUYAMA, K. SAKAI, Y. GOTO & V. S. HUBBARD. 1991. Am. J. Clin. Nutr. **51:** 991-993.
2. WELSCH, C. W. 1992. Cancer Res. **52**(Suppl.): 2040s-2048s.
3. NORRHEIM, L., H. SØRENSEN, K. GAUTVIK, J. BREMER & Ø. SPYDEVOLD. 1990. Biochem. Biophys. Acta **1051:** 319-323.
4. GONZALEZ, M. J., R. A. SCHEMMEL, L. DUGAN, J. I. GRAY & C. W. WELSCH. 1993. Lipids **28**(9): 827-832.
5. BÉGIN, M. E. & G. ELLS. 1992. Lipids **27**(2): 147-149.
6. DEMOZ, A., A. SVARDAL & R. K. BERGE. 1993. Biochem. Pharmacol. **45**(1): 257-259.

The X-Ray Structure and Biophysical Studies of a 15-Lipoxygenase[a]

J. C. BOYINGTON,[b] B. J. GAFFNEY,[c,d] L. M. AMZEL,[b]
K. S. DOCTOR,[e] D. V. MAVROPHILIPOS,[c]
Z. V. MAVROPHILIPOS,[c] A. COLOM,[c] AND
S. M. YUAN[c]

[b]*Biophysics and Biophysical Chemistry Department*
Johns Hopkins University School of Medicine
Baltimore, Maryland 21205

[c]*Chemistry Department*
[e]*Department of Biophysics*
Johns Hopkins University
Baltimore, Maryland 21218

The structure and stability of lipoxygenases have been characterized by solving the x-ray structure of soybean lipoxygenase-1 and by comparing electron paramagnetic resonance (EPR) spectral and scanning calorimetry data from soybean lipoxygenase and rabbit reticulocyte lipoxygenase. The iron chemistry and thermal stability of soybean and reticulocyte 15-lipoxygenases are similar.

X-RAY STRUCTURE

The x-ray structure[1] of soybean lipoxygenase-1, an arachidonic acid 15-lipoxygenase, shows that the catalytic iron is contained in the center of the larger of two domains. This domain of approximately 693 residues is a bundle of 22 α-helices and two β-sheet regions. It contains two cavities leading from the surface of the protein to the iron atom. One cavity, wide at the protein surface and narrow near iron, could provide a path for oxygen to reach the iron center. The other is long and bent, consistent with a role in fatty acid binding. The iron ligands are within the large domain: the ε-nitrogens of three histidines (residues 499, 504, and 690) and the carboxyl of the C-terminus, Ile 839. The smaller domain at the N-terminus is folded in a β-barrel. Plant and animal lipoxygenases differ in length at the N-terminus; the entire β-barrel domain and associated loops are apparently lacking in animal lipoxygenases. FIGURE 1 shows a series of tracings of the backbone of lipoxygenase-1 in the overall structure of the protein. The backbone of the long helix is shown

[a]This work was supported by National Institutes of Health Grant GM 36232.
[d]To whom correspondence should be addressed.

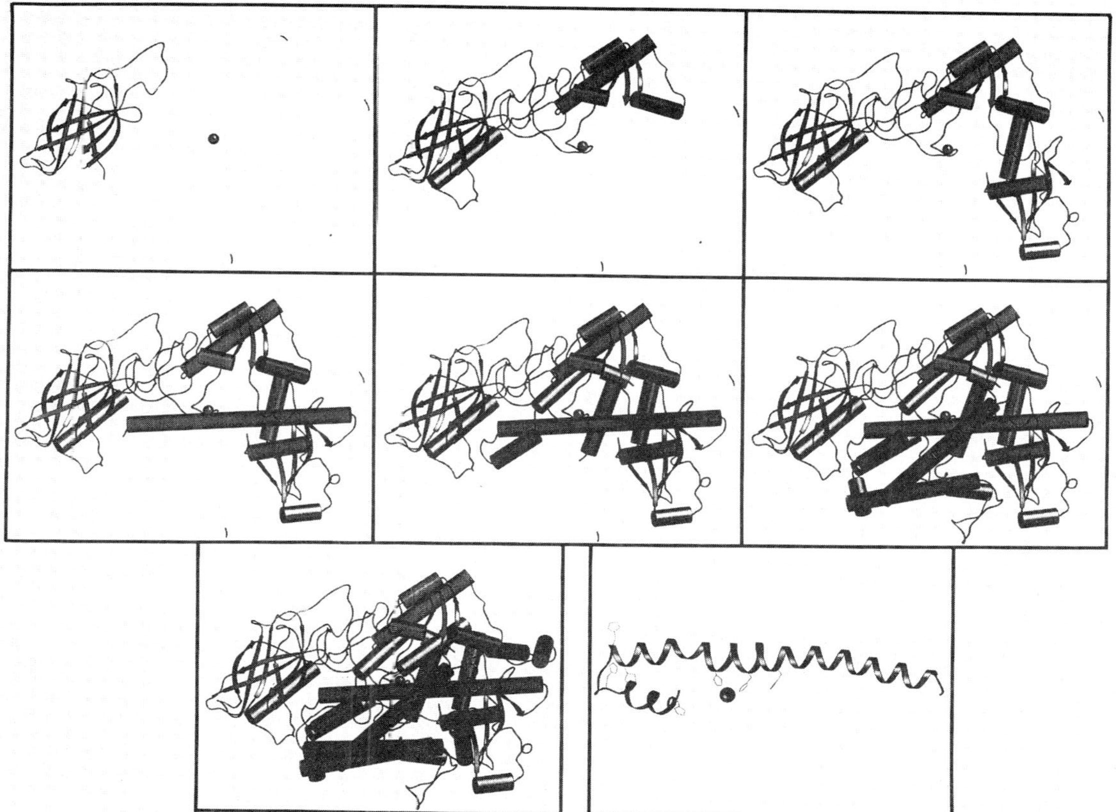

FIGURE 1. This sequence of figures shows a tracing of the lipoxygenase backbone with structured regions (β-sheets and helices) displayed. The sequence of displayed areas progresses in order from the N-terminus to the carboxyl end. The *sphere* in the middle is the iron atom. The sequence goes from upper left to upper right and then in this order from top to bottom. Only the N-terminal β-barrel is highlighted in the **upper left figure;** in the fourth figure (**middle left**), the long helix containing two histidine-iron ligands has been added; and the **next-to-last figure** contains all the secondary structural elements. The long helix and the region of conserved histidines in lipoxygenase sequences (residues 474–531 in L-1) are shown in the **lower right figure.** This figure has been rotated from the orientation in the rest of the figure by 90 degrees about an axis along the long helix.

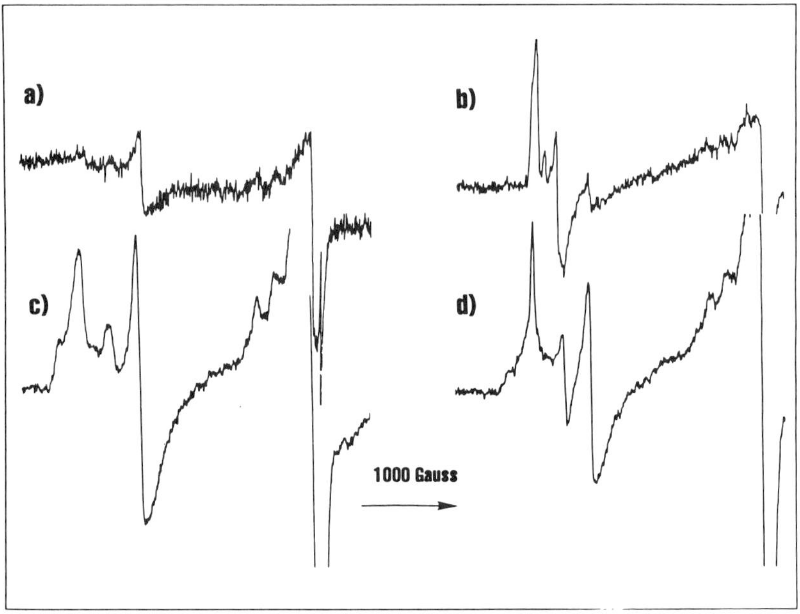

FIGURE 2. Electron paramagnetic resonance spectra of lipoxygenase purified by anion exchange chromatography and HPLC from rabbit reticulocytes are shown in **a, c, and d.** Native lipoxygenase has no significant spectrum except for a small background signal at g′ = 4.3 and background copper at g′ = 2. Addition of excess linoleic acid 13-hydroperoxide to the sample used for **a** gives spectrum **c.** In **c,** both ferric iron at low field and a large copper signal are evident. Addition of 10 mM 4-methyl pyridine (4-picoline) to the sample in **c** gives the spectrum **d.** Because of the heterogeneity in iron environment indicated in **c,** samples were examined by EPR at several stages in purification. The sample giving spectrum **b** was a crude preparation containing a green impurity close to lipoxygenase in SDS gel mobility (probably catalase); this sample was not subjected to hydroperoxide oxidation.

separately, highlighting the π helix region near the iron and the region rich in histidines.

ELECTRON PARAMAGNETIC RESONANCE

Axelrod showed that soybean lipoxygenase-1 contains a single ferrous iron atom that is converted to the ferric form when oxidized by one equivalent of the hydroperoxide product of lipoxygenase-1 action on linoleic acid.[3] We[2] and others[4] have shown that the electron paramagnetic resonance line shapes of lipoxygenases are a very sensitive function of anions and other ligands in the buffer. The EPR spectra in FIGURE 2 show that the iron center in rabbit reticulocyte lipoxygenase is similar to that in soybean lipoxygenase-1 in that the native enzyme appears to be ferrous (FIG. 2a) and gives an EPR signal only after oxidation with hydroperoxide (FIG. 2c). This ferric center is sensitive to the presence of ligands: FIGURE 2d is obtained after addition of 4-methyl pyridine to the sample giving FIGURE 2c. However, the reticulocyte lipoxygenase EPR spectra indicate more heterogeneity in the iron center than the spectra of the soybean enzyme. FIGURE 2b shows the EPR spectrum of the major impurity that is difficult to separate from reticulocyte lipoxygenase. This impurity appears to be catalase and is absent from gels of the samples giving spectra a, c, and d; but traces of this or of free heme in lipoxygenase samples cannot be ruled out completely.

SCANNING CALORIMETRY

The role of the second, N-terminal domain in thermal stability of the three isozymes of soybean lipoxygenase has also been studied by scanning calorimetry.[5] These three isozymes differ most in sequence in the N-terminal region. They each undergo a thermal transition that is modeled as a two-step process: a reversible unfolding followed by an irreversible kinetically limited denaturation. Reticulocyte lipoxygenase also exhibits similar thermal behavior, with a transition in the same range as the soybean isozymes.

REFERENCES

1. BOYINGTON, J. C., B. J. GAFFNEY & L. M. AMZEL. 1993. Science **260:** 1482-1486.
2. GAFFNEY, B. J., D. V. MAVROPHILIPOS & K. S. DOCTOR. 1993. Biophys. J. **64:** 773-783.
3. PISTORIUS, E. K., B. AXELROD & G. PALMER. 1976. J. Biol. Chem. **251:** 7144-7148.
4. CHASTEEN, N. D., J. K. GRADY, K. I. SKOREY, K. J. NEDEN, D. RIENDEAU & M. D. PERCIVAL. 1993. Biochemistry **32:** 9763-9771.
5. GAFFNEY, B. J., J. M. STURTEVANT, S. M. YUAN, D. M. LANG & E. DAGDIGIAN. 1994. Biophys. J. **66:** A178.

Linoleic Acid 8R-Dioxygenase and Hydroperoxide Isomerase of the Fungus *Gaeumannomyces graminis*

Mechanism of Catalysis and Inhibition

I. D. BRODOWSKY, L. Y. ZHANG,[a] E. H. OLIW, AND M. HAMBERG[a]

Department of Pharmaceutical Biosciences
Uppsala University
Uppsala, Sweden

[a]*Department of Physiological Chemistry*
Karolinska Institutet
Stockholm, Sweden

Supernatants of the fungus *Gaeumannomyces graminis* sequentially and efficiently converted linoleic acid to 8R-hydroperoxylinoleic acid and then to 7S,8S-dihydroxylinoleic acid and 8R-hydroxylinoleic acid (FIG. 1).[1,2] The reaction was catalyzed by

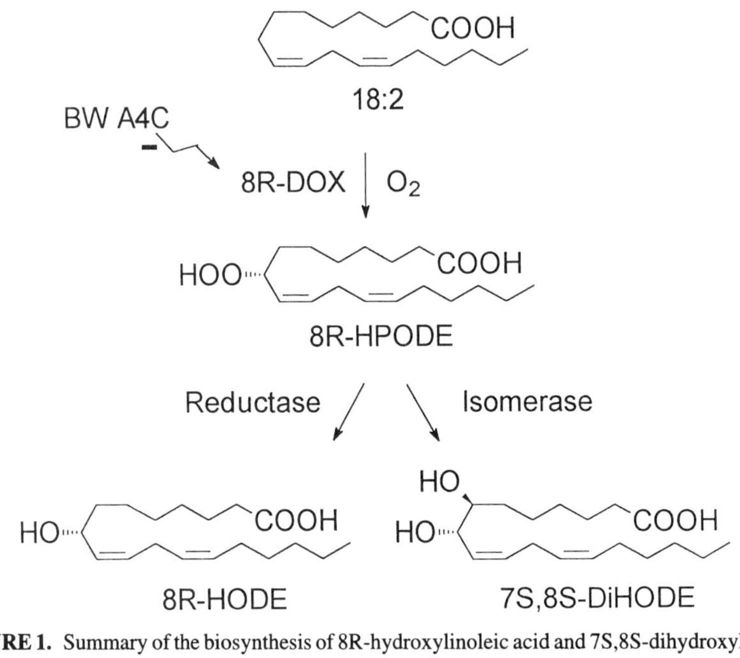

FIGURE 1. Summary of the biosynthesis of 8R-hydroxylinoleic acid and 7S,8S-dihydroxylinoleic acid by 8R-dioxygenase and hydroperoxide isomerase.

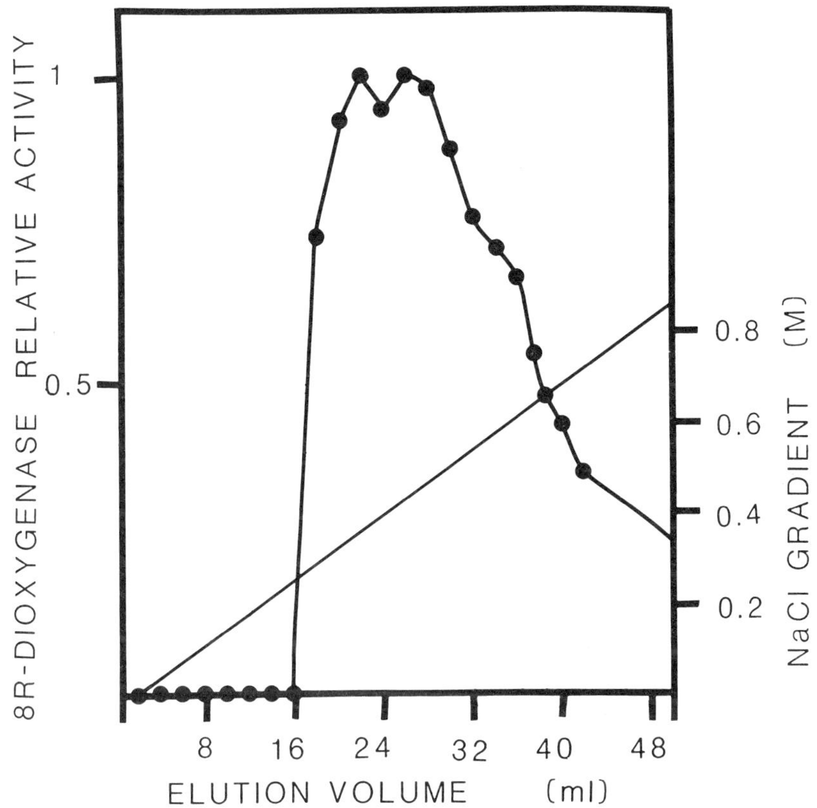

FIGURE 2. Ion exchange chromatogram of a supernatant of *G. graminis* in a 50-mM potassium phosphate buffer, pH 7.3, with sorbitol (0.75 M), EDTA (1 mM), and EGTA (1 mM). 8R-DOX activity eluted with 0.3 M NaCl (TLC assay).

the 8R-dioxygenase (8R-DOX) and the hydroperoxide isomerase. The main product, 8R-hydroxylinoleic acid, could also be isolated from incubations of supernatants of *Laetisaria arvalis* with [1-^{14}C]linoleic acid.[3] In addition, microsomes from rat liver and NADPH formed 8-hydroxylinoleic acid.[4]

8R-DOX was inhibited by typical lipoxygenase inhibitors. Supernatant was preincubated with inhibitor for 20 min at 4 °C and incubated for 15 min at 4 °C with [1-^{14}C]linoleic acid in the presence of GSH-peroxidase and GSH. The data was interpolated from duplicate or triplicate incubations at different inhibitor concentrations and evaluated by HPLC analysis. Formation of 8R-hydroxylinoleic acid was inhibited by BW A4C, zileuton, and linoleate-hydroxamic acid with IC_{50}-values of 0.2 μM, 3-10 μM, and 20 μM, respectively. Nordihydroguaiaretic acid and eicosatetraynoic acid were less potent with IC_{50}-values of 0.09 mM and 0.15 mM, respectively.

The mechanism of hydrogen abstraction by the 8R-DOX and the isomerase was determined by GC-MS analyses of the products of incubation of four different stereospecifically deuterium labeled linoleic acids.[5] The 8R-DOX oxygenated [8S-^{2}H]linoleic acid so that most of the deuterium label was lost in 8R-hydroxylinoleic acid, while the deuterium label of [8R-^{2}H]linoleic acid was retained.[5] Further experiments showed that 7S,8S-dihydroxylinoleic acid formed from [7S-^{2}H]linoleic acid had lost most of the deuterium label, while the deuterium label of [7R-^{2}H]linoleic acid was completely retained in the diol.[5] 8R-DOX thus abstracted the pro-S hydrogen at C8 of linoleic acid and inserted O_2 antarafacially, while the isomerase abstracted the pro-S hydrogen at C7 from 8R-hydroperoxylinoleic acid and inserted oxygen from the hydroperoxy group suprafacially.

Ammonium sulfate precipitations, ion exchange chromatography, and gel filtration experiments were performed with supernatant. These experiments indicated that the 8R-dioxygenase and isomerase activities reside in different enzymes (FIG. 2). Incubations with partially purified 8R-DOX yielded 8R-hydroxylinoleic acid as the main product.

We conclude that 8R-DOX has a similar inhibition profile and course of oxygenation to lipoxygenases and that the 8R-DOX and isomerase activities can be separated.

REFERENCES

1. BRODOWSKY, I. D. & E. H. OLIW. 1992. Biochim. Biophys. Acta **1124:** 59-65.
2. BRODOWSKY, I. D., M. HAMBERG & E. H. OLIW. 1992. J. Biol. Chem. **267**(21): 14738-14745.
3. BRODOWSKY, I. D. & E. H. OLIW. 1993. Biochim. Biophys. Acta **1168:** 68-72.
4. OLIW, E. H., I. D. BRODOWSKY, L. HÖRNSTEN & M. HAMBERG. 1993. Arch. Biochem. Biophys. **300**(1): 434-439.
5. HAMBERG, M., L. Y. ZHANG, I. D. BRODOWSKY & E. H. OLIW. 1993. Arch. Biochem. Biophys. **303**(1): 77-80.

Lipoxygenases in Human, Monkey, and Bovine Corneal Epithelia[a]

MARIA LIMINGA, ANETTE VON MALMBORG,[b] AND ERNST OLIW

Department of Pharmaceutical Biosciences
Biomedical Centre
Uppsala, Sweden

[b]*Kabi Pharmacia Ophthalmics*
Uppsala, Sweden

A 12-lipoxygenase has previously been identified in rabbit corneal epithelium.[1] The object of this study was to investigate lipoxygenases of human, monkey, and bovine corneal epithelia.

The bovine epithelium contained a microsomal arachidonate 12-lipoxygenase and a cytosolic 15-lipoxygenase.[2] 12S-Hydroxyeicosatetraenoic acid (12S-HETE) and 15S-hydroxyeicosatetraenoic acid (15S-HETE) were identified by GC-MS and chiral HPLC.[2] BW A4C, an acetohydroxamic acid lipoxygenase inhibitor, reduced the biosynthesis of 12S-HETE and 15S-HETE by over 90% at 10 μM.[2] Bovine corneal epithelial 12-lipoxygenase metabolized ^{14}C-labeled linoleic acid poorly in comparison with [^{14}C]arachidonic acid.

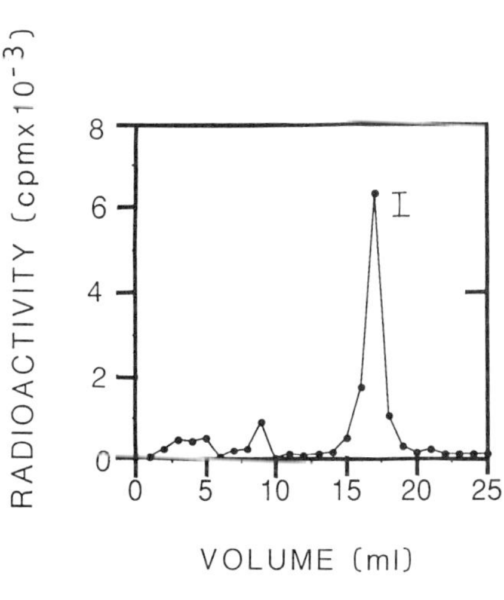

FIGURE 1. Reverse phase HPLC analyses of metabolites formed from [^{14}C]arachidonic acid by human corneal epithelial cells. Human corneal epithelial cells were isolated and homogenized, and the low-speed supernatant was incubated with 10μM [^{14}C]arachidonic acid (37 °C, 30 min). After extractive isolation the material was analyzed by RP-HPLC. The RP-HPLC column (ODS 5 μm, 150 × 4.6 mm) was eluted with methanol/water/acetic acid (75:25:0.01, v/v) at 1 ml/min/fraction. The radioactivity was determined. The peak of radiolabeled material (marked I) coeluted with 15S-HETE.

[a]This work was supported by grants from Kabi Pharmacia Ophthalmics, Uppsala, and the Swedish Medical Research Council (Number 6523).

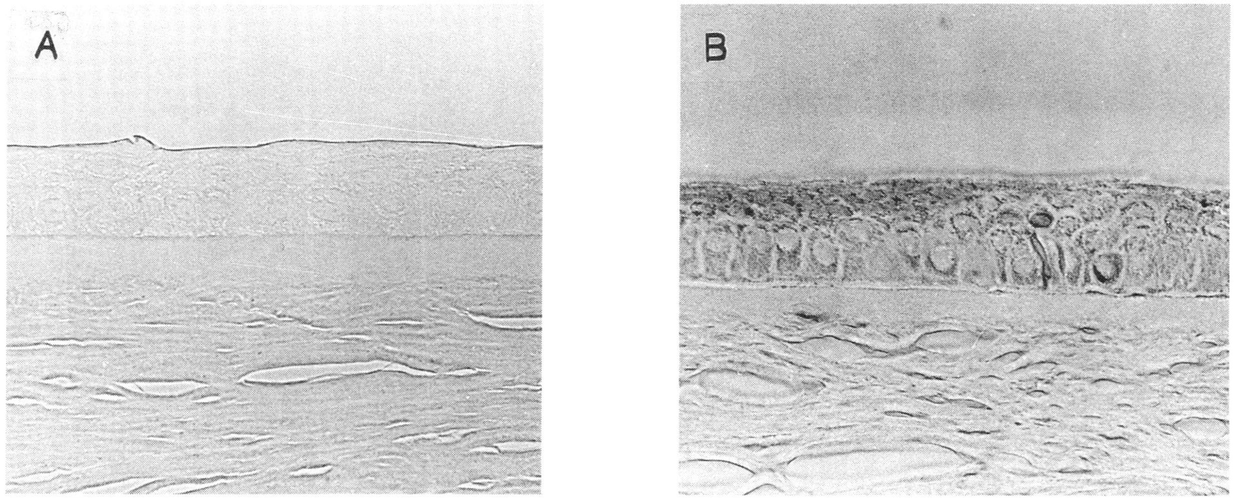

FIGURE 2. Immunostaining of human corneal epithelium with a polyclonal antibody against human recombinant reticulocyte 15-lipoxygenase. (**A**) Control. (**B**) Staining with the primary antibody. Human eyes with a postmortem time less than 24 hours were fixed in 4% formaldehyde. The eyes were embedded in paraffin and cut into 6-mm-thick sections and fixed to glass slides. Unspecific binding sites were blocked with bovine serum albumin. The slides were incubated with the primary antibody, and the controls with normal rabbit serum for 60 min at 20 °C. Biotinylated goat anti-rabbit IgG was used as secondary antibody. An anti-biotin complex reagent was used for detection. Magnification: 125×.

The cytosolic fraction of corneal epithelial cells of the monkey *Macaca fascicularis* (n = 12) converted [^{14}C]arachidonic acid to two major products. The products were identified as 12S-HETE and 15S-HETE by reverse phase HPLC, straight phase HPLC, and chiral HPLC. No lipoxygenase activity could be detected in the microsomal fraction.

The cytosol of human corneal epithelium converted [^{14}C]arachidonic acid to one prominent metabolite. The product cochromatographed with 15S-HETE on reverse phase HPLC (FIG. 1), straight phase HPLC, and chiral HPLC. GC-MS analysis of the hydrogenated product (Me_3Si ether methyl ester derivatives) showed strong and characteristic signals, *inter alia*, at *m/z* 383, 367, 343, 314, 173, and 73. The C value was 22.0. The metabolite was identified as 15S-HETE. Esculetin, a lipoxygenase inhibitor, reduced the 15S-HETE formation by approximately 70% at 20 μM. Immunohistochemical analysis of human cornea using a polyclonal antibody against human recombinant reticulocyte 15-lipoxygenase[3] gave positive staining of the epithelial cells, while the stroma was negative (FIG. 2). The results provide biochemical, pharmacological, and immunohistological evidence of a 15-lipoxygenase in human corneal epithelium. The biological effects of 15S-HETE in human corneal epithelium may therefore merit further investigation.

REFERENCES

1. HURST, J. S., M. BALAZY, H. E. P. BAZAN & N. G. BAZAN. 1991. The epithelium, endothelium, and stroma of the rabbit cornea generate (12S)-hydroxyeicosatetraenoic acid as the main lipoxygenase metabolite in response to injury. J. Biol. Chem. **266**(11): 6726-6730.
2. LIMINGA, M., L. HÖRNSTEN, H. W. SPRECHER & E. H. OLIW. 1994. Arachidonate 15-lipoxygenase in human corneal epithelium and 12- and 15-lipoxygenases in bovine corneal epithelium: Comparison with other bovine 12-lipoxygenases. Biochim. Biophys. Acta. **1210:** 288-296.
3. SIGAL, E., D. GRUNBERGER, E. HIGHLAND, C. GROSS, R. A. F. DIXON & C. S. CRAIK. 1990. Expression of cloned human reticulocyte 15-lipoxygenase and immunological evidence that 15-lipoxygenase of different cell types are related. J. Biol. Chem. **265**(9): 5113-5120.

Inhibition of Platelet Eicosanoid Metabolism by the Bacterial Phenazine Derivative Pyocyanin

MICHAEL MULLER, KRYSTYNA SZTELMA, AND TANIA C. SORRELL

Centre for Infectious Diseases and Microbiology
University of Sydney
Westmead Hospital
Westmead, NSW, 2145, Australia

Eicosanoid metabolites of arachidonic acid play a key role in modulating the host inflammatory response. In particular, eicosanoids have significant actions in regulating platelet aggregation and neutrophil infiltration at sites of vascular lesions. Platelets

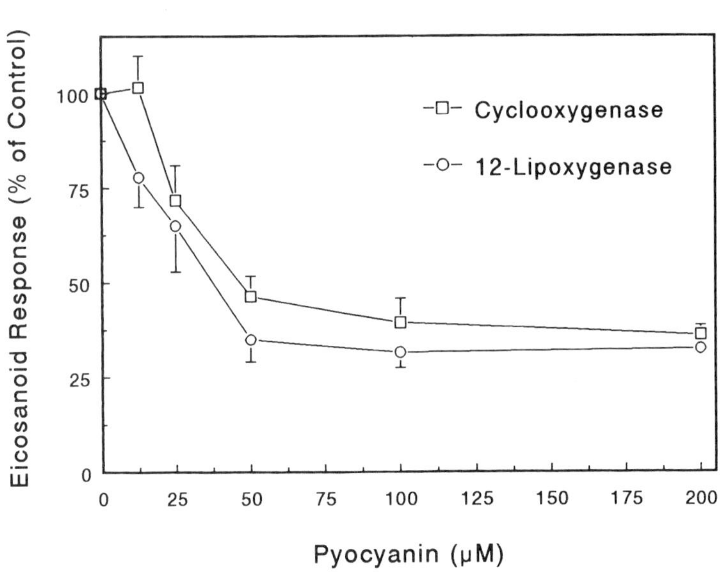

FIGURE 1. Inhibition of human platelet cyclooxygenase and 12-lipoxygenase activity by pyocyanin. Platelets were prelabeled for 60 min with [^{3}H]arachidonic acid and, after washing, were suspended at 3×10^8/ml in Hank's balanced salt solution containing 0.01% bovine serum albumin. The platelets were exposed to pyocyanin (0–200 μM) for 20 min followed by stimulation for 5 min with A23187 (5 μM) at 37 °C. Pyocyanin was prepared as previously described.[6] Eicosanoids were extracted by the method of Bligh and Dyer[7] and separated by thin-layer chromatography on silica gel 60 plates with a developing solvent of chloroform:methanol:acetic acid:water (90:8:1:0.8). Bands were scraped and radioactivity determined by scintillation counting. Results represent the mean ± SEM of three separate experiments.

TABLE 1. Effect of Antioxidants on Restoring Platelet Eicosanoid Activity[a]

	TxB_2	HHT	12-HETE
Control	28.6 ± 6.1	48.9 ± 10.6	31.6 ± 4.1
SOD	91.1 ± 5.7	92.2 ± 11.2	72.2 ± 10.6
Catalase	142.8 ± 8.9	147.1 ± 8.7	112.2 ± 15.7
BHT	91.3 ± 9.9	120.2 ± 9.5	83.7 ± 6.0
Vitamin E	65.8 ± 7.9	77.9 ± 4.0*	55.4 ± 3.6

[a]Effect of antioxidants on pyocyanin-mediated platelet eicosanoid inhibition. Platelets were prelabeled with [^{3}H]arachidonic acid as described in the legend to FIGURE 1. Prior to exposure to pyocyanin (100 μM), platelets were treated with the indicated antioxidants for 5 min. Butylated hydroxytoluene (BHT) and vitamin E were used (at a final concentration of 10 μM) as well as superoxide dismutase (30 U/ml) and catalase (60 U/ml). Each agent was added 5 min prior to the addition of pyocyanin. Eicosanoids were extracted and quantified as described for FIGURE 1. Results are expressed as the percent of eicosanoids available after pyocyanin exposure with respect to control values ± antioxidant treatment, and represent the mean ± SEM of three separate experiments. All values were significantly different from control values except where indicated. Significance was determined using the Student *t*-test; * = not significant.

have been implicated in the pathogenesis of lung disease due to enhanced adhesion to pulmonary endothelium[1] and are known to participate in the transcellular metabolism of arachidonic acid and its derivatives.[2,3] The opportunistic bacterial pathogen *Pseudomonas aeruginosa* produces the phenazine derivative pyocyanin (1-hydroxy-5-methylphenazinium hydroxide inner salt). Pyocyanin occurs in the sol phase of sputa from patients with cystic fibrosis and bronchiectasis in amounts up to 27.3 μg/ml.[4] Previously, we have identified pyocyanin as an inhibitor of human neutrophil 5-lipoxygenase activity[5] and leukotriene B_4 omega-oxidation.[6] However, the effect of pyocyanin on enzymes of the cyclooxygenase and other lipoxygenase pathways has not yet been established. In the present work we investigated the effects of pyocyanin on eicosanoid production by platelets.

Pyocyanin dose-dependently inhibited production of thromboxane A_2 (TxA_2), measured as thromboxane B_2 (TxB_2), 12-hydroxyheptadecatrienoic acid (HHT), and 12-hydroxyeicosatetraenoic acid (12-HETE), from human platelets that had been prelabeled with [^{3}H]arachidonic acid ([^{3}H]AA) and stimulated with the calcium ionophore A23187 (FIG. 1). However, concurrent addition of [^{3}H]AA and A23187 to pyocyanin-treated platelets showed no evidence of inhibition of eicosanoid production (data not shown). This result excluded a direct inhibitory effect by pyocyanin on platelet cyclooxygenase and 12-lipoxygenase enzymes. Prelabeled platelets treated with the antioxidants BHT, vitamin E, or superoxide dismutase for 5 min prior to the addition of pyocyanin (100 μM) partially restored production of TxB_2, and HHT and 12-HETE levels, while catalase provided greater protection (TABLE 1).

These data indicate that pyocyanin, reduced by platelets, reacts with molecular oxygen to generate the reactive oxygen species superoxide and hydrogen peroxide. These species inhibit the metabolism of endogenous arachidonic acid, possibly by oxidative changes to platelet membrane components; however, the ability of platelets to produce eicosanoids by the metabolism of exogenous arachidonic acid remains unaffected.

As eicosanoids exert potent regulatory roles at inflammatory sites, changes in eicosanoid production may alter the course of the inflammatory reaction. Oxidant stress induced by bacterial pyocyanin may have a significant role in the pathogenesis of chronic inflammatory lung disease due to *P. aeruginosa,* particularly in cystic fibrosis.

REFERENCES

1. Durham, S. K., M. A. Horan, A. Brouwer, R. J. Barelds & D. L. Knook. 1989. J. Pathol. **157:** 339-345.
2. Marcus, A. J. 1984. J. Allergy Clin. Immunol. **74:** 338-342.
3. McCulloch, R. K., K. D. Croft & R. Vandongen. 1992. Biochim. Biophys. Acta **1133:** 142-146.
4. Wilson, R., D. A. Sykes, D. Watson, A. Rutman, G. W. Taylor & P. J. Cole. 1988. Infect. Immun. **56:** 2515-2517.
5. Muller, M. & T. C. Sorrell. 1991. Infect. Immun. **59:** 3316-3318.
6. Muller, M. & T. C. Sorrell. 1992. Infect. Immun. **60:** 2536-2540.
7. Bligh, E. G. & W. J. Dyer. 1959. Can. J. Biochem. **37:** 911-917.

Effect of Quercetin on Prostaglandin A_1–Induced Heat Shock Response in Human Cells[a]

GIULIANO ELIA[b] AND M. GABRIELLA SANTORO[b,c,d]

[b]*Institute of Experimental Medicine*
National Council of Research
00137 Rome, Italy

[c]*Department of Experimental Medicine*
University of L'Aquila
67100 L'Aquila, Italy

The exposure of eukaryotic as well as prokaryotic cells to temperatures a few degrees above the physiological level triggers the synthesis of a particular set of proteins called stress proteins or heat shock proteins (HSP). HSP synthesis represents an essential, finely regulated, and highly conserved cellular response to adverse environmental conditions including exposure to transition metal ions or sulfhydrylic reagents, glucose deprivation, hypoxia, and virus infection, as well as heat shock.[1] Besides their defensive activity, stress proteins participate in essential metabolic processes, including protein folding and intracellular translocation, and control of cell cycle progression and differentiation; their synthesis can be induced by specific bioactive molecules, including hemin and prostaglandins.[2] It has recently been shown that induction of HSP70 synthesis by prostaglandins of the A type is modulated by cycloheximide-sensitive activation of the heat shock transcription factor (HSF) from its control non-DNA binding state to a transcriptionally active form, which binds to the heat shock element (HSE) in the heat shock gene promoter.[3]

Induction of HSP70 synthesis has been associated with inhibition of cell proliferation after treatment with PGA_1.[2] However, no direct evidence regarding a causal relationship between HSPs and control of cell proliferation has been achieved to date, also due to the lack of specific inhibitors of HSP synthesis. Since it has been recently reported that the flavonoid quercetin, an *in vitro* inhibitor of 5-lipoxygenase and cyclooxygenase, inhibits HSP70 synthesis induction by heat shock in a human colon carcinoma cell line,[4] we have investigated the effect of quercetin on PGA_1-induced HSP70 synthesis.

K562 human erythroleukemic cells, grown in RPMI 1640 medium supplemented with 10% fetal calf serum and antibiotics, were treated with PGA_1 (6 or 10 μg/ml) or ethanol diluent at 37 °C, or were subjected to heat shock (HS, 45 °C for 20 min)

[a] This work was supported in part by the Italian National Council of Research, Progetti Finalizzati "ACRO," and "FATMA."

[d] Address for correspondence: M. Gabriella Santoro, Institute of Experimental Medicine, Viale K. Marx, 15/43, 00137 Rome, Italy.

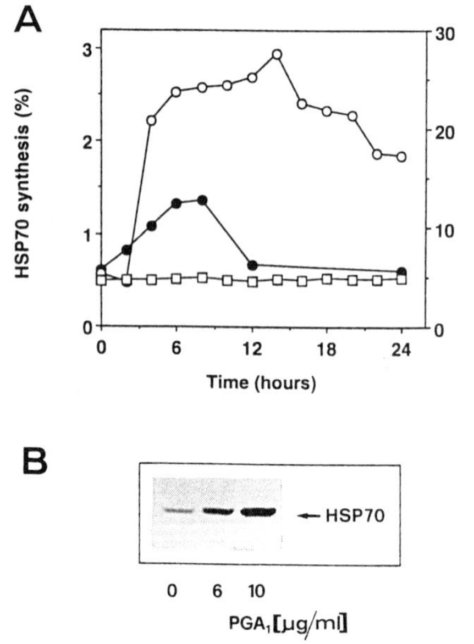

FIGURE 1. Kinetics of HSP70 synthesis after PGA_1-induction. (**A**) K562 cells were treated with 6 μg/ml or 10 μg/ml PGA_1 and labeled with L-[^{35}S]methionine (5 μCi/10^6 cells, 1-h pulse) at different times after treatment. Samples containing an equal amount of radioactivity were processed for quantitative densitometric analysis after SDS-PAGE separation and autoradiography. Levels of HSP70 synthesis after treatment with 6 μg/ml (*solid circles*) or 10 μg/ml (*open circles*) PGA_1 are shown. Levels of actin in samples treated with 10 μg/ml PGA_1 (*squares*) are shown as control. (**B**) Identification of HSP70 by immunoblot analysis.

in the presence or in the absence of 30 μM quercetin (3, 3′, 4′, 5, 7-pentahydroxyflavone). At this concentration, quercetin did not significantly inhibit K562 protein synthesis for at least 24 h in untreated cells. At different times after treatment, K562 cells were labeled with L-[^{35}S]methionine (5 μCi/10^6 cells) and processed for SDS-PAGE, autoradiography, and densitometric analysis as described.[2] Quantitative evaluation of proteins was determined using the BioRad 1-D Analyst software. For identification of HSP70, equal amounts of protein from PGA_1-treated samples were separated by SDS-PAGE and processed for immunoblot analysis, using a monoclonal anti-72/73kDa HSP antibody from HeLa cells, as described.[2]

As shown in FIGURE 1, both the kinetics and the extent of HSP70 synthesis vary in relation to PGA_1 concentration. Treatment with quercetin inhibited HSP70 synthesis in the first few hours after induction by PGA_1-treatment (FIG. 2A,B). This block was exerted at the transcriptional level, as shown by Northern blot analysis of RNA extracted from K562 cells treated with PGA_1 in the presence or the absence of quercetin (data not shown).

However, analysis of the kinetics of HSP70 synthesis after PGA_1 treatment revealed that the presence of quercetin did not reduce, but only delayed, HSP70 synthesis. In fact, in quercetin-treated cells HSP70 synthesis started 2-3 h later than in the PGA_1-treated control, and continued up to 20 h after treatment at a much higher level than control (FIG. 2C,D). Similarly, HSP70 synthesis in HS-treated cells was delayed in the presence of quercetin, but it continued at much higher levels for at least 3 h longer than in control cells.

Finally, quercetin did not affect PGA_1 antiproliferative activity. On the contrary, quercetin alone at nontoxic concentrations was found to inhibit K562 cell growth by more than 90%. While the mechanism of quercetin antiproliferative activity and its

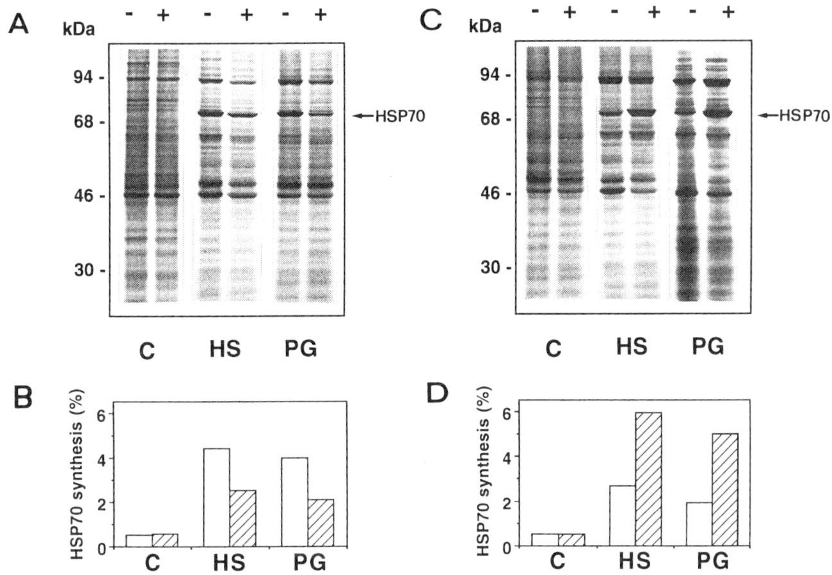

FIGURE 2. Time-dependent modulation of HSP70 synthesis by quercetin in PGA_1-treated cells. K562 cells treated with PGA_1 (10 μg/ml) in the presence (+) or in the absence (–) of quercetin (30 μM) were labeled with L-[^{35}S]methionine (5 μCi/10^6 cells, 1-h pulse) at 2 h (**A,B**) or at 20 h (**C,D**) after treatment. The effect of quercetin on HSP70 synthesis at 2 h (**A,B**) or at 8 h (**C,D**) after heat shock (45 °C, 20 min) is also shown. Samples containing an equal amount of radioactivity were processed for SDS-PAGE analysis and autoradiography (**A,C**). Percent of HSP70 synthesis was determined by quantitative densitometric analysis (**B,D**). **C** = Control; **PG** = PGA_1; **HS** = heat shock.

relation to 5-lipoxygenase inhibition remains to be elucidated, these results indicate that the described interaction of quercetin with the complex constituted by the *cis*-acting HSE and the *trans*-acting cytoplasmic HSF[4] is limited in time, and the heat shock response is restored in a few hours and actually enhanced after exposure to the drug.

REFERENCES

1. Lindquist, S. & E. A. Craig. 1988. Annu. Rev. Genet. **22:** 631-677.
2. Santoro, M. G., E. Garaci & C. Amici. 1989. Proc. Natl. Acad. Sci. USA **86:** 8407-8411.
3. Amici, C., L. Sistonen, M. G. Santoro & R. I. Morimoto. 1992. Proc. Natl. Acad. Sci. USA **89:** 6227-6231.
4. Hosokawa, N., K. Hirayoshi, H. Kudo, H. Takechi, A. Aoike, K. Kawai & K. Nagata. 1992. Mol. Cell. Biol. **12:** 3490-3498.

Induction of a 32-kDa Stress Protein by Prostaglandin A_1 in Cultured Murine Cells[a]

ANTONIO ROSSI[b] AND
M. GABRIELLA SANTORO[b,c,d]

[b]*Institute of Experimental Medicine*
National Council of Research
00137 Rome, Italy

[c]*Department of Experimental Medicine*
University of L'Aquila
67100 L'Aquila, Italy

Prostaglandins of the A type (PGA) induce the expression of a 70-kDa heat shock protein (HSP70) in a large variety of monkey, canine, porcine, and human cells (reviewed in Ref. 1). Induction of HSP70 has been associated with a cytoprotective effect of PGA_1 after virus infection or thermal injury.[2,3] In eukaryotic cells HSP70 proteins, including the constitutively expressed HSC70 and the major inducible HSP70, are encoded by a large multigene family.[4] The activation of heat shock gene transcription is mediated by a transcription factor known as heat shock factor (HSF), which binds to heat shock elements (HSE) in the promoter region of the heat shock genes.[4] The presence of multiple HSFs has now been described in several species, including human. In mouse cells two different types of heat shock factors (mHSF1 and mHSF2) have been described that show only 38% identity overall and have different functional properties. mHSF1 displays inducible DNA-binding activity like the HSF present in *Drosophila* and vertebrate cells, whereas mHSF2 is constitutively bound to DNA similarly to *Saccaromyces cerevisiae* and loses DNA binding activity by incubation at elevated temperatures.[5] Because of the different type of regulation of the HS response in murine cells as compared to human cells, we investigated the effect of PGA_1 on the heat shock response in C2C12 mouse myoblasts and other murine cell lines.

C2C12 mouse myoblasts (kindly supplied by S. Alemà, Institute of Cell Biology, CNR, Rome, Italy) were maintained in Dulbecco's modified Eagle's medium supplemented with 15% fetal calf serum, 2 mM glutamine, and antibiotics. Cells were treated with PGA_1 (30 μM) or control diluent at 37 °C, or were subjected to a 20-min 45 °C heat shock, as control for HSP70 induction. Cells were labeled with L-

[a]This work was supported by grants from the Italian National Research Council, Progetti Finalizzati "A.C.R.O.," and "F.A.T.M.A."

[d]Address for correspondence: M. Gabriella Santoro, Institute of Experimental Medicine, CNR, Viale K. Marx 15/43, 00137 Rome, Italy.

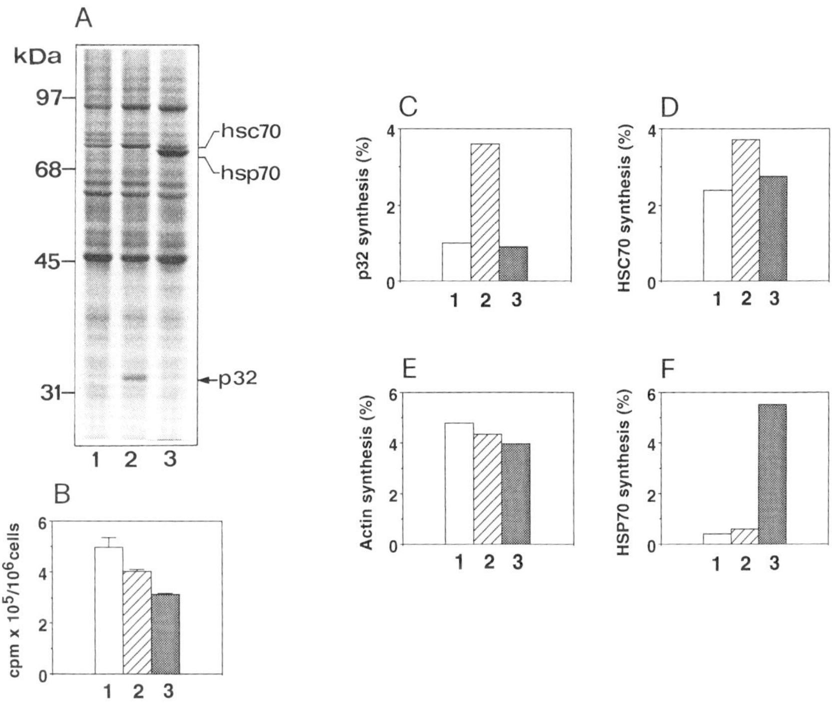

FIGURE 1. Effect of PGA_1 on HSP70 and p32 synthesis. Control (**lane 1**), PGA_1-treated (**lane 2**), or heat shock-treated cells (**lane 3**) were labeled with L-[^{35}S]methionine (12 μCi/10^6 cells, 7-h pulse) soon after treatment, and the radioactivity incorporated into TCA-insoluble material was determined as described.[2] Total protein synthesis, measured as L-[^{35}S]methionine incorporation into TCA-insoluble material, is shown in **B.** Samples containing equal amounts of radioactivity were processed for SDS-PAGE and autoradiography (**A**). After densitometric analysis of autoradiograms, quantitative evaluation of p32 (**C**), HSC70 (**D**), and HSP70 (**F**) synthesis was determined using the BioRad 1-D Analyst software, and expressed as percentage of total protein synthesis. Actin synthesis (**E**) was determined as a control.

[^{35}S]methionine soon after heat treatment or addition of PGA_1 for the following 7 h, and samples containing equal amounts of radioactivity were processed for SDS-PAGE, autoradiography, and densitometric analysis as described.[2] Heat treatment resulted in approximately 40% inhibition of total protein synthesis and induction of HSP70 synthesis (FIG. 1A,B,F). PGA_1 only slightly (~15%) inhibited total protein synthesis (FIG. 1B), and, while it moderately increased HSC70 synthesis, it was not able to induce HSP70 synthesis in these cells (FIG. 1A,D,F). HSC70 and HSP70 proteins were identified by immunoblot analysis (FIG. 2). Equal amounts of protein for each sample were separated by SDS-PAGE and processed for immunoblot analysis as previously described,[2] using monoclonal anti-HSP70 antibodies 3A3 (3A3 mAbs, kindly provided by R. Morimoto, Northwestern University, Evanston, IL), which recognize both the 73-kDa constitutive HSC and the 72-kDa inducible HSP.

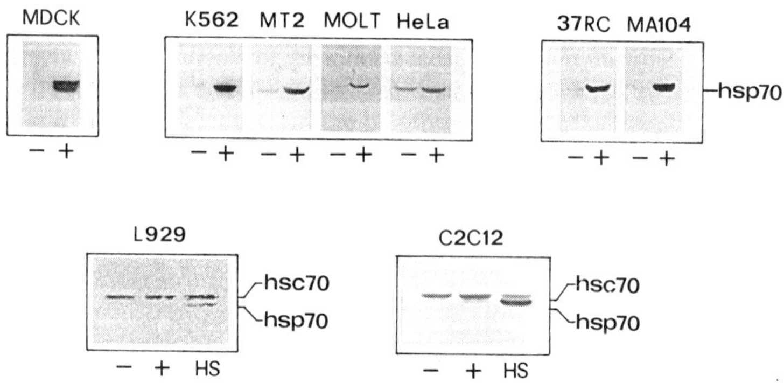

FIGURE 2. Induction of HSP70 by PGA_1 in several mammalian cell lines. Canine (MDCK), human (K562, MT2, MOLT, HeLa), or monkey (37RC, MA104) cells were treated with 30 μM PGA_1 (+) or control diluent (–) for 24 h; samples containing equal amounts of protein were processed for immunoblot analysis, using anti-HSP72/73 mAbs from HeLa cells. L929 murine fibroblasts and C2C12 murine myoblasts were treated with 30 μM PGA_1 (+) or control diluent (–), or were heat shocked at 45 °C for 20 min (**HS**); 12 h after treatment, samples containing equal amounts of protein were processed for immunoblot analysis using 3A3 mAbs.

The lack of induction of HSP70 synthesis was not dependent on the concentration used. In fact, when PGA_1 was tested at concentrations ranging from 6 to 60 μM, no induction of HSP70 synthesis was detected, even at the highest dose, which caused a 45% decrease in total protein synthesis; HSC70 synthesis was found to increase dose-dependently (data not shown). The lack of induction of HSP70 synthesis after PGA_1 treatment appeared to be a general characteristic of murine cells, since, apart from C2C12 myoblasts, L929 mouse fibroblasts (FIG. 2) and several other murine cell lines, including B16 melanoma, N2A neuroblastoma, and Friend erythroleukemic cells (data not shown), were unable to respond to PGA_1 treatment by synthesizing HSP70. As previously reported, PGA_1 was instead able to induce HSP70 synthesis in a variety of human (K562 erythroleukemia, MT2 leukemia, MOLT-4 T-cell lymphoma, HeLa cells), monkey (37RC and MA104 cells), and canine (MDCK cells) cell lines (FIG. 2). These results indicate that murine cells behave differently from cells of other mammalian species in their response to PGA_1. The lack of response of murine cells to HSP70 induction by PGA_1 could be related to a peculiar control mechanism of HSP70 gene expression by the murine heat shock transcriptional factors mHSF1 and mHSF2. The ability of PGA_1 to stimulate HSC70 synthesis without inducing HSP70 could be a useful tool in the study of murine HSF activation and function.

While it was not able to induce HSP70 synthesis, PGA_1 was found to dramatically induce the synthesis of a 32-kDa protein (p32) in C2C12 cells (FIG. 1 A, C). The p32 protein has been identified as heme oxygenase (HO) by two-dimensional gel electrophoresis and Northern blot analysis (manuscript in preparation). PGA_1 was found to induce heme oxygenase expression more efficiently than other classical inducers (i.e., sodium arsenite), at concentrations that were nontoxic to the cells and did not inhibit cell protein synthesis. Moreover, induction was dependent on the cell type, and myoblastic cells appeared to be extremely sensitive to PGA_1. Heme oxygenase plays a key role in the physiological heme catabolism in mammals, catalyzing the oxidative cleavage of heme. Since heme oxygenase has been shown to be a low molecular weight stress protein induced in mammalian cells by oxidant stress and is considered to be part of a general inducible antioxidant defense pathway,[6] these results suggest the possibility that arachidonic acid metabolites could be involved in a protective mechanism during oxidative injury.

REFERENCES

1. Santoro, M. G., E. Garaci & C. Amici. 1990. Adv. Prostaglandins Thromboxane Leukotriene Res. **21:** 867-874.
2. Amici, C. & M. G. Santoro. 1991. J. Gen. Virol. **72:** 1877-1885.
3. Amici, C., A. Palamara & M. G. Santoro. 1993. Exp. Cell Res. **207:** 230-234.
4. Morimoto, R. I. 1992. J. Biol. Chem. **267:** 21987-21990.
5. Sarge, K. D., V. Zimarino, K. Holm, C. Wu & R. I. Morimoto. 1991. Genes Dev. **5:** 1902-1911.
6. Applegate, L. A., P. Luscher & R. M. Tyrrell. 1991. Cancer Res. **51:** 974-978.

Subject Index

Index of Contributors